The Chemistry and Biology of Winemaking

Dedication

For Bix, Winston and The Oaf.

The Chemistry and Biology of Winemaking

Ian Hornsey
Founder Partner and Launch Head Brewer, Nethergate Brewery

RSCPublishing

ISBN-13: 978-0-85404-266-1

A catalogue record for this book is available from the British Library

Published by The Royal Society of Chemistry,
Thomas Graham House, Science Park, Milton Road,
Cambridge CB4 0WF, UK

Registered Charity Number 207890

For further information see our web site at www.rsc.org

Preface

There are good reasons for believing that wine has played a prominent role in the civilisation of mankind; indeed, the ancient Greeks regarded anyone who did not imbibe wine as a barbarian. We may presume that the drink would have played no small part in the lives of great men such as Archimedes, Pythagoras and Pliny the Elder, and, from a much later era, we know that much of Pasteur's eminence can be attributed to his studies on 'the diseases of wine'.

This little book aims to show that there is a lot more to viticulture and viniculture than is normally considered to be the case. It is neither merely intended to be a condensed version of some of the excellent, extensive treatises already available, nor, because of the limitations of space, does it include some of the more fashionable topics, such as wine tasting. Rather, it is hoped that the book will introduce the reader to the ways in which the science of oenology, in its broadest sense, has contributed to the well-being of mankind, and to intimate how some of the organisms involved might play an important role in his future.

First and foremost, due emphasis has been given to the yeast and its ability to undergo ethanolic fermentation. Yeasts were active in the environment long before humans came on to the scene and since the advent of man on the planet these microbes have, almost certainly, been an ever-present component of his diet. Over time, yeasts have learned to manipulate human cells just as effectively as humans have exploited the unique properties of yeast cells. Yeasts are the most important organisms in biotechnology, and research with the organism is about to extend the boundaries of functional genomics. The fundamentals of genome variation in yeast are the

v

same as those in humans, and so much useful information can be gleaned from comparative analysis. If we look further into the future, the first complete 'wiring diagram' of a cell will probably be for a yeast. By organising genetic information into a wiring diagram, the known functions and interactions of genes can be used to generate a computer model of a living cell. Such diagrams have the potential to solve any biological problem, including cellular approaches for the treatment of disease. From a functional point of view, yeasts may be considered to be tiny 'factories', capable of producing essentials such as proteins, vitamins and foodstuffs generally.

In a slightly different context, the worldwide movement against the use of chemicals in the environment, and about our foodstuffs, has meant that man is ever on the lookout for novel means for biological control methods for pests, diseases and contaminants. Killer yeasts, which are dealt with here, show some promise as agents of biological control.

Like it or not, genetically modified organisms are going to play a major role in our future well-being. The bacterium causing crown gall disease of the vine and some other plants has contributed greatly to the way in which plant genomes can be modified, and has taught us much about the infection process. An insight into the relationship between the bacterium and the vine, and the biochemistry involved, is presented here.

Finally, lactic acid bacteria, some of which are responsible for a secondary wine fermentation, are some of the most versatile microbes known to man, and, as I hope this book indicates, are destined to contribute significantly to biotechnology in the future.

Thus, I hope that the slightly different slant of certain sections in this book will make it useful to students and others readers, with a wide range of chemical and biological interests, as well as being of benefit to aficionados with a more perfunctory interest in wine.

Contents

Introduction

Unless otherwise stated, the term 'wine' is applied here to the product made by the alcoholic (ethanolic) fermentation of sound grape juice by yeast (*Saccharomyces cerevisiae*), followed by an appropriate period of ageing. Indeed, for a definition we need to go no further than the *Oxford English Dictionary*, which contains the ultra-concise entry: 'The fermented juice of the grape used as a beverage'. Someone once said that: 'wine is a mixture of chemistry, biology and psychology', and there is certainly a considerable contribution from the latter discipline involved in winemaking. For example, few wine consumers have a preconceived idea of grape flavour, something that allows the winemaker a high degree of stylistic licence, as long as the wine is balanced, and has well-integrated flavours and aromas. Conversely, if a wine is made from strawberries, then the consumer expects it to taste of strawberries! Wine has always fascinated man because of its complexity, and because it is an ever-changing entity. We now know that in the wine *milieu* there is an interaction between more than 500 substances, all of which contribute towards the ultimate flavour, aroma and structure of the drink, yet it is very difficult to assess the importance of any one of them in isolation.

In terms of organic chemistry, wine is a complex mixture of a large number of compounds including carbohydrates, alcohols, aldehydes, esters, acids, proteins and vitamins. It is also home to a number of polyhydroxy aromatic compounds, such as tannins, anthocyanins and flavonols, which contribute hugely to colour and taste. The basic raw material for a wine fermentation is a fermentable sugar, such as fructose or sucrose, rather than the less soluble, non-fermentable starch, which is the raw material for most beers.

1

Indeed, of all the fruits that would have been gathered by our ancient forefathers, only the grape stores carbohydrates predominantly as soluble sugars. All of the major sources of nutrients in grapes are, in fact, in a form that may be readily metabolised by yeast, and many historians would aver that the fermentation of grape juice was the *inevitable* result of ancient man storing grapes for out-of-season use, an event that presupposes that he had ceased to function as a hunter-gatherer. Indeed, to take matters further, winemaking as a frequent cultural practice could only result from a sedentary life-style, and cannot possibly pre-date the evolution of agriculture. A nomadic life-style is incompatible with the need to accumulate enough grapes to turn into significant quantities of wine. If the first alcoholic beverages were the result of serendipity, then it is far easier to envisage how wine may have resulted from grapes, than it is to foresee how barley (say) might have given rise to beer; one simply had to gather the fruit together, damage it slightly and leave it! Even this was not always necessary, because, under the appropriate environmental conditions the grape will undergo fermentation while still on the vine, a fact that would surely not have gone unnoticed by the more observant of early humans. They might not have known what was occurring to the fruit, but they would certainly have observed the muscular dys-functioning of certain frugivorous animals that had fed on such material (one has visions, here, of birds falling off of their perches!). The appropriate research would have been carried out, and, from then on, there was no looking back. The notion that 'man first met grape' during the Palaeolithic has led some workers to postulate about the 'discovery' of wine, something known as the 'Palaeolithic Hypothesis'.

The production of beer, of course, necessitates that the starch, from whichever source it comes (normally grain), is first broken down into fermentable sugars. This is effected, either by enzymes produced naturally by the germinating grain (called malting, if carried out artificially), by exposure to saliva (chewing) or by addition of enzymes produced extraneously by microbes. None of this is necessary in a wine fermentation. In addition, unlike cereal crops, the surface of the mature cultivated grape develops a microflora containing a high number of acid-tolerant yeast cells (especially *Saccharomyces cerevisiae*), which encourage spontaneous fermentation once the grape skin is ruptured. It is of interest that the wine yeast, which has until recently has been *S. cerevisiae* var. *ellipsoideus*, is apparently not an organism indigenous to the microflora of the grape skin. The natural habitat of the ancestral forms of this fungus was the sap exudate of oak trees, and it has been proposed, not entirely unreasonably, that inoculation of the grape surface either arose from vines growing on oak trees, or from the simultaneous harvesting of grapes and acorns! Once under way, fermentation of grape juice proceeds apace, and soon liberates copious quantities of ethanol, which, together with the natural low pH of the juice (principally due to tartaric acid), severely limits the growth of other (unwanted) grape skin microorganisms. The presence of considerable quantities of tartaric acid in the mature grape is seen as one of its distinguishing biochemical characters, and, in some instances, grapes actually represent a commercial source of this chemical. The 'cream of tartar', used in baking is, in effect, the crystalline form of the acid that is formed during fermentation.

The results of the first major chemical analyses of ethanolic fermentation were published between 1789 and 1815, but it was not until the monumental work of Pasteur, during the latter half of the 19th century, that scientists fully appreciated the role of yeasts in such fermentations, and, since that time, *S. cerevisiae* has become one of the most studied organisms on the planet! Since the 19th century, our understanding of wine, wine composition and wine transformations has evolved immensely as a result of advances in biochemistry and microbiology. As in brewery fermentations, air is initially necessary in order to permit synthesis of some vital, cellular components of yeast, but prolonged exposure to air results in acetic acid (vinegar) as an end product, rather than ethanol. Exclusion of air would have presented a major problem to the ancients, a fact that was somewhat ameliorated by the fact that vinegar, itself, was a valuable commodity for them, facilitating the production of some forms of pottery, and permitting the preservation of perishable foodstuffs (*i.e.* pickling). Although the likelihood of grape juice being fermented into wine appears to be a much more logical procedure than obtaining beer from cereal grain, there seems to be little doubt that during the conversion of the hunter–gatherer to the sedentary agriculturalist, cereals became a part of domestic life way before the grape did. In addition to the ease of production of wine, as compared to beer, wine is also much more stable (storable) than beer, principally because of its higher alcohol content and greater acidity.

Since the ancients firmly believed that wine was the invention of the gods, it is unsurprising that this intoxicating beverage would become a prominent feature of many of their cults. Offerings of wine were given to the deities, and the drink played a central role in religious ceremonies. Copious quantities of wine were brought to temples, and consumed at lavish banquets held in honour of the gods, and at funerary feasts. In Mesopotamia, wine was linked to their version of the creation, as can be seen from the Babylonian epic, '*Enūma eliš*', written to glorify the important god Marduk. According to the legend, Marduk was appointed 'king of gods' during a banquet, while all of the other gods were under the influence of wine. The symbolic power of the grapevine was manifested by the Babylonian goddess Geshtin-an-na, whose name literally translates as 'grapevine of heaven'. In ancient Egypt, as well, wine was the drink of the gods, and earthly worthies would make offerings of wine to placate them. Beer was used in similar fashion, but, by the 2nd Dynasty, wine seemed to be at the top of the list of liquid offerings to the gods. From an early period, bunches of grapes and jars of wine became basic funerary gifts in the tombs of royalty, and other high-ranking officials. According to the Classical writers, the gift of the vine was attributable to the god Osiris, ruler of the world of the dead, and the association of this deity with wine dates from very early times. The Greek historian Herodotus compared Osiris to Dionysus. The popularity of wine was given a boost by the cult of the Greek god Dionysus (Bacchus to the Romans), the son of Zeus, and god of wine, fertility and vegetation.

The cult and myth of Dionysus are the richest in the Greek pantheon, and encompass the earliest of the cult rituals. It has often been claimed that Dionysus was a late arrival among the Greek gods, appearing in the epics somewhere between the 8th and 7th centuries BC, or, maybe, even as late as the 5th century BC, and that his cult origins seem to point to the north (Thrace and Macedonia), to Asia Minor or to Asia itself. Evidence now suggests, however, that Dionysus was already known in Greece before 1200 BC. What appears to be the first reference to Dionysus forms part of an early Greek (Linear B) inscription found during excavations at the Mycenean palace at Pylos (the home of Nestor in

Homer's *Iliad*). One of the recovered tablets bears the preserved personal name *di-wo-ni-so-jo*, which has been interpreted as referring to the deity Dionysus, rather than some other, mortal, name. There are two (incompatible) stories of Dionysus' birth, one obviously a Theban story, the other Orphic in origin. The Theban legend has it that Dionysus' mother was Semele, the daughter of King Cadmus of Thebes. The intimate relationship with Semele made Zeus' wife, Hera, so jealous that she arranged for Semele to be confronted by Zeus, which, since she was mortal, would cause her to burn to death. Although Semele was six months pregnant at her death, Hermes managed to save the embryo from her womb, and implant it in Zeus' thigh, from which Dionysus was born 3 months later. Thus, Dionysus was born from the thigh of his own father. The second story records that Dionysus was the son of Zeus and Persephone. Numerous locations have been suggested as Dionysus' birthplace, including (alphabetically): Dracanum, Icarus, Naxos and Mount Nysa, the latter, according to Homer, being 'in the barbarous land of Thrace'. There has always been much debate among scholars as to the true origin of Dionysus, some have argued that both he and his cult originated somewhere in the Levant and from there spread westwards to the rest of the Mediterranean world. Such a notion is given credence by the mid-6th century BC Greek potter and vase painter Exekias, creator, in 540 BC, of a kylix depicting on its inner surface Dionysus reclining in a ship whose mast has sprouted vines and bunches of grapes, while around the ship play a school of dolphins (Figure 1). The illustration depicts an event from Homer's *Hymn to Dionysus*. The story goes as follows: One day, the god, who was on the Island of Icarus, was captured by Tyrrhenian pirates, who had originally agreed to allow him passage to Naxos, but decided to hold him to ransom instead, and, since he was handsome, tried to seduce him. Suddenly, flutes were heard; ivy and grapevines fouled the oars and sails, and wild beasts appeared on the deck (lions, panthers and bears). The pirates jumped into the sea, and were turned into dolphins. One of them was put into the sky as a constellation (Delphinus) as a warning to sailors to behave!

Just as there are several myths relating to his birth, the place of origin of Dionysus is shrouded in mystery, as well. As already intimated, there are numerous tales of the god travelling throughout the Greek world, having arrived somewhere from the east. As

Figure 1 *Dionysus sailing the Mediterranean, after he has miraculously grown a vine up his ship's mast and transformed his assailants into dolphins. Painting by Exekias inside a Greek drinking vessel (kylix), mid-6th century BC*

knowledge of the world to the east of Greece expanded with Alexander's campaigns to India, so did the regions from which Dionysus was supposed to have arrived in Greece. In Hellenistic and Roman times, for example, he is often depicted as having arrived in triumph from India, accompanied by exotic animals. Most authorities, however, would consider him to be an Anatolian import and, from an historical perspective, the Thracian people, from whose land he was said to have emerged, founded the kingdom of Phrygia in Asia Minor, where they were known in the *Iliad* as 'growers of the vine' (iii. 185). In fact, throughout the classical period, the inhabitants of Asia Minor were considered adherents of strange ecstatic cults, and very un-Greek in spirit. Some authorities (*e.g.* Stanislawski, 1975)[1] maintain that knowledge of Dionysus would have been impossible without any knowledge of the vine, while Graves[2] stated quite categorically that: 'The main clue to Dionysus' mystic history is the spread of the vine culture over Europe, Asia and North Africa'. The name 'Bacchus' came into use in ancient Greece during the 5th century BC, and came to represent the god of wine. Bacchus translates as 'the screamer' in ancient Anatolian, and relates to the loud cries that accompanied the frenzied worship at the celebratory 'bacchanalia'.

As a deity, Dionysus is full of contradictions, which is hardly surprising, given the circumstances of his birth. On one hand, he is seen as being full of benevolence, alleviating the woes of mankind; on the other, he is highly vindictive, punishing all those who oppose him, and/or his cult. Although he originated as a minor fertility symbol among the Anatolians, he became transformed into a complex deity representing wine in the context of religious and political protest. These protests invariably took the form of festivals of unrestricted drinking of alcohol, leading to ecstatic revelries, and orgies, all fuelled by drunkenness, that were intentionally designed to transgress normal patterns of behaviour. Contrary to popular belief, Dionysus did not necessarily want his adherents to get intoxicated, he wanted them to drink wine so that they could relax and be joyful. The cult spread widely throughout the Mediterranean world and was hugely responsible for creating demand for wine among the population at large. In addition to the joyous aspects of Dionysus, he also had a dark side to his personality, which was associated with the Underworld. This aspect is concerned with Dionysus' role as a god of fertility, who dies, and is resurrected in accordance with the annual cycle of the seasons. It is also has something to do with Dionysus' unique birth, or, rather, 're-birth' in the body of his father. Dionysus is usually depicted in statues holding a decanter of wine, and a bunch of grapes. According to Dayagi-Mendels,[3] it is not without possibility that this mythical tale of death and resurrection may have, centuries later, had some influence on Christian beliefs concerning the life of Jesus. Thus, the legacy of the close relationship of wine to divinity has outlived the polytheism of Greece and Rome, and can be seen today in the rather more staid rituals of Christianity.

The grapevine and wine are central subjects of the Bible. The Land of Canaan flowed not only with milk and honey, but with wine, as well. The hills of Judea were ideal sites for vineyards, and so the Hebrews came to regard wine as a gift and a part of everyday life, something for which they gave praise to God (*Psalms*, 104:15). While drinking wine to 'gladden the heart' was acceptable, over-indulgence most definitely was not (*Proverbs*, 20:1). Similar sentiments are to be found in the New Testament, where Christ made frequent use of the growing vine in his parables, and even referred to himself as 'The True Vine.' Wine was, of course, the subject of one of the best known miracles in the New Testament – the changing of water into wine at the wedding in Cana of Galilee (*John*, 2:1–11), and, as the 'Blood of Christ' it is the central theme at the Last Supper. Wine is one of the two elements of the Eucharist, the holiest sacrament of the Christian churches, and is the central act of Christian worship practised by almost all denominations of Christians. According to the synoptic gospels, Jesus gave wine to the disciples at the end of the Last Supper, saying: 'Drink ye all of it: for this is my blood of the New Testament, which is shed for many for the remission of sins.' (*Matthew*, 26:27–29). The Eucharist is understood by Christians to commemorate the death and resurrection of Jesus Christ and to mediate communion with God and community among the worshippers. The theology of the Eucharist varies widely among the different Christian denominations. The Orthodox and Roman Catholic Christians understand the presence of Christ concretely, whereas the Protestant churches, to varying degrees, find transubstantiation out of harmony with their Biblical interpretation, and view the Eucharist in symbolic terms. The vessel used in the celebration of the Eucharist is the chalice and, from the beginning of Christianity, special rites of consecration attended its use. The most famous example of this type of vessel is the Great Chalice of Antioch, which bears the earliest known portraits of Christ and his disciples, and dates from the 4th–5th century AD.

Some scholars question the interpretation of several Greek, Hebrew and Chaldean words that have, hitherto, been indiscriminately translated as 'wine', whereas they might be describing some non-intoxicant, such as unfermented grape juice, or by-products of it. To illustrate the potential problems involved, let us briefly consider grape juice concentrate. For centuries, fresh juice (must) has been boiled down to a thick syrup, like molasses, and then stored in jars. This was used by the ancients as a substitute for honey, and was eaten by spreading on bread. It was also used to sweeten or preserve wine and fruit, and it was taken as a drink after addition of water. Cato, Columella and Pliny the Elder all describe how *mustum* was boiled down to concentrate its sugar content. If it was reduced to half of its original volume the product was known as *defrutum*, but if the grape juice was boiled down to one-third of its bulk, it yielded a thickened syrup with the finest of flavours, called *sapa* ('best wine'). Pliny referred to these unfermented products as 'a product of art, not of nature', and it has been suggested that it was from *sapa* that Christ miraculously produced the 'wine' at the wedding at Cana, which was, therefore, an unfermented product.

When all things are considered, it seems as though intoxication was undoubtedly the reason why, almost from its inception, wine and religion were inextricably bound together – wine served as a convenient means of contacting the gods; the more that was drunk, the closer one could get to them!

Wine is a source of energy, and also contains sufficient nutrients to enable it to be classed as a food. Louis Grivetti[4] speaks of wine as 'the food with two faces', maintaining that no other food can claim to have such a unique dichotomy, because it is praised when consumed in moderation, and condemned when consumed in excess. He elaborates by relating some of the laudatory expressions that have been coined for the drink, such as: 'a chemical symphony'; 'bottled poetry' and 'captured sunshine', while also documenting some less flattering sobriquets, such as: 'the opener of graves'; 'the destroyer of homes' and 'the quencher of hopes'.

Generally speaking, three basic types of wine are produced: (1) still, or table wines with an alcoholic content in the range of 7–13%; (2) sparkling wines produced from a secondary fermentation, whereby the CO_2 is deliberately trapped in the liquid phase; and (3) fortified wines, which result from the addition of spirits to still wines in order to raise the alcoholic content to around 20%. Apart from their use in viniculture, grapes can be eaten in fresh form as fruit, dried as raisins or sultanas, or pressed for their juice. Throughout the history of 'civilised' man, it is the stability (*i.e.* storability) of wine, and of raisins, that has made the grapevine an important source of food at any time of year.

To finish this section, let me quote what I understand to be an old Spanish proverb: 'Wine has only two defects: if you add water to it, you ruin it; if you do not add water to it, it ruins you.'

REFERENCES

1. D. Stanislawski, *Geographical Rev.*, 1975, **65**, 427.
2. R. Graves, *The Greek Myths I*, Penguin, London, 1988.
3. M. Dayagi-Mendels, *Drink and Be Merry: Wine and Beer in Ancient Times*, The Israel Museum, Jerusalem, 1999.
4. L.E. Grivetti, Wine: The Food with Two Faces, in P.E. McGovern, S.J. Fleming and S.H. Katz, (eds), *The Origins and Ancient History of Wine*, Gordon & Breach, Amsterdam, 1996.

CHAPTER 1

The History of Wine

1.1 THE PREHISTORY OF WINE

Homo sapiens would have probably encountered *Vitis vinifera* subsp. *sylvestris* (Figure 1) at a very early date, when the first groups of humans migrated from East Africa some 2 million years ago. The most likely venues for making the acquaintance were the upland areas of eastern Turkey, northern Syria or western Iran, or, maybe, the hilly lands of Palestine and Israel, or the Transjordanian highlands. Significantly, most eastern Mediterranean myths pinpoint the origin of viniculture to somewhere in north-eastern Asia Minor. In theory, however, the event could have occurred anywhere within the distribution range of the wild grapevine (Figure 2). As we have already intimated, Palaeolithic man was probably the first to become familiar with wine, purely by the accidental 'spoilage' of stored, or over-ripe grapes. Wine may, of course, have been the result of unsuccessful attempts to store grape juice, which is a particularly unstable beverage. Because our Stone Age forebears did not dwell in permanent, year-round settlements, they had no opportunity to investigate what was actually happening when grapes/grape juice spoiled (fermented), and so they were unable to learn about and perfect the process. It was not until man adopted a sedentary way of life, and had developed the need for a continuous supply of wine, that he put his mind to a rudimentary form of viticulture. From what we know at present, the Neolithic period in the Near East, somewhere between 8500 and 4000 BC would seem to provide all prerequisites for the intentional

Figure 1   *Wild grapevine* Vitis vinifera *subsp.* sylvestris *(after Zapriagaeva[1])*

manufacture of wine. Not only was mankind flirting with agriculture and beginning to lead a settled existence, but was starting to manufacture items such as pottery vessels, essential items if liquids were to be stored for any period of time.

The world's earliest known pottery, from Japan, dates to the 11th millennium BC, but it seems to have been 'invented' independently in many places throughout Anatolia, Mesopotamia and the Levant almost simultaneously around 7000–6000 BC. The earliest pots were simple cups, vases and dishes, probably originally sun-dried. Later, more sophisticated vessels were fired in kilns. Although probably used by nomadic peoples, pottery was a real benefit for those with a sedentary life style. With the domestication of plants and animals, clay pots provided a sound way to store

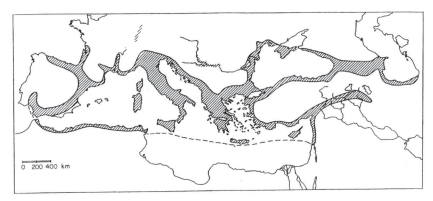

Figure 2   *The distribution range (shaded area) of* V. vinifera *subsp.* sylvestris.
*Note that towards the east the wild vine extends beyond the boundaries of
this map, and reappears at a few places in Turkmenistan and Tadzhikis-
tan (After Zohary & Spiegel-Roy[10])*
(Reproduced by kind permission of the American Association for the
Advancement of Science)

liquids (*e.g.* milk, fruit juice), and, most importantly, the where-
withal for cooking (heating and boiling) and fermenting. The
inventor of the first clay pot is unknown, but its creation was most
likely attributable to women who carried most of the burden for
domestic chores. The ability to heat food, and to ferment, opened
up a whole new world of potential food sources that were safer,
tastier and easier on the digestive tract.

If we are looking for a single site for the origin of the domestic
grapevine (the Noah 'hypothesis'), then, according to the archaeological, archaeobotanical and historical information that we have
at present, some northern mountainous region of the Near East
would appear to be the most likely ones. With the ingenuity of
Homo sapiens, and the vast natural range of the wild vine, it is
tempting to suggest that there might have been multiple domesti-
cations of the plant in different places, and at slightly different
times. This is not impossible, but all the available evidence does
seem to point to an upland site in the northern part of the Near
East.

As we shall see, the earliest confirmed traces of wine have been
found in sediments in a pottery jar from the Neolithic village of
Hajji Firuz in north-eastern Iran. Relatively little is known about
wine making at Neolithic sites further north and at higher altitudes

in the Taurus and Caucasus mountains, where the wild subspecies of *V. vinifera* thrives today. Domesticated grape seeds have been recovered from Chokh in the Dagestan mountains of the eastern Caucasus, dating from the beginning of the 6th millennium BC, and from Shomutepe and Shulaveri along the Kura River in Transcaucasia, dating from the 5th through early 4th millennia BC. Should confirmatory residue analysis from pottery fragments from these areas ever be possible, then we may be able to locate the ultimate origins of viticulture. Unfortunately, few Neolithic sites in this vast region have, so far, been excavated. It has long been thought that the domestication of *V. vinifera* occurred in Transcaucasia, or in neighbouring Anatolia, around 4000 BC. The region between and below the Black Sea and the Caspian Sea, where the grapevine is indigenous, represents the closest the vine comes to the Near Eastern origins of plant and animal agriculture. As Jackson[2] says: 'Therefore it seems reasonable to assume that grape domestication may have first occurred in the Transcaucasian Near East.' Certainly, as far as Vavilov's 'centres of genetic diversity' are concerned, it is in this Transcaucasian region that wild grapevines are both the most abundant and most variable. Before this, of course, wild vines would have been used as a source of grape juice, and, as is the case for most domesticated plants, they would have required constant attention from man in order to give him satisfactory yields. The notion that viticulture and viniculture originated somewhere in the Near East, and was disseminated from there, is supported by the fact that there is a remarkable similarity in most Indo-European languages in the words for 'vine' and 'wine', indeed, Renfrew[3] goes further and maintains that the spread of agriculture into Europe resulted from the dispersal of people speaking Proto-Indo-European languages. Conversely, there is little resemblance between the words for 'grape' in the same group of languages, and this is taken to signify that the appreciation of, and use of the grape occurred a long time before the advent of viticulture and viniculture, and the above-mentioned dispersal of Indo-European languages into Europe. In contrast, Semitic languages, that evolved in regions where the vine was not indigenous, have often adopted words for 'vine' and 'grape', which are very similar to those for 'wine', and are thought to be related to *woi-no*, an ancestral term for wine. According to some authorities, this lexical similarity implies that knowledge

relating to grapes, viticulture and viniculture was obtained concurrently. If the above Noah hypothesis is correct, then it is
possible that from Transcaucasia the vine (and wine) became
transplanted into the Indian sub-continent, and then back to the
Mediterranean basin. Support for such a notion comes from the
fact that, in Sanskrit, the oldest of the Indo-European family of
languages, 'vena' translates as 'favourite'.

1.2 DISSEMINATION OF VITICULTURE

In simple terms, it is still commonly believed that domesticated
vines were carried westwards by humans, along with the shift of
agriculture generally, into the Mediterranean Basin, and from
thence into different parts of Europe by Greeks and Roman
colonists. The Greeks are credited with the transport of cultivated
vines into Italy, and the Romans took them to Spain, France and
Germany. It should be stressed, however, that movements such as
these are not as certain as parts of the historical record might
suggest, for changes in *V. vinifera* seed and pollen morphology
indicate that domestication was occurring in some parts of Europe
before the agricultural revolution supposedly reached that continent! Work by Stevenson,[4] for example, has indicated that an
extensive system of viticulture was already in existence in southern
Spain several centuries before colonisation by the Phoenicians.
Until many wild grapevine populations were annihilated by 'foreign' pests and diseases in the mid-19th century, the subspecies
sylvestris remained abundant throughout its natural range, from
Spain to Turkmenistan. Large populations of wild vines still exist
in certain regions within their indigenous range, such as south-
western Russia, where they were afforded protection from the
abovementioned scourges, and have been used for winemaking
over several millennia. It is thought that these locally occurring
and adapted vines might be the progenitors of most of today's
European cultivars, and it is notable that the Pinot group of
cultivars, for instance, possesses many traits resembling *V. vinifera*
subsp. *sylvestris*.

From whatever its exact place of origin, the domesticated vine
seems to have diffused in two directions. One route was to Assyria,
and thence to the Mesopotamian city states of Kish and Ur, and, in

later times, Babylon, where wine, because of its rarity, was mostly regarded as a drink for the upper classes, and the priesthood. The hot, dry climate, and soil salinity in Mesopotamia were not conducive to growing vines, and so most of the wine consumed by the civilisations there was imported from further north, from Assyria, for example, where there was higher rainfall. From Mesopotamia, under the influence of man, wine and the vine reached the Jordan Valley, somewhere around 4000 BC. The vine never grew in this part of the world, and from there it reached ancient Egypt, where, again, the grapevine is certainly not indigenous, but where there were certain areas favourable for the growth of *Vitis*. Vines became established in the Nile Delta by 3000 BC., at the beginning of the Early Dynastic period, and, by the New Kingdom times a very sophisticated form of viticulture was being practised. Viticulture followed the Nile upstream into Nubia, where it flourished until the advent of Islam. By 1000 BC wine drinking was widespread over much of the Near East, and was practised by, among many others, the Hebrews, the Canaanites, and the closely related Phoenicians, even if it was largely restricted to the higher strata in society.

The second route carried wine across Anatolia to the Aegean, where, by 2200 BC, it was being enjoyed by the Minoan civilisation on Crete, and in Mycenaea, where viticulture and viniculture became highly specialised, and wine was an important commercial commodity. Of great significance is the fact that wine (and vines) became an integral part of Greek culture, and, wherever they established a colony around the Mediterranean, wine drinking, viticulture and viniculture were sure to follow. Indeed, Sicily and southern Italy became very important vine-growing areas to the Greeks. Another important site was Massilia (now Marseilles), from where the Greek methods of viticulture and viniculture spread inland, following the valley of the River Rhône. The suggested major diffusion routes of viticulture in southwest Asia and Europe are shown in Figure 3.

1.3   THE EARLIEST CHEMICAL EVIDENCE FOR ANCIENT WINE

In 1996, McGovern et al.[6] reported on the re-investigation of some residues found on the inner surfaces of pottery sherds, the analysis

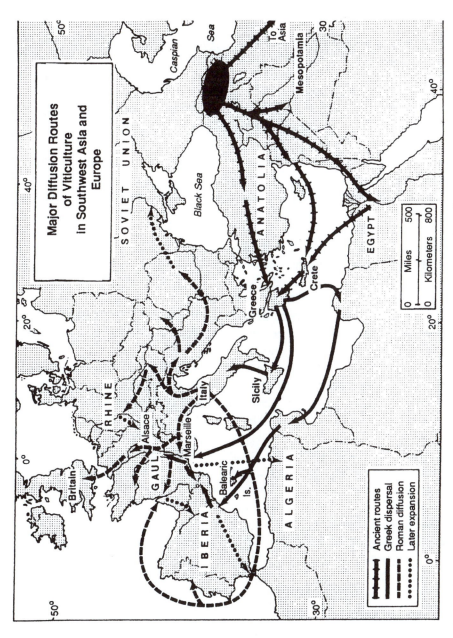

Figure 3 *Likely diffusion routes of viticulture in southwest Asia and Europe (after de Blij[5])* (Reproduced by kind permission of Professor H.J. de Blij)

of which showed, beyond doubt, that the original jars had contained wine. The sherds had been recovered in 1968 from Hajji Firuz Tepe, a Neolithic site south-west of Lake Urmia, in the northern Zagros Mountains. The site is on the eastern fringes of the 'Fertile Crescent'. Wild *Vitis* still grows in this region, and pollen cores taken from the deposits of Lake Urmia showed that it grew there during Neolithic times. The site was excavated as part of the University of Pennsylvania Museum's Hasanlu project. The jars had been found embedded in the earthen floor of the kitchen of a mud-brick building dated to *ca.* 5400–5000 BC. During the original excavations, a yellow residue was noted on the inside lower half of a jar fragment. At the time, the deposit was assumed to be from some sort of dairy product, even though chemical analysis yielded no positive results. After a gap of 25 years, some sherds (one with a reddish residue) were re-examined using more sophisticated methods. Results showed that the deposits contained tartaric acid, calcium tartrate and the oleoresin of the terebinth tree (*Pistacia atlantica* Desf.). Tartaric acid occurs naturally in large amounts only in grapes, and was converted into its insoluble calcium salt in the calcareous environment of the site. *P. atlantica* grows abundantly throughout the Near East, and was widely used as a medicine and a wine preservative in antiquity. Judging by their long, narrow necks, the fact that the residues were confined to their bottom halves, and the presence of clay stoppers of approximately the same diameter as the necks, it was evident that these jars once contained some sort of liquid, and had been sealed. All the evidence supported the conclusion that the Hajji Firuz jars originally contained resinated wine, the like of which has been found in other, more recent contexts in the ancient Near East and ancient Egypt. If all the six jars found contained wine, then it was estimated that there would have been around 14 gallons in all, quite a significant production for household use. As McGovern commented,[7] 'If the same pattern of usage were established across the whole of the site's Neolithic stratum, only part of which was excavated, one might conclude that the grapevine had already come into cultivation.' He also tentatively asked the question (somewhat tongue-in-cheek) whether the difference in colour (yellow and red) of some of the deposits might have represented early attempts at making white and red wine.

The first published chemical evidence for ancient Near Eastern wine[8] emanated from excavations carried out on the Period V site at the Late Chalcolithic village of Godin Tepe, in the Kangavar Valley in west-central Iran, during the years 1967–1973. The site is located high in the Zagros Mountains, and Period V dates from 3500 to 2900 BC, which is contemporary with the Late Uruk period (as the Late Chalcolithic in lowland Greater Meso-potamia is known) in southern Mesopotamia, and sits along-side the 'High Road', or the 'Great Khorasan Road', which later on became part of the famous 'Silk Road', leading from Lower Mesopotamia to Iran. Godin Tepe controlled the most important east–west route through the Zagros Mountains between Baghdad and Hamadan, and may have been a Sumerian or Elamite trading post. Certainly, Godin Tepe was well positioned to parti-cipate in trade, and, especially, to protect the trade route, being situated, as it was, some 2 km above the alluvial low-lands. It was adjacent to the eastern edge of Lower Mesopotamia, where some of the world's earliest literate cultures had formed themselves into city-states, such as Lagash, Ur, Uruk and Kish. In addition, the Elamite capital of Susa (modern Shush, in Iran) was some 300 km to the south, and the proto-Elamites at this time were already well on their way towards developing an urban way of life. The city-states in the Tigris–Euphrates valley were based on the irrigation culture of cereals, dates, figs and other plants.

From the evidence given by imported items, these adjacent lowland cultures appear to have been in contact with several other parts of the Near East, such as Anatolia, Egypt and Transcaucasia. There was much evidence of an import trade in precious commo-dities, such as gold, copper and lapis lazuli, and even in everyday requirements, such as wood and stone, which were not readily available in the lowlands. Many resources, essential to urban lowland life, were available in the Zagros Mountains, and so Godin Tepe was ideally placed to take account of, and assimilate developments occurring elsewhere in the world.

In 1988, some years after the original excavations at Godin Tepe had been completed, a reddish residue found on some pottery sherds from unusually shaped jars, was re-examined, using infrared (IR) spectroscopy. The spectra obtained clearly showed the pre-dominant presence of tartaric acid, and its salts, principal compo-nents of grapes, and it was presumed that the jars had once contained wine, which had evaporated and left a residue. The wine jars proved to be of a type not found elsewhere, or, at least, imported from a region not explored archaeologically (or not reported). The jars, which had been stored on their sides, and had been stoppered, were unique in several respects, all of which were consistent with them being used to store liquid. Firstly, a rope design had been applied as two inverted 'U' shapes along opposite exterior sides of each vessel (termed *rope-appliqué*). The specific rope pattern on the jars implied that actual rope had once been laid underneath, in an effort to stabilise them once they had been laid on their sides. The red, organic deposits on the interior of the one whole vessel available for examination were confined to the base and to the side wall directly opposite to that which had the rope decoration on its outer surface – exactly the position that one would expect if precipitation of sediment had occurred inside the jar after it had been placed on its side. Secondly, the jars had tall necks, and relatively narrow mouths, compared to other known jars of this period. Thirdly, the interior of one of the sherds appeared to be lined with a slip (a compact, fired fine layer of clay), which would have helped to make the vessel resistant to leakage. All in all, these jars, with their elongated, narrow necks, were ideally suited for storing, transporting and pouring wine. The question now arose as to whether the wine was produced locally, and probably destined for export, or whether it was imported from elsewhere. As Badler and co-workers reported,[8] it is quite possible that the wild Eurasian grapevine grew around Godin Tepe in antiquity, since grapevines are common today throughout parts of the Zagros Mountains, especially some 500 km north of the village. The ecology and climate of the Godin Tepe region, with a perennial supply of water provided by its river, the Gamas Ab, is well suited to growing wild, or domesticated *Vitis*, and there have been occasional reports of the vine growing there. Moister, milder conditions in ancient times might have permitted viticulture during the Late Uruk period. Excavations from Period V yielded no unequivocal evidence of grapes, but grape seeds dating to the much later Period II (1st millennium BC) have been identified. In the words of Badler *et al.*:

> Given the beneficial upland environment for grape cultivation, and the evident contacts that existed with lowland horticultural areas, it is quite conceivable that viticulture and wine trade already existed at

Godin Tepe and other key sites in the hill country in the Late Uruk period.

Also found with the jars was an unusually large funnel, and circular lid, which might have been used as a strainer of some sort (*e.g.* for extracting juice from grapes), and might be indicative of viniculture. If, as seems the most likely scenario, the wine was indeed imported, rather than being indigenously produced, then the most likely source was Transcaucasia, where Godin Tepe had contacts by the 4th millennium BC. Another possibility is that the wine came from slightly further west, in south-eastern Anatolia. The wild grapevine grows prolifically along the northern and southern boundaries of Anatolia, and it is known that consumption of fresh grapes, or raisins, extends at least back to the Neolithic period, as exemplified by the 9th millennium BC site at Çayönü. Cultivation and processing of *Vitis* may date to as early as the late-4th millennium BC in southern Turkey, in, for example, Kurban Höyük.

The hot, dry climate in most of Mesopotamia made it difficult to cultivate the grapevine. Vineyards were mostly to be found in the northern, hilly, region. There is little evidence of vineyards prior to the 2nd millennium BC, but one example comes via documents from the time of King Gudea of Lagash (2125–2110 BC), that make reference to irrigated terraces, protected by trees, for planting Vines. Records from the royal palace at Mari, dating to the 18th century BC have yielded clay tablets containing information on viticulture and viniculture.

1.4   THE HOLY LAND

According to *Genesis*, 9:20, one of the first things that Noah did after emerging from the Ark was to plant a vineyard: 'And Noah began to be an husbandman, and he planted a vineyard'. As we know, he then went on to enthusiastically sample the products of the vine, with inevitable results. The Bible also mentions the various steps necessary to care for a vineyard (*Isiah*, 5:1–6), and, as well as other things, prohibits the crossing of strains in a vineyard, lest 'the fruit of thy vineyard be defiled' (*Deuteronomy*, 22:9). Even in antiquity, two main ways of growing vines were recognised: along the ground, and upwardly trained (mentioned in

Ezekiel, 17:6–8). According to the Mishneh Torah, the preferred wine for use in the temple was from vines that grew along the ground: 'From whence did they bring the wine? They may not bring it from (grapes grown on) trellised vines, but only from vines growing from the ground' (*Menahot*, 8:6). The kings of Judah and Israel are said to have owned vast vineyards and huge stores of wine. King David's vineyards were so numerous that he had to appoint special officials to supervise them; one 'over the vineyards' and the other 'over the produce in the vineyards for the wine cellars' (*1 Chronicles*, 27: 27). The land that is now Israel lies in one of the oldest wine-producing areas in the world, there being evidence of viticulture in Jericho, Lachish and Arad, during the period 3500–3000 BC. In antiquity, some of the wines shipped to Egypt were apparently so bad that they had to be flavoured with honey, pepper or juniper berries, in order to make them more palatable.

The Moslem conquest of 636 brought a halt to commercial viticulture for around 1200 years, and during the period of Ottoman rule, from 1516 to 1917, Moslem law prevented Jews and Christians from manufacturing all but 'house wine', which enabled non-Moslems to use wine during their religious ceremonies. It was not until 1870 that Jews began to make wine again on any scale, when Baron Edmond de Rothschild imported some French grape varieties. The wines produced in those days were mostly red, sweet and unsophisticated, such that, in 1875, when the then British Prime Minister, Benjamin Disraeli, was given a bottle of kosher red from Palestine, he commented that it tasted 'not so much like wine, but more like what I expect to receive from my doctor as a remedy for a bad winter cough'.

The modern wine industry in Israel owes its existence to the importation of French cultivars by Baron de Rothschild. Prohibition of the consumption of wine under Islamic law had the effect of fostering the cultivation of table varieties in the Middle East and North Africa. This, in turn, led to the introduction of table grapes into Spain, from where they spread to France and Italy, and eventually to the New World. The growing of table grapes also spread to the Balkans during the Ottoman occupation, but the Turks, unlike the Arabs, allowed their Christian and Jewish subjects to cultivate and use wine grapes.

1.5   EGYPT

As James[9] so forcefully put it:

> A great deal of nonsense has been written about the origins of the vine
> and of wine in Egypt, much of it based on the fanciful ideas purveyed
> by classical and post-classical writers, whose authority in other mat-
> ters of early Egyptian culture is known to be wholly unreliable.

For a start, the vine is not a component of the native flora of
Egypt, and, secondly, until effective irrigation techniques were
developed, the vine remained rare in ancient Egypt, especially in
Upper Egypt where the climate was hot and dry. From the
evidence that we have at present, it seems most likely that the
domesticated grape was introduced into Egypt from the Levant
at least as early as the Predynastic period (Zohary and Spiegel-
Roy[10]; Zohary and Hopf[11]). Archaeobotanical evidence for the
grapevine in ancient Egypt manifests itself in many forms, and
includes both charred and desiccated whole grapes (or fragments), grape seeds, stems, leaves and woody material. At present, the earliest finds of grape seeds from Egypt are from the
Predynastic settlements of Tell Ibrahim Awad and Tell el-Fara'in
(Buto) in the Nile Delta, and, also from Tomb U-j at Umm el-
Qa'ab at Abydos, on the middle Nile. The numerous grape seeds
from the latter site, which was the burial chamber of King
Scorpion I (McGovern et al., 1997[12]), are, to date, the earliest
to be ascribed to winemaking from Egypt (ca. 3150 BC), and they
were found inside storage jars, which had patently been used to
store and transport wine. It should be stressed, at this point, that
presence of the grape does not necessarily indicate viniculture.
The Abydos find consisted of three rooms stacked with around
700 wine jars. McGovern's team calculated that if each jar had
been full, then there would have been some 1200 gallons of wine
at the site. Of the 207 jars that were recovered, 47 of them (ca.
23%) contained grape seeds, and several contained completely
preserved grapes. The seeds were most closely related to the
domesticated V. vinifera subsp. vinifera. Eleven of the jars (ca.
5.5%) contained the remains of sliced figs, which had been
perforated and strung together. It is surmised that the figs were
suspended in the liquid in order to impart sweetness, and the

addition of whole grapes may have served the same purpose. This appears to be the sole example of figs being used as an additive in this way. In addition to the jars, many clay sealings with rim and string impressions were recovered. The sealings had probably been pressed onto leather or cloth covers and tied over the jar mouths with string, an arrangement that would have been particularly convenient for storing young alcoholic beverages from which fermentation gases needed to escape. In addition, most of the Abydos tomb vessels were bottle-shaped, with narrow mouths, which would have been easy to stopper, and, therefore, well-suited for long-distance travel.

The volatile constituents of the jars had long since disappeared, leaving yellowish deposits. Chemical analysis of these deposits confirmed the presence of tartaric acid, which strongly suggested that the liquid had been wine. Analysis also showed the presence of aromatic hydrocarbons derived from the resin of the terebinth tree. The presence of terebinth resin is consistent with the ancient practice of resinating wines in order to prevent oxidation (to vinegar) during storage.

McGovern *et al.* speculated as to where the wine might have originated from, in view of the fact that the wild grapevine never grew in ancient Egypt, and that the domesticated subspecies was not transplanted to the Nile Delta before the end of the 4th millennium BC. The clay sources of the jars were characterised by neutron activation analysis (NAA), and comparison of results with a data base showed that the pottery originated in more than one place, but that none of the jars were of Egyptian origin. Of the jars that could be clearly identified by their clay composition, it was apparent that they had originated in southern Levant, an area where there is also archaeobotanical evidence for grapevine cultivation. It was, therefore, presumed that the wine from Abydos came from this region. Interestingly, the clay seals were made of Nile alluvial clay, suggesting that the jars were sealed on entering Egypt (somewhere in the Nile Delta), or at the tomb itself. Archaeological investigation has established that the use of the overland trade route between the southern Levant (the lowlands of Israel, the Palestinian uplands, the Jordan Valley, and Transjordan) and the eastern Nile Delta – 'the Ways of Horus' as it was known to the ancient Egyptians – intensified during Early Bronze Age 1. There were two-way exchanges in both goods

Figure 4   *First-dynasty seal impressions, with wine press in top right hand corner (after Kaplony[13])*

and technologies and, during the latter stages of this period, the Egyptians took control of the route, and established a number of trading stations in the southern Levantine coastal plain. The most important of these centres was at 'En Besor' (in modern Palestine).

A two-stage process in Early Bronze Age 1 interactions between ancient Egypt and the southern Levant may account for the presence of the Abydos wine jars and the start of an indigenous Egyptian winemaking industry in the wide alluvial plains of the Nile Delta soon after. In the first phase, increasing Egyptian demand stimulated the trade in horticultural products, particularly grapes. Once a sufficient market for wine had been established in Egypt, the transplantation to, and cultivation of grapevines in the Nile Delta could be accomplished, and this would have been quickly followed by vinification itself, probably under the auspices of foreign specialists. The earliest indications so far, for ancient Egyptian wine itself are from the very start of the historic period, the beginning of the 1st Dynasty (*ca.* 3000 BC). The first appearance of the hieroglyph for what is generally considered to be a wine press (Figure 4) occurs at around this time, and the ancient Egyptian term for wine (*irp*) is known from the 2nd Dynasty onwards. Representations of vineyards date as far back as the Old Kingdom, the 3rd millennium BC.

Figure 5   *Winemaking scenes in the tomb of Amenemhet (after Newberry[14])*

A thriving royal winemaking industry had probably become established by the 1st and 2nd Dynasties (*ca.* 3050–2700 BC), but certainly by the 3rd Dynasty, which was the beginning of the Old Kingdom (*ca.* 2700 BC). That much can be confirmed by results of organic analyses carried out on residues from thousands of jars found in the tombs of the pharaohs at Abydos and Saqqara. Over a period of time, the Egyptian elite were able to call upon a variety of locally produced wines that could satisfy both daily needs and religious purposes. New Kingdom (1550—1069 BC) tomb reliefs and paintings depict elaborate storehouses and wine cellars in the palaces and temples at Amarna and Thebes, with row upon row of amphorae awaiting consumption during royal celebrations. The bulk of the information regarding Pharaonic wine production comes from the artistic and linguistic records of the New Kingdom, particularly from Thebes, where there are 42 tombs in the necropolis illustrating the winemaking process. Most tomb-paintings show grapes being harvested and pressed, perhaps the most notable being those in the tomb of Nakht at Thebes. Figure 5 shows Middle Kingdom winemaking scenes in the tomb of Amenemhet at Beni Hasan (dated about 2050–2000 BC).

From the evidence available in the pictorial record generally, the juice from crushed grapes was collected in vats and fermented. When partly fermented, the juice/young wine was decanted into amphorae and left to mature. It might then be filtered, and have spices and/or honey added before being transported to its destination in amphorae. The shoulders of amphorae (and, sometimes, their mud stoppers) are frequently inscribed with hieroglyphics listing the pharaoh's regnal year, the variety of grape, the vineyard, its owner and the person responsible for production. Such information is the forerunner of the modern wine bottle label, and enables us to pinpoint the location of certain vineyards.

It is predominantly the multiple funerary uses of wine (*e.g.* the presence of wine jars in tombs) that have helped to preserve the record of ancient Egyptian viniculture that comes to us today. Other finds of grape remains include those from the 1st Dynasty graves at Abydos and Nagada; from the 3rd Dynasty Step Pyramid of Djoser at Saqqara; from 12th Dynasty Kahun; from 13th Dynasty Memphis and from Second Intermediate Period Tell el-Dab'a. Several 18th Dynasty sites have also yielded grape remains: these include Amarna, Deir el-Medina, Memphis and the tomb of Tutankhamun. There are numerous post-Pharaonic finds of grape. In Egypt, viticulture and wine consumption became much more widespread during the Ptolemaic period, mainly due to the presence of a large number of Greek immigrants, but also due to improved irrigation techniques at that time.

As Powell[15] has discussed, in both ancient Egypt and Mesopotamia, wine was reserved for the elite, and for special occasions. Grapes and raisins were always more expensive than barley or dates. During the Ramesside period (19th–20th Dynasties; New Kingdom) at Deir el-Medina, for example, wine was five to ten times more expensive than beer.[16] Palmer[17] has queried whether the population at large in ancient Egypt, and other parts of the Near East would have actually preferred wine to beer if they had obtained greater access to it. No entry on winemaking in ancient Egypt would be complete without referring the reader to Mary Ann Murray's[18] superb chapter on the subject.

1.6 GREECE

The precise chronology of the arrival of viniculture in Greece is not documented. The earliest evidence for the wild grape appears in the Franchthi Cave, in the Argolid, dating to *ca.* 11,000 BC, but there is no inference that they would have been used for making wine. Allowing for the difficulty experienced in differentiating the cultivated form from its wild cousin, grape cultivation appears to have arisen near Sitagroi, in eastern Macedonia, during the Late Neolithic period, or at least by 2800 BC, but the origins of viticulture in

Greece are also uncertain. Probably the best evidence for it dates to
ca. 2200–2000 BC, and comes from Lerna, in southern Greece.
During the Early Bronze Age finds of domesticated *Vitis* are
widespread throughout the Greek mainland, in the Cyclades, and
on the island of Crete. Whether these grapes were vinified, eaten
raw, or turned into raisins, is not precisely known. The earliest
archaeological evidence for viticulture comes from the Early
Minoan II site (*ca.* 2170 BC) of Myrtos, on Crete, which yielded
a pithos containing crushed grape skins and pips in its base.[19]
Another early site is the Early Helladic (*ca.* 2100 BC) Aghios
Kosmas in Attica. The earliest evidence of the practicalities of
winemaking in Greece lies in a stone foot press at Vathipetro, a
Minoan villa on Crete, dated to 1600 BC. The sophistication of this
site suggests that Minoan wine production had been underway
there for some time. The Myceneans (1600–1150 BC) were also
familiar with viticulture, and grape seeds have been found at
various sites, such as Tiryns and Mycenae, and, in general, by
the Late Bronze Age (*ca.* 1550–1150), viticulture was well established among both the Aegean and Levantine cultures. Some
authorities feel that viticulture in Greece may in some way be
connected to the expansion of the Hittites in Anatolia, around
about 3000 BC, and the resultant upheavals that gave rise to
westward movements of certain elements of the population (in
effect, refugees) who eventually occupied Crete and the Aegean
islands, and spawned the Minoan civilisation (2200–1400 BC).
Decoded Linear B tablets from the Minoan site at Knossos in
Crete have revealed a thriving trade-driven economy, which was
subsequently to have considerable influence on the culture and
commerce of ancient Greece.[20] It is thought that viticulture and
viniculture, once having been introduced to Crete, went from there
to the Peloponessus. Certainly, by the end of the Minoan civilisation, shortly after 1500 BC, winemaking was common throughout
mainland Greece and the Aegean. Archaeological finds on the
Greek mainland demonstrate a close relationship with the Mycen-
aean culture, which met its demise around 1100 BC. Other candidates for bringing the wine culture to Greece include the
Phoenicians, but one should not preclude the possibility that it
was introduced directly into the Balkan Peninsula via the Bosporus, and thence southwards onto the Greek mainland. The latter
proposal becomes somewhat plausible when one appreciates that

Homer, in the *Iliad*, reported that the Thracians had supplied the Greeks with wine during the Trojan War (1300 BC).

The demise of the Mycenean culture is believed to have resulted in a short period of economic depression, and cultural sterility, in Crete, and on the mainland. This was followed by a gradual (1050–900 BC) recovery, both in arts and technology (including the emergence of iron-working), in which peoples from the Aegean islands and mainland Greece began to colonise parts of Asia Minor. Trade routes were re-established, and, during the period from 900 to 700 BC, Greece underwent major cultural, political and economic changes. As a result, urbanisation commenced, and, with the adoption of a Semitic alphabet, the written word re-emerged. It was during this period that the Homeric works are thought to have been recorded, including the *Hymn to Dionysus*. By the end of this period, there is little doubt that wine had assumed a religious significance in Greece, and become an integral part of Greek temporal culture. In Classical Greece, drinking practices were revolutionised and drinking wine spread to all levels of society. To meet increased demand, numerous vineyards were planted and significant advances were made in viticultural techniques.

All this was made possible because detailed information, on the cultivation of grapes and the manufacture of wine, was made available by the Greek writer Theophrastus (371–287 BC). Theophrastus was responsible for two major botanical works. In his first treatise, *Historia de Plantis* (Enquiry into Plants), he had described some of the diseases of the vine, and was observant enough to note that rust on vines occurred most often in areas exposed to dew, and in windless valleys. In his second major work, *De Causis Plantarium* (Of Plants, an Explanation), he explains that the successful growth of the vine, and the quality of the grapes, depends largely on soil conditions. He also discussed conditions for planting vines, methods of grafting, pruning techniques, *etc*. Theophrastus was born as Tyrtamos, at Eremos, on the island of Lesbos, and went to Athens to study under Aristotle, who was 12 years his senior, at the Lyceum. It was Aristotle who changed his name to Theophrastus (divine speaker) because of his oratorical capabilities. He succeeded Aristotle as head of the Lyceum, serving in that capacity for 35 years.

Prior to the era of people like Theophrastus, the pre-classical Greek poets were often wont to wax lyrical about their native wine. Nearly 3000 years ago, the poet Hesiod extolled the virtues of Greek wine made from a grape variety known as 'Biblio', and told of a heavy, sweet wine made from dried grapes. Homer paid tribute to wine so many times that the Latin poet Virgil dubbed him 'vinosus Homerus'. Virgil also remarked that it would be easier to count the grains of sand on a beach than the number of Greek wine varieties!

During the period 750–550 BC, the Greek city states were established, and the requirements of an expanding population were satisfied by colonial expansion along the Mediterranean and Black Sea coasts. Some of the areas colonised were ideally suited to viticulture, which boosted the wine-based segment of the economy. Essential 'foreign' commodities, such as grain and timber, were traded for home-produced wine and olive oil. By the time that Philip had seized power, and crowned himself king in Macedonia, in 359 BC, Greek colonies were to be found on the Iberian Peninsula, in southern France, southern Italy, northern Africa, Asia Minor, Georgia and other parts of southern Russia, all of which were to become important wine-producing areas. In some instances, vestiges of the influence of Greek colonisation, notably viticulture, may still be discerned. This is particularly the case for the wine industries of Georgia and the Crimea, and, especially, the important Russian winemaking region of Krasnodar Krai, on the Black Sea. The tradition of viniculture in this area began with the founding of the ancient Greek polis of Phanagoria *ca.* 542 BC. The Greek settlers found that the climatic and edaphic characters of the region were propitious for viticulture (the area is located in latitude 45°N, the same as Bordeaux). Phanagoria was supposedly situated on an 'island', called the Isle of Phanagor by the ancients, which is actually the Taman Peninsula, between the Black Sea and the Sea of Azov. Before the Greeks inhabited the region, the land had already been inhabited by some agrarian tribes, but, after its establishment, agriculture, viticulture and viniculture formed the economic basis of Phanagoria. Initially, Chiosan wine from the Mediterranean was imported into the polis, with bread, meat, fish and leather being exported in exchange, and the earliest finds of grape seeds (possibly suggesting winemaking) come from the 5th century BC.

By the second half of the 4th century BC, viniculture in Phanagoria can be confirmed, and the industry was important enough for images of bunches of grapes to be minted onto some coins of the day. Phanagoria remained an important city until the end of the 4th century AD, when the Huns encroached upon the Bosporus, and was later to be devastated by Muslim invaders in 1486. Viticulture recommenced under the aegis of Catherine the Great, and in the 19th century winemaking became commercialised. Like many other wine-producing areas within the old USSR, viniculture on the Taman Peninsula suffered as a result of the state measures against alcoholism in 1985, but there has been a resurgence of late, including the formation of the 'Fanagoria Wine Company' (дом вина "ФАНАГОРИЯ"), near Krasnodar Krai.

Echoes of Greek culture, emanating from their colonisation, may be found in southern Italy today, particularly in the dialect, Grecanico, spoken in parts of Apulia and Calabria. Even the name 'Italy' is thought to have originated in the Greek language. The first Greek settlers in southern Italy were so impressed with the way that their vines flourished there, that they called their new colony *Oenotria* (the wineland).

The wine trade in ancient Greece was extensive, and well organised, and wine travelled wherever ships sailed. The ancient Greeks are credited with invoking what might be described as the first system of 'appellation' designation, in order to assure the origins of their better products. What some authorities have designated the 'first golden age of wine' came to an end with the disintegration of Magna Graecia during the Peloponnesian Wars. By the time that Athens fell to the Romans in 86 BC, the Greeks had laid the foundations for advanced viticulture throughout a vast expanse of, what was to become, the Western World. The ancient Greeks can be credited with: a technical mastery of large-scale wine production; the elevation of wine to a deeply rooted cultural phenomenon; and the development of a sophisticated (wine-stimulated) system of commerce, all of which have had a profound effect on Western concepts of culture. After the end of Magna Graecia, the hegemony for the spread of viticulture and viniculture was thrust principally on the Romans, who, while lacking the aestheticism of the ancient Greeks, were sufficiently appreciative of Greek culture to adopt many facets of it, including their alphabet, some of their deities, and their reverence for wine.

In relation to the latter point, the ancient Greeks and Romans had plenty to say about the characteristics and the pleasures of wine, perhaps the most oft-quoted being '*in vino veritas*', 'the truth is in wine', which appears in Plato's *Symposium*, written *ca.* 380 BC. Such a statement indicates that, at the time of Socrates, to which the *Symposium* refers, they could not distinguish philosophy from wine. In the same era, Hippocrates wrote: 'Wine is an appropriate article for mankind, both for the healthy body and for the ailing man', and in Homer's *Odyssey*, we are treated to: 'Wine can of their wits the wise beguile, make the sage frolic, and the serious smile'.

Although the Greeks fared well, politically, under Roman rule, the state of their wine industry fell somewhat into decline until Constantinople became the seat of Roman power in her Eastern Empire. That was in the 4th century AD (324, to be precise), when Emperor Constantine I decided to build his capital on the ancient Greek colony of Byzantium. As the centre of the vast Byzantine Empire, Constantinople became one of the wealthiest and most important cities on earth, and the centre of the Eastern Christian world. The Eastern Roman Empire paid more heed to Classical culture and values, than the Rome-based Western Empire, which was overrun by barbarians during the 4th and 5th centuries. Constantinople became capital of the Roman Empire, and of a realm that would become the first Christian empire. The city became an important trading centre, and, in particular, there was a flourishing wine trade by the end of the first millennium; and the Greek population were, once again, able to capitalise on their viticultural assets. Greece, itself, suffered from Slavic incursions from the north around 650 AD, which had a negative effect on viticulture over much of the mainland, because traditional grape-growing areas were abandoned as unsafe. Greece was reclaimed by Byzantium in 1260, and viticulture gradually regained its former prominence. By the late Middle Ages, however, viticulture fell victim to feudal land management under Byzantine control, and wine quality suffered as a result. This meant that, like elsewhere in Europe, the monasteries became sanctuaries of wine production.

By the 13th century, Greece became a target of territorial expansion from the west, first by the Franks, and then by the Venetians. The main Venetian influence was to stimulate the trade in wine, and the effect of this was felt mainly in the Ionian islands, Crete and parts of the Aegean islands. Venetian markets were to be found as far north as the British Isles, where malvasia (malmsey), a highly prized strong, sweet wine which was a predecessor to the fortified wines of Iberia. Malvasia was originally produced at Monemvasiá, on the eastern coast of the Peloponessos, but was later on produced in Crete and some of the Aegean islands. The production of malvasia survived the Ottoman occupation, and trade with northern Europe, in wine, lasted until the end of the 19th century, and left an indelible influence on the palates of its inhabitants, many of who developed a penchant for sweet, strong wines. The following century saw the Ottomans invading Greece from the east, the first notable conquest being Thessaloniki (Salonika), in 1430. Over the next 200 years the invaders took control of all of Greece, the culmination being the fall of Crete in 1669. The imposition of Moslem rule in Greece had the predicted effects on viticulture, although the downturn in wine production was mostly due to over-taxation, rather than any religiously imposed ban on alcohol. Only the monasteries benefited from any tax-relief on wine.

1.7   ROMANS

According to Petronius's *Satyricon*, written in the 1st century AD, *vita vinum est*; 'wine is life', and the ancient Romans took him at his word. The Romans inherited viticulture from two possible sources: from Greek settlers in southern Italy (Calabria) and Sicily, and from the Etruscans, who came to Tuscany from the eastern Mediterranean around 800 BC. Etruscan vineyards extended into northern Italy, and their wine reached as far north as Gaul. Who was actually responsible remains a matter for debate. Pliny the Elder, favours the 'Greek connection', as can be seen from his encyclopaedic *Historia Naturalis*, which records that Eumolpus of Athens was responsible for Teaching the Romans viticulture and arboriculture. Viticulture was relatively unimportant to the Romans during the early years of the Republic, when maintaining and expanding the empire had precedence over everything else, and, as in other cultures, viticulture was associated principally with wealth. The first Roman wines were rather coarse and crudely made, and those of better quality had to be imported from Greece, and her empire. Famed for their high-quality wines were the islands of Khios, Naxos, Lesbos, Rhodes, Lemnos, Thasos and Cyprus, where the rolling hills supported highly successful vineyards. It was not until the last few centuries BC that vineyards became commonplace, and vinification techniques improved somewhat. Legend has it that wine became so common at around this time, that during the Second Punic War (218–210 BC), Hannibal and his troops bathed their horses in it! By the middle of the 2nd century BC, with the defeat of the Etruscans, the Carthaginians and Philip of Macedonia, Rome controlled the Mediterranean, and there was now sufficient wealth, and the markets, to invest in vineyards. The vine, and knowledge of viticulture were disseminated with nearly every Roman conquest. The increasing importance of viticulture and viniculture prompted some classical writers to exhaustively document the subjects. The earliest work on wine and agriculture was a 28-volume treatise, written by Mago, in Punic, during the 4th century BC. After the destruction of Carthage in 146 BC, it was translated into Latin, at the behest of the Roman senate, and became the foundation for all subsequent Roman writings on viticulture. Mago's original has long been lost. The first strictly Roman survey of the subject was provided by Cato the Censor (Marcus Porcius Cato), the second century BC statesman, who, ironically, insisted upon the demise of Carthage, and who, around 160 BC, penned *De Agri Cultura*, which also happens to be the oldest surviving prose work in Latin. In the work, Cato explains the correct way to run an estate, and gives details of the equipment and finance necessary for successful viticulture. *De Agri Cultura* was obviously an influential work, for, around two centuries later, Pliny reported that, by 154 BC, viniculture in Italy was *sans pareil*. That same year saw the cultivation of vines prohibited beyond the Alps, and, for the first two centuries BC, wine was exported to the colonies, especially Gaul, with which the wine trade was so extensive, because of its inhabitants predilection for drinking wine unmixed, and without moderation. Wine sent to Gaul was exchanged for slaves, whose labour was required to cultivate the large estate vineyards. In Cato's time, most commercially produced wine emanated from large, slave-based, villa estates. In 37 BC, Marcus Terentius Varro (116–27 BC) wrote *Res Rusticae* (country matters), a manual on farming, which, although less detailed than Cato's work, pays more attention to wine itself, rather than dealing solely with viticulture. For instance, Varro mentions that some grapes only produce wines that must be drunk within 12 months, before they become too bitter, while others, such as Falernian, yield wines that mature with age, and increase in value accordingly. From much the same era was Publius Vergilius Maro (Virgil), the great poet, and a leading naturalist of his time (70–19 BC). Best known for his epic poem, the *Aeneid*, in our context he is notable for his four books on farming, called *Georgics*, in which he transformed nature and commonplace agricultural activities, including viticulture, into verse. Many of Virgil's observations were not original, and he 'borrowed' from several previously published Greek and Roman works.

A little later, during the 1st century AD, we read that Roman viticulture was experiencing something of a boom, new varieties of vine were being raised and transplanted across the Roman Empire, and planting experiments showed that results could be improved by matching grape varieties to growing conditions. Using suitable varieties, the Romans even took viticulture as far north as Britannia. It was probably not by coincidence that this same period witnessed the publication of two major treatises, both of which had sections devoted to viticulture. The first was the twelve volume *De Re Rustica* (On Agriculture) published around AD 65 by Columella (Lucius Junius Moderatus Columella), who was born around the beginning of the first millennium AD, and hailed from Hispania Baetica (modern Cadiz). At some time, he moved to Italy and lived there most of his adult life; he died there around the 60s AD. The work covered all aspects of Roman agriculture, with Books III and IV, and half of V, being devoted to wine and the vine. Columella considers viticulture and viniculture comprehensively, including the choice of site for a vineyard, methods of planting and training, propagation and the economics of the whole operation. In particular, it was noted that the performance of two of the popular varieties of the day, *Eugenian*, and *Allobrogian*, were very much influenced by the location of the vineyard. As part of his treatment of wine, Columella provides an extensive catalogue of the 50-odd grape varieties available to Roman growers in the 1st century AD. The author was particularly interested in the production of quality wines, going into great length about the subject of ageing, and his catalogue of grape varieties is arranged according to quality, rather than size of yield: (1) varieties producing the highest quality wines; (2) those producing wines of lower quality, but which may still be aged; (3) varieties yielding large quantities of *vin ordinaire*. Columella went to extraordinary lengths to extol the virtues of the Aminnean vine, repeatedly commending the quality of wine that it produced. Many growers of the day regarded this variety as inevitably producing small yields, something rejected by Columella, who proved that it could be fruitful. Columella was convinced that the Aminnean vine was capable of giving yields to match any of the new varieties that were being introduced throughout the Roman Empire, and he provided impressive figures to back his point. Some authorities feel that the figures are too impressive, and have dismissed them accordingly. The Aminnean vine was brought to Italy by Greek colonists who first settled at Cumae, near the Bay of Naples. Columella was certainly more concerned with the quality of wines than many growers of his time, but he was also extremely aware that vineyards had to remain competitive, in terms of yields. Thus, as is the case today, Columella tried to balance quality with quantity, a feat that was to be attempted by Pliny a few years later. He also warned against investing in vineyards without carefully examining all aspects of the business beforehand, and pointed out that lack of regular maintenance of a vineyard invariably resulted in a profuse crop of substandard fruit. *De Re Rustica* is the most comprehensive surviving Latin work on the subject, and many of the grapevine-related recommendations therein are pertinent to viticulturalists today.

Around a decade (AD 77) after Columella published his masterpiece, Pliny the Elder (Gaius Plinius Secundus) completed his 37 volume *opus magnum*, *Historia Naturalis*. As it turned out, the work appeared just 2 years before his death, while he was watching the eruption at Mount Vesuvius. In Book XIV, Pliny reviews the history of wine, viticulture and vinification, and laments the production of cheap wines, which were popular at that time, and the loss of quality vintages. In all, *Historia Naturalis* contains references to 91 grape varieties, 50 kinds of premium wine, 38 foreign wines and 18 sweet wines. It should be stressed that the works of both Columella and Pliny were written against a background of ignorance and confusion, as far as the populace were concerned. While some Romans used vine-growing as a route to fame and fortune, others faced financial hardship. Growing vines at this time was not a guaranteed way to 'make a million'. For the ones who failed, it was largely because they were hampered by poor vineyards, and/or had received inadequate training in cultivating grapevines. This last point was forcibly made by Pliny, who maintained that he, personally, knew of very few skilled viticulturalists and viniculturalists. *De Re Rustica*, and *Historia Naturalis* were intended to eradicate viticultural ignorance, and to enable the grower to achieve a happy balance between quality and quantity.

In these agricultural works, particular attention was paid to methods of supporting growing grapevines. Varro commented upon the Spanish method of training vines at a low level, as compared to the Italian custom of training them on higher props. Upright poles were called *pedamenta*, while cross-pieces were known as *iuga*. In some Asian vineyards, it was noted that branches were allowed to lie upon the ground, and were raised up on forked sticks when the grape clusters began to develop. According to Varro, this was the most economical way to cultivate vines, but the harvest could be damaged by mice and foxes! Columella provided much additional detail, particularly with reference to the age of the vine. He recommends that the young vine should be trained to rest on cross-pieces, rather than hang on an upright pole, so that it is better protected against wind damage. After its third season, the vine is provided with a more robust support, which is placed one foot away from the base of the vine. The height to which the vines should be trained on supports varied between four and seven feet, depending on vineyard conditions. In the more humid locations, where winds tended to be light, vines were trained higher, whereas in poorer soils, on hot slopes, or in areas subject to storms, vines needed to be trained closer to the ground. Of note in *De Re Rustica*, was a section devoted to the *arbustum*, which is a plantation of vines supported on living trees, rather than an artificial lattice. By Pliny's time, *arbusta* seem to have disappeared, for he recognised five methods of training vines: (1) trained with the branches lying over the ground. Such vines yielded a large quantity of fruit, albeit of inferior quality. Grapes, being low-lying, were protected from winds. This method was the one favoured by the Greeks; (2) vines growing upward, but without support (*i.e.* free-standing). By pruning the lower branches, vines can be made to grow upward, making the grapes easier to harvest. Such vines are depicted in the famous Lachish[*] relief that embellished the walls of the royal palace of Sennacherib at Nineveh, dated 700 BC; (3) vines trained on an upright post; (4) trained on an upright with a cross bar; or (5) trained over bars in a rectangular formation (*Hist. Nat.* 17, 164).

---

[*]A fortified city in Judean hill country, second only to Jerusalem in regional importance. Sennacherib invaded Judea and destroyed Lachish in 701 B.C. Accounts of the Assyrian invasion are found in the Bible. Assyrian artists carved a depiction of the siege in relief to decorate Sennacherib's palace at Nineveh.

Pliny reports that, historically, the best wine had been Caecuban, which came from Latium, but this could no longer be produced, because the vineyards had become neglected, and had been grubbed out by Nero for the construction of a canal! In Pliny's time, the best wine was considered to be Falernian, grown on the slopes of Mount Falernus on the border between Latium and Campania. Next in esteem were the wines of the Alban Hills, south-east of Rome. Falernian was an exceptional first-growth wine, made from the Aminnean grape, and of which Pliny recognised three types: Caucinian, made from grapes grown on the higher slopes of the mountain; Faustian, made from grapes grown midway down (grown on the estate of Faustus, son of the dictator Sulla); and plain Falerian, made from grapes grown on the lower slopes. Because of the amount of care that went into its production, Faustian wine was the most highly regarded. Legend has it that the Falernian of 121 BC was the best ever recorded vintage. Some of the best vineyards in Italy were along the Campanian coast, around Pompeii, many wealthy Romans owning villas and vineyards there. In particular, the soils around Mt. Vesuvius were especially favourable for growing vines, and nearly every house in Pompeii had its own vineyard, the produce of which went to the city's 30-odd wineries. The eruption of Vesuvius in AD 79 caused much human misery, and destroyed some of the best vineyards in Italy, but, the smothering of Pompeii by volcanic ash, preserved a unique and vivid picture of Roman life at that time. Around 200 taverns (*taberna*) have been identified at Pompeii, with many of them still exhibiting tariffs for the wine supplied. Prices varied from one to four *asses* per *sextarius* (pint), according to wine quality (a loaf of bread, at that time, cost two *asses*), and on one wall, there is an inscription that reads: 'For one *as* you can drink wine; for two, you can drink the best; for four you can drink Falernian'. The usual tavern drink, however, was a red wine, not more than a year old, drawn from an amphora placed on the counter, and consumed from earthenware mugs. Wine was almost invariably mixed with water before being drunk, and drinking undiluted wine, called *merum*, was either viewed as a provincial and barbarian habit, or, as a drink reserved for the gods. From what we understand, Roman wines were dense, bitter and more alcoholic than found today, and, almost always, aged. The Romans usually mixed one part of wine to two parts of water, and hot water, or even seawater could be used. This contrasts somewhat with the Greeks, who also diluted their wine, but with three, or four, parts of water. According to Pliny, seawater was added to 'enliven the wine's smoothness'. Diluting wine in ancient times served two purposes: firstly, it turned it into a thirst-quenching drink that could be consumed in large quantities; and, secondly, the presence of alcohol made the water safer to drink, an important consideration in the growing cities of the Greek and Roman empires, where potable water was at a premium. Snow was sometimes added to wine in order to cool it! In the wake of the disaster at Pompeii, and probably partly as a result of the works of Columella and Pliny too, new vineyards were planted all over Italy, even fields that had traditionally been used to grow grain, were planted with vines. This inevitably led to the market becoming flooded with wine, which, in turn, led to a downturn in prices. There was even a danger of wine becoming cheaper than water! The situation became so serious, that, in AD 92, the emperor Domitian issued an edict proscribing the plantation of any new vineyards in Italy, and ordering the uprooting of at least half of the vineyards in the provinces. This measure was partly designed to protect Italy's domestic wine industry, and partly to ensure an adequate supply of grain for the population, and it was not to be rescinded until almost two centuries later.

The Romans gave the refined art of wine-tasting to the world. The sommeliers of the day were known as *haustores*, who refrained from eating spiced food before tasting, and never swallowed the wine under scrutiny. Wines were separated according to their taste, and their colour, examples being: *vinum dulce* (sweet); *vinum molle* (soft); *vinum album* (white) and *vinum sanguineum* (blood-red). Like most ancient civilisations, the Romans were obsessed with wines that had been aged for a long time. Perhaps this was to mask some of the inherent flavours resulting from their mode of production. Falernian wine, for example, was to be matured for 10 years before drinking was recommended, and was proclaimed to be excellent if left for 30 years. It was not uncommon for wine to be aged much longer than this, and an instance is recorded in the *Satyricon*, by Petronius the Arbiter (*ca.* AD 27–66). Book two of this classic contains the *Dinner of Trimalchio*, at which honied wine was drunk,

to be followed by 100-year-old Opimian Falernian, after which
Trimalchio utters the immortal words:

> Ah me! To think that wine lives longer than poor little man. Let's fill
> 'em up! There's life in wine and this is the real Opimian.

Thus, the practice of ageing was fundamental to Roman wine-
producers, although it followed a different pattern from that employed
in modern times. While today wine is bottled after a period of ageing in
cask, in olden days the young wine was sealed straightaway in amphorae, with ageing being promoted by the action of smoke and heat (said,
by some, to be a rudimentary form of pasteurisation). Even as far back
as Homeric times, the Greeks appreciated that wine improved with age,
for, as we learn from the *Odyssey*, Nestor calls for wine that had been in
a jar for 11 years when he was about to entertain Telemachus.

In the absence of distillation techniques, wine was the strongest
drink of the Romans, and Falernian, with an alcohol content of
some 15–16%, was as full-bodied as any. It was a white wine, which,
after storage for 10–20 years, assumed an amber hue. Vintage wines
could be stored for such periods of time because of the development
of amphorae, which were large (having a capacity of approximately
seven gallons), tapered, two-handled, stoppered jars, with a narrow
neck (Figure 6). The Romans also introduced large, staved, wooden
barrels (casks) in which wine could be aged.

The winemaking methodology in Greek and Roman times, consisted of grape harvest (usually September), followed by retrieval of
grape juice, which was initially effected manually, by means of
treading. Once treading was complete, the remainder of the juice
would be removed by means of a press. Juice (must) was then
fermented in large vats for around 6 months. There were, of course,
no means available for controlling, or terminating fermentations. The
young wine was then sieved through a cloth to remove any gross
material, before being placed in amphorae, for storage, and/or trans-
portation. Most scholars would agree that, by 500 BC, the Greeks
had turned winemaking into a science, causing a high demand for the
product, and that this methodology was passed on to the Romans.

The consumption of wine became a cultural institution for people of
all classes in the ancient world, and the semi-ritualistic drinking of
wine can be traced back to the Homeric times of the mid-8th century
BC, when warriors would feast and drink together. Later on, in the
Archaic period (700–480 BC) the elite would hold 'symposia' (meaning

Figure 6 *Amphorae found in debris from the destruction of Athens by the Romans in 86 BC. The jars lean against the terrace parapet of the Stoa of Attalos, and, from left to right, they are from Rhodes, Knossos, Chios and Rome. Behind them, one can see the Market Hill of Athens and the Temple of Hephaistos*
(Reproduced by kind permission of Princeton University Press)

'drinking together'), where, after dining, political discussions would ensue. Such events would often be followed by some sort of sexual activity; nothing changes! The symposium reached its zenith during the Classical period in Greece (480–323 BC), when it became a forum for intellectual dialogues by such men as Plato and Xenophon. By the mid-4th century BC, most middle-class homes in Greece had a separate room for a symposium (the 'androne'), and the pursuit survived the Hellenistic period (323–30 BC) to be passed on to the Romans.

The symposium has been said to have provided the spawning-ground for democracy in ancient Greece, being the aristocratic forum in which influential men could exchange ideas. With the exception of courtesans, and the like, women were not allowed to participate in symposia. Outside of formal occasions, such as symposia, wine was available to the Greek masses in establishments such as bars (*kapeleion*) and street kiosks. Similarly, in Rome, there were public bars (*popinae*) and taverns, which were more sophisticated. Again, it was frowned upon for civilised women to enter these institutions. The Greek (and, to some extent, the Roman) tradition of political association through drinking was not governed purely by social class, since wine consumption was coupled with group activity, be it political, intellectual, military or even sexual. Drinking wine in ancient Greek and Roman societies was clearly seen as a civilised activity.

1.7.1 Resinated Wine

Pliny the Elder (AD 23–79), the renowned Roman scientific en cyclopaedist, devoted a fair proportion of Book XIV of his *Historia naturalis* to the problem of preventing wine turning into vinegar. According to Pliny, the best way to prevent 'wine disease' was to add the resin of certain trees: pine, cedar, frankinsense, myrrh or terebinth, to the wine. Pliny also recommended the use of terebinth resin for human chest ailments, and other disorders, commenting: 'When warmed it is used as an embrocation to relieve pain in the limbs'. The thinking was that, if the resin was capable of naturally preventing wound infection in trees, it should be effective in humans, and should certainly be able to preserve wine! Similar sentiments were expressed by another 1st century AD writer, Columella, who in *De re rustica* (XII. 18ff.) wrote about a *medicamentum* made of myrrh, terebinth resin, pitch and various spices. From what we have already said, it is fairly evident that the people of Hajji Firuz must have known enough about the preservative (and medicinal?) properties of tree resins to have resinated their Neolithic wine. It should be noted, however, that, since in an upland region like Hajji Firuz, the wild grapevine can use the terebinth tree as a natural support, and that grapes and the resin are produced at roughly the same time, it is not without possibility that the mixing of wine and resin might have occurred accidentally. In the 21st century, Greece is now the only country where resinated wine, *retsina*, is still produced with any regularity. In *retsina* production, the resins normally used originate from the native Aleppo pine (*Pinus halepensis*), or the sandarac tree (*Tetraclinis articulata*), which is imported from North Africa. Pliny has the following entry for *retsina*:

> The method of flavouring wine is to sprinkle it in its raw state with pitch during the first fermentation – which takes nine days at the outside – so that the wine may acquire the smell of pitch and some hint of sharp flavour. Some authorities think that a more effective way is to use unrefined flower of resin which enlivens the smoothness of the wine.

1.8   POST-ROMAN EUROPE

Under the influence of Rome, grape-growing spread throughout much of Europe, including the Rhine valley, and into Germany, and, by AD 300, viticulture and viniculture were being practised from the Atlantic shores to the valley of the Danube.

With the fall of the western Roman Empire, the European wine trade was disrupted, and commercial viticulture and viniculture went into decline, but by this time, wine was being produced in all the Latin countries of the world, especially so in Transalpine Gaul (modern France). The early Middle Ages (AD 500–1000) saw the dissolution of many large estates, and the establishment of smaller enterprises, often in association with the monasteries, who were to become the custodians of the art of winemaking at this time. The depression of the wine trade was short-lived, for the spread of Christianity to northern Europe, with the concomitant need for sacramental wine, led to the establishment of a completely new international trade network in the commodity. As drinking wine once more became an accepted part of life, so the trade outgrew its religious associations, and, by the end of the Middle Ages, it was a firmly established custom over much of mainland Europe. Religious establishments were also responsible for a resurgence in the study of agronomy, something that was stimulated by Pier de Crescenzi's *Ruralium commodorum libri duodecim*, the fourth of the 12 books being dedicated to viticulture. Born in Bologna in 1228, de Crescenzi mentions many types of grape in his work, some of which can be related to our modern varieties, notably the white Trebbiano (Tribiana) from Tuscany, which is still with us, and the red Nubiola, from Asti, probably the ancestor of the well-known Nebbiolo, grown in Piedmont. A little later, we have *De naturali vinorum historia*, written in 1596 by Andrea Bacci (1524–1600), sometime doctor to Pope Sixtus V, and professor of botany in Rome from 1567 until his death in 1600. Bacci's fundamental work contained information about wines from Italy, Spain, France and other European countries, and was supplemented by another work, *Ditirambo di Bacco in Toscana*, by the doctor and naturalist, Francesco Redi of Arezzo (1626–1698). The book, which was, in essence, a catalogue of Tuscan wines, was notable for containing a tirade against other forms of drink, such as tea, coffee, cider and beer.

European viticulture expanded steadily from the 16th to the 19th century, despite a series of catastrophes: the Thirty Years War (1618–1648), which left much of central Europe in ruins, and decimated the vineyards of the Palatinate; the frost of 1709, which almost eradicated the most northerly vineyards of France and Germany; and the phylloxera epidemic, which hit France in 1868, and then spread through all the major vine-growing areas of Europe. The history of viticulture and wine in most of today's wine-producing countries makes fascinating reading, and it is impossible to do justice to it here. Accordingly, a couple of examples are presented here.

Germany has been a wine-producing area since Roman times, even though it is generally an unfavourable region for growing grapes. It is only on well-drained soils with favourable exposures that grapes can attain maturity, and then only in warmer seasons. As a result, it is the slopes along the rivers that provide the best areas for vine and grape growth. The origins of viticulture in Germany can be traced back to the 1st century AD, the earliest vineyards being on the left bank of the Rhine. Plantings probably spread to the Mosel around the 3rd century. The vine advanced further in the Middle Ages, mainly through the monasteries in particular, an example being in the Rheingau, where Benedictines founded an abbey, which later became the Schloss Johannisberg. Kloster Eberbach was established by Cistercians in 1135. The planting of vines reached a high point in the 15th century, when the area under vine was four times larger than it is today. This included Alsace, which was the most highly esteemed region during that period. The most important early variety was probably one called Elbling, but Silvaner, Muskat, Traminer, Spätburgunder and Trollinger are also known from early times. Riesling arrived relatively late, and is first reliably documented in the Rheingau in 1435 and in the Mosel not much later. Different varieties were generally mixed within a vineyard, rather than separately cultivated. A serious crisis developed around the 17th century, when prices fell, due to massive overproduction, and competition from beer. The Thirty Years War, which ended in 1648, also had a significant effect, and resulted in Alsace becoming a French province. In the wake of all the troubles, the industry recovered and quality improved as land unsuitable for viticulture was returned to other uses. There was a rationalisation of varieties, with Riesling replacing less useful varieties, and this was often carried out by decree from political and clerical authorities. The term 'Cabinet' was first used in 1712 by the Kloster Eberbach to indicate wines of superior quality. In 1720 the first monoculture of Riesling was planted at Schloss Johannisberg. Noble rot was discovered a little later, and Kloster Eberbach produced a successful wine from botrytised grapes in 1753. The invention of Spätlese is generally dated at 1775, when the harvest at Schloss Johannisberg was delayed by accident, resulting in a late harvest of largely rotten grapes. The wines made from these grapes became a legend.

In the 19th century, in the wake of the French occupation, most of the church's wine estates were secularised, and this resulted in much technological progress, such as the invention of the Oechsle must weight scale, which helped to improve winemaking tremendously. In many ways, German wine entered a golden age. The great estates of the Rheinpfalz and Mosel-Saar-Ruwer rose to fame, alongside the Rheingau. At the height of its prestige, Rhine wine generally sold at prices above those of first growth Bordeaux. The Mosel's first Trockenbeerenauslese was made by the Thanisch estate from the Bernkasteler Doctor vineyard in 1921, and created something of a 'Doctor cult'. Generally, however, times were not easy during the deterioration of the political and economic situation in the early 20th century, and phylloxera, as it did in many countries, added to these troubles. The worst blow to German wine since the 17th century came with the Nazis, when World War II eventually devastated Germany's wine regions, along with much of the rest of Europe. Most authorities would agree that it is along the Rhine, from where it leaves Switzerland nearly up to Bonn, and along many of the Rhine tributaries, that is most important for producing quality wines; areas such as Rheinpfalz (Palatinate), Rheinhessen and Rheingau.

The products of the Austrian wine industry are still, largely, a well-kept secret, even though wine has been consumed in Austria for over 2800 years. Evidence of this comes from the discovery of *V. vinifera* seeds in an ancient grave at Zagersdorf (near Eisenstadt), dating back to the Hallstatt period, approximately 700 BC. The Celts who settled the area *ca.* 400 BC, apparently appreciated that grapes were not just for eating! Next came the Romans, who are known to have planted vines near Vienna around 300 AD. In addition, vine cultivation became widespread in Noricum, a Roman province in the eastern Alps, between what is now Hungary, Switzerland and Italy. The wines were deemed to be of such quality that they were to be imported directly back to Rome. The area best known in Austria is Carnuntum, where many Roman ruins, including an amphitheatre, can be found. Carnuntum still produces good dry white wines today. The wines themselves were often 'improved' by the addition of honey, pepper or spices. As an indication of their value, locals were prepared to exchange a young slave for an amphora of wine!

When the Romans fled, their place was taken over by barbarian hoards sweeping across from the Hungarian Plains towards Italy, and many plantings were totally destroyed. The defeat of the Magyars in 955 opened up the Danubian provinces and Styria to whole new plantations of vines, under the auspices of representatives of Bavarian monasteries, who followed Charlemagne's armies of liberation. The first monastery was Göttweig, perched high on the cliffs opposite Krems. The monastery is still there today. This opened up the entire Wachau region for new vines and great wine production similar to the Rhein and Mossel regions of Germany. By the end of the Middle Ages over 42 monasteries possessed vineyards in and around Krems along with daughter houses along the banks of the Danube. Emperor Joseph II dissolved many of these in 1792.

As with the Wachau region, it was the monks who proved the greatest innovators in the vineyard. The Cistercian order from Burgundy's Côte d'Or founded the monastery near Güssin in 1157. They brought with them the Burundian Pinot Gris or Rülander. The locals named the grape, the Grey Monk. Pinot Gris, and this is still one of the great Austrian grapes from Burgenland. It was the Hungarian King Imre who created the reputation of the wine towns of Rust and Ödenburg (which today is Sopran, in Hungary), when in 1203 he granted the Cistercians of the Abbey of Heiligenkreuz, a strip of land bordering the northern shores of the Neusiedlersee. The monastery is still producing wine today.

By the 11th and 12th centuries, Austria was exporting wine all over Europe. Barrels of Viennese and Krems wine found their way not only to the neighbouring lands of Bohemia, Moravia and Hungary, but also to northern Germany, the Baltic States and even the British Isles. Wine production was so valuable that laws enacted in 1352 called for the cutting off of ears for grape stealers and amputation of a hand for workers' laziness during the vintage.

It was about this time that the village of Rust in Burgenland was enjoying a reputation for its wine, which would bring it fame and fortune over the next three centuries. By 1470 the wine of Rust had gained such a reputation that the King of Hungary granted the town a market, and by 1542, the wine had become such a valuable commodity that the King granted a Markenschutz (trademark) allowing barrels to be branded with the letter R. The corks of wines produced in Rust are still branded with an R today. In 1681 the town was ennobled as an Imperial Free City, but disaster came the following year when a plague of locusts ravaged the vineyards, and continued their destructive activities for the next 10 years. In 1683 there was an even more famous plague on the rampage, the Turks on their way to lay siege to Vienna. For a long period, Rust was situated in Hungary, and it was not until 1921 that it was returned to Austria.

For long periods, two wines, the Ausbruch of Rust and the Tokay of Hungary vied with each other for recognition as supreme products. They were produced by different methods, but both depend on the noble rot, *Botrytis cinerea*, for their sweetness. Unfortunately Tokay wines went through a long period of difficulty as Hungary fell under the Sultan's rule during the first Turkish siege of Vienna in 1529. But in the 18th century Tokay wines would re-emerge with as a grand a style, while the Ausbruch style almost disappeared.

Disaster hit the Austrian industry during the second half of the 19th century. In Rust, the Neusiedlersee (the large lake responsible for the mist and the noble rot) dried up between 1865 and 1871. Greed had caused many vintners to start cutting corners resulting in poorer quality. And probably worst of all for the entire country, vineyards were ravaged by *Peronospora* (downy mildew), oidium (powdery mildew) and phylloxera (first identified in Kloisterneuburg in 1872). This latter disaster laid waste to over 27,000 acres in over 120 areas of the province of Lower Austria. In Austria, as elsewhere in Europe, the solution to the problem was to graft the European vines onto American rootstock. In some places, however, particularly in the poorer areas, growers simply replaced their *V. vinifera* plants with *V. labrusca*, with the result that the local population became used to the characteristic flavours and aromas imparted by the American vines.

The Thirty Years War with Sweden, the increased consumption of beer, taxes (which were called ungeld – no money) and the

French Revolution and its aftermath, all took their toll on the Austrian wine industry. One notable victim was the great monastery at Kloisterneuburg, where Napoleon blew up the cellars containing vintages that went back hundred of years. Only the great tun (barrel) of the monastery was to survive, and this can still be seen today. History repeated itself in 1945 when Russian soldiers drank the massive monastery dry! The French invasions of 1805 and 1809 laid waste to the vineyards. The Battle of Wagram, which sealed Austria's defeat, was actually fought over one of the choicest vineyards. By the time of Waterloo, wine consumption had significantly dropped due to unavailable manpower (due to conscription into Napoleon's armies) and poor harvests. After Waterloo, Austrian winemakers found ways to improve their wines. The first wine nurseries were established in 1817. During this century the Austrian wine industry, like that in Germany, flourished with both technological advances and much improved quality. Pinot Noir and the white Grüner Veltliner cultivars were both introduced. The second quarter of the century saw the launch of the great wave of popularity for sparkling wines made in imitation of the great wines of Champagne. The south German Robert Schlumberger founded the largest sekt house in Austria, Schlumbeger in 1842. All Austrian sparkling wines are referred to as sekt.

Most people will have heard of the Austrian wine scandal of 1985, when most of the industry there was under the control of 'middlemen', some of who resorted to adding diethylene glycol to wine to sweeten it for the German market. The scam was discovered when one of them claimed a tax deduction for the antifreeze! Fortunately, no injuries were reported, but in spite of quick criminal prosecutions, the international sales of Austrian wines plummeted, and many producers went bankrupt. The scandal has resulted in a completely new set of wine laws in Austria, which are now deemed to be the toughest compliances for vintners in the world. Today's vintners are determined to lay the antifreeze jokes to rest, and take their place again among the world leaders. The quality of Austrian wine has risen dramatically over the past 20 years, and especially in the last five as Grüner Veltliner has taken on the world's top Chardonnays in major blind tasting competitions. The story of the wines of Austria has just been written by Blom.[21]

Hungary has always been an important European viticultural nation, with many thousands of acres under the vine at various times, the first vineyards being seemingly planted by the Romans. The word for 'wine' in Hungarian is *bor*, which is almost unique among European languages in that it not evolved from the Sanskrit *vena*. From the chronicles of the Arab writer Ibn Rostah, we are led to believe that the Hungarians have been familiar with wine since the 5th century AD. Wine growing was so important at one time that, in the Founding Document of the Abbey of Pannonhalma, Stephen the First (1001–1038) ranks grapes above all other crops that were to be handed over as tithe, a form of taxation in those days. Contemporary records indicate that King Béla the Third (1173–1198) distributed large quantities of wine among the Crusaders as they were passing eastwards through Hungary, and this is seen as representing the early days of wine trading in the region. Largely because of circumstances forced upon it, the Hungarian wine industry has experienced a series of 'peaks' and 'troughs' over the years. Like the indigenous population, viticulture and wine-making suffered greatly from the invasion by the Tartars in 1241, and the situation did not really recover until the dynastic accession of the House of Anjou in the mid-14th century. During the mid-16th century, most of the country was invaded by the Turks who remained there for around 150 years, turning the countryside into a veritable desert, and it was not until liberation in 1686 that viticulture and viniculture could be revived. By the 18th and 19th centuries, grapes were grown over almost all of the country, with wines such as Tokaji Aszú being much sought after by European royalty. By the early 19th century the Hungarian wine industry had reached an unprecedented size, with reported exports of 4 million barrels. The industry is now recovering from the state-enforced restrictions placed on it after World War II, and, once again, Hungarian wine is held in high esteem. The wines from the Tokaj-Hegyalgi region, in the north-east of the country were given special recognition as far back as 1770, a century before Bordeaux received similar treatment! In 2002, UNESCO designated this region as a World Heritage Site, partly on the basis of the fact that 'the Tokaji wine region represents a distinct viticultural tradition that has existed for at least a thousand years and which has survived intact up to the present'. The area covers some 6000 hectares and covers a total of 28 towns and villages, and is dominated by the extinct

Tokaj volcano. The book recently written by Rohály and Mészáros,[22] however, shows that there is more to Hungarian wine than just Bull's Blood and Tokaj.

1.9 THE NEW WORLD

It was to the Europeans that fell the responsibility of distributing the domesticated grapevine, and associated technologies, to the New World. Vines were first taken to Australia as part of the cargo of first fleet in 1788, and several small vineyards were immediately set up. Thus, viticulture and viniculture are as old as white settlement on the continent. When the thousand-odd settlers formed a penal settle-ment at Port Jackson, it was imperative that they became self-sufficient as quickly as possible. By 1803 these early settlers were reading an article in the Sydney Gazette, Australia's first newspaper, on a 'Method of preparing a piece of land for the purpose of forming a vineyard' (translated from the French).

An influential figure in the fledging Australian wine sector was viticulturist James Busby. Busby had lived near Bordeaux in France before emigrating to Australia in 1824 and running an agricultural school that specialised in viticulture. Busby considered the vine specially suited to the colony and wine a possible staple. To promote the industry he published *A Treatise on the Culture of the Vine and the Art of Making Wine*, in 1825, which was a translation of extracts from standard French texts on the subject. The book was largely scientific and impractical and had little or no impact, so in order to encourage small settlers to plant vineyards he wrote the much more practical, *A Manual of Plain Directions for Planting and Cultivating Vineyards and for Making Wine in New South Wales* (Sydney, 1830) which sold for the grand sum of 3s 6d. He also distributed over 20,000 cuttings to interested persons.

In 1830, Busby took the first cask of wine made from the school's vineyard to England where it was pronounced by the palates of the time as 'very promising'. In 1831, Busby undertook a 3-month tour of Spain and France and returned with a collection of vine cuttings and started the first source beds in Sydney's Botanic Gardens, along with duplicate plantings in Victoria and South Australia. All of today's well-known cultivars, such as Shiraz, Cabernet Sauvig-non and Muscat, originated from these stocks, and, by the 1850s, large areas of vineyard had been established in Victoria, New South Wales and South Australia.

Another important early pioneer was Gregory Blaxland (1778–1853), native of Kent and free settler. Blaxland arrived in Sydney in 1806 via Madeira and the Cape of Good Hope and bought 450 acres at Brush Farm (Eastwood) where over the next 25 years he conducted many experiments with crops, grasses and viticulture. He selected wine varieties from old gardens and vineyards that had been planted by both seed and cutting, and initiating the Madeira practice of training the young vines and the Cape practice of trenching the ground to keep the roots moist in summer. In November 1819, he prepared one of the seminal documents in the early history of the Australian wine industry entitled *A Statement on the Progress of the Culture of the Vine*, which accurately described the disease anthracnose, called by him 'disease or blight' and summarised his experiments to date.

The quickening of immigration to Australia in the 1840s led to a rapid expansion of the industry and several of today's famous vineyards and wineries in New South Wales and South Australia had their origins in this decade. Dr. Henry John Lindemann (1811–1881), ex-naval surgeon and native of Surrey, who had visited the wine regions of France and Germany, settled at Gresford on the Paterson River where he planted 'Cawarra' in 1843. The same year, in order to further encourage the industry and wean the drinker away from the 'ardent spirits' the New South Wales government legislated to enable vignerons to sell wine in smaller quantities than before and waived the requirement for the necessity of taking out a publican's licence. As a result the number of acres doubled to over 1000 between 1843 and 1850 when the production of wine exceeded 100,000 gallons for the first time.

In South Australia the first vineyards were in what are now Adelaide's inner suburbs but it was not long before they spread to other parts of the colony. John Reynell (1809–1873) planted 500 cuttings from Tasmania near the present township of Reynella in 1841, George Anstey planted 2000 cuttings from Camden Park at Highercombe in 1843, Dr Christopher Rawson Penfold (1811–1870), believing that wine was a useful medicament, planted his first vineyard at Magill at the foot of the Mount Lofty Ranges in 1844 and other Plantings were made by Dr A.C. Kelly near Morphett Vale, J.E. Peak at Clarendon and by the Jesuit Fathers at Sevenhill College near Clare. German settlers pioneered the Barossa Valley where Johann Gramp (1819–1903) planted the first vines at Jacob's Creek in 1847, Samuel Hoffman settled at Tanunda in 1848, and Samuel Smith (1812–1889) planted Yalumba in 1849. By the late 1840s South Australian wines were making their appearance at dinners given in London by promoters and friends of the new colony. Government regulations, however, were strict especially on distillers and this checked expansion of the industry a little. When one vigneron sent a case of wine to Queen Victoria and got a medal from Prince Albert in 1846, Mount Barker magistrates fined him 10 pounds for making wines without a licence!

Swiss and later German settlers contributed much to the establishment of viticulture in Victoria. Charles Joseph La Trobe, superintendent of the Port Phillip District, had spent several years in Neuchatel Canton in Switzerland and this connection led to the emigration of Swiss families who planted vineyards at Geelong and Lilydale. The first vines around Geelong were planted at Pollock's Ford by David Louis Pettavel and F. Brequet, and by John Belperroud in the Barrabool Hills in 1842; they were soon followed by 11 settlers from Neuchatel who planted the vine around their homes in the Barrabool Hills and in the late 1840s by German immigrants who settled at Germantown (Grovedale). By 1850 there were over 160 acres of vineyards in Victoria. Space prohibits documentation of the numerous other enterprises set up on the continent in what might be considered to be the 'Golden Age', 1850–1870.

For the majority of the 1800s and into the mid-1900s, the Australian wine industry primarily serviced a steadily growing domestic demand for wine, with only occasional forays into export markets. In the mid-1980s the Australian wine sector turned its gaze outwards, spurred on by changing domestic tastes that encouraged the development of high quality red and white wines. Today, wine exports are a multi-billion-dollar enterprise that continue to grow as more consumers around the world enjoy the quality and diversity of Australian wine. Much has now been written about the Australian wine industry, and Halliday,[23] Allen,[24] Iland and Gago[25] and Faith[26] can all be recommended to those with an acute interest.

In 1001, when Leif Ericsson landed on the northern coast of what we now call Newfoundland, one of the first things that he noticed was the proliferation of wild grapes. Accordingly, he named the place 'Vinland'. Coming from a beer/mead-drinking culture, he may not have realised the significance of these plants, but he thought that he knew what they were. When the Spanish, whose beverage of choice was wine, arrived in the Americas, they were surprised to find that the natives, living, as they did, in a land filled with vines, did not drink wine. Fermented drinks were made from corn and agave, but no wine. Reports of winemaking in North America appear in early Spanish, Dutch, French and English colonial records dating back to the early 16th century. As early as 1609, Jamestown colonists were doing their best to make wine from the local, indigenous grapes. These early settlers called the *lambrusca* grape the 'fox grape' because of its heavy, musky scent. America is home to more species of grapevines than any other continent in the world, but the first European settlers found it impossible to establish vineyards using European grape varieties. This was caused by a variety of vine diseases in North America to which the European varieties had never been exposed. The native grapes were tried but proved to be too high in acid, too low in sugar and had distinctive, unfamiliar tastes. Frustrated by the failure of the Jamestown settlement to produce drinkable wine, the British government of the day enacted a law that required the settlers to 'plant and maintain ten new vines each year until they have attained the art and experience of dressing a vineyard'. Faced with continuing failure, the Virginians were sent a group of French wine experts, together with vine cuttings from Europe, but the resultant vineyards failed, and for over 200 years, new pioneers to the Eastern United States tried cultivating European grape varieties with no success. For refreshment, the colonists were constrained to drink cider, rum, whiskey and imported Madeira. Failure was repeated up and down the Atlantic coast, and was abandoned as a bad job by the end of the 17th century. The failure to produce decent wine lasted up to, and through, the American War of Independence, and, in the early years of the new Republic, luminaries such as Thomas Jefferson determined to improve American winemaking. The Americans had become hard drinkers, with whiskey being their main tipple, but Jefferson saw wine as a far more 'democratic' drink. He was convinced that his home state of Virginia had a climate that was ideal for viticulture, and he set up his own vineyards there. Jefferson encouraged others, and although he ultimately failed as a vintner, his enthusiasm helped to 'kick-start' the American industry. By the early 19th century, vine breeding programmes were under way, and, in 1823, the first really successful hybrid, Catawba, emerged. Bred by one John Adlum, Catawba at last produced a 'drinkable' wine.

If viticulture in the eastern states was fraught with difficulties, the situation on the West Coast was vastly different. In northern Mexico and the Baja, Spanish colonists/missionaries had been successfully cultivating European vines for many years, and, in 1769, some Franciscans moved north, and settled in California, where they found the environment extremely hospitable for growing grapes to make wine for sacramental purposes. It had a Mediterranean-like environment, and was mostly free of the pests and diseases prevalent in the east. Within a decade, missionaries as far north as San Francisco were producing wine for church consumption. After America took California from Mexico in 1848, the year before gold was discovered, European immigrants brought winemaking techniques with them, and a commercial industry emerged. One of the earliest Californian enterprises was the still extant Buena Vista vineyard, set up by the Hungarian, Count Agoston Haraszthy. Haraszthy, born in 1812, had travelled widely in America to find a suitable base, and, in 1857, planted 25 acres of vines near the town of Somona. He dug tunnels into the limestone hillside, toured Europe to gather the vine cuttings that were to be the basis of the Californian industry, and is generally considered to be the father of modern viticulture in California (see Figure 7). During the 1870s, the Californian industry was largely centred around San Francisco Bay, and became the largest wine-producing area in the USA.

There are so many stories relating to the early days of American viticulture that it difficult to know where to start. Let us suffice with the story of just one of its architects. When John James Dufour, a Swiss immigrant fleeing Napoleon's armies, set foot in America in 1796, there was no American wine industry. He had been sent by his family to scout the best possible place to start a Swiss colony devoted to winemaking. He travelled through the Mid Atlantic states and found nothing that represented a successful vineyard. He then crossed the Appalachian Mountains, descended the Ohio River, and eventually settled near Lexington, Kentucky where he founded a vineyard funded by the sale of shares to the

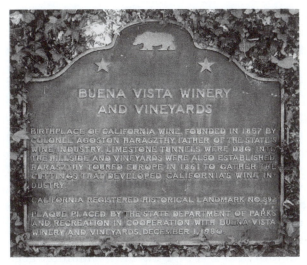

Figure 7   *Plaque in remembrance of Haraszthy*

wealthy citizens of that city. Dufour planted the vineyard in 1799 with mostly European varieties. The vines grew well for a couple of years, but soon failed, following the same path of decline that others had witnessed over the past two centuries. Despite the failure, he noticed that one variety, the Cape, seemed to do better than all the others. When the investors lost interest, Dufour sought out a new sight for the Swiss colony that was on its way from Europe. He purchased land in the newly surveyed Indiana Territory north of the Ohio River. He took cuttings of the Cape grape to plant at the new site that would later become Vevay, Indiana. The Cape grapes planted at Vevay proved to be the basis for the first successful wine production in the United States. Today we know why the Cape was successful. It was not a true European variety, but a cross of a wild native grape and a European grape, making it hardy enough to survive in North America while also exhibiting useful winemaking characteristics.

The Swiss of Vevay sold wine to merchants in Louisville, Cincinnati, Vincennes and St. Louis. For the first time, American grown wines were available to the public. The patriotic enthusiasm of the war of 1812 spurred the sale of 'Vevay' wine. New Harmony, Indiana was also a site of wine production during this time. But trouble was around the corner. Starting in 1818 the speculative bubble in land prices that had developed on the frontier burst and agricultural prices began to plummet. Soon it was possible to buy a gallon of whiskey for the same price as a bottle of wine. By the 1830s the vineyards were all but gone. The agricultural magazines of the time no longer singled out Vevay as the centre of the American wine industry. Cincinnati became the wine capital of the country from about 1840 until the end of the Civil War. Both banks of the Ohio River were lined with vineyards, earning the nickname the 'Rhineland of America.' It is in the 19th century that viticulture is trailed from the Atlantic coast to the Ohio and Missouri valleys, and then westwards to California.

When the 18th Amendment, which prohibited the production and sale of alcoholic beverages in America, was finally ratified, winemakers were not spared, even though wine was considered, by many in authority, as less of a threat. Prohibition began in 1920 and resulted in the bankruptcy of many winemakers, but a loophole in the law permitted the home production of 200 gallons per annum. Some Californian growers managed to survive, and sold grape concentrate, conveniently packaged together with yeast and implicit instructions on how not to make wine!! Other winemakers survived by making 'medicinal', or sacramental wines. The Beaulieu Vineyard of Rutherford, in the Napa Valley, survived, and even prospered because they were contracted to supply altar wine for the Archdiocese of San Francisco. After World War II, winemaking became big business in the USA. Today, wine is made in every state except Alaska and Wyoming. As is the case in Europe, some states have specialised in vinifying certain grape cultivars; Oregon, for example, specialising in Pinot Noir, and New York majoring with Riesling. The history of viticulture and viniculture in North America is the subject of a highly recommended, two-part masterpiece by Thomas Pinney.[27,28]

Spanish colonists to the Americas (New Spain) needed wine, and, in 1524, Hernán Cortés, the Governor, ordered vines from Europe to be planted in, and around, the settlement that was to become Mexico City. Because of the climate, the scheme was unsuccessful, and the Spanish Government abandoned the plan and moved further south. Here, viticulture was far more successful, and by the middle of the 16th century, thriving industries existed in Peru, Chile and Argentina. By the start of the 17th century, South American wine was being exported back to mainland Europe.

Wine and winemaking in South America has a longer history than anywhere else in the New World. In the mid-16th century, Jesuit missionaries brought viticulture to South America, and found the climate of much of Chile, in particular, extremely equable for growing grapes, even if the natives were not too keen on their presence. Rich soils, rainy winters and warm, dry summers proved to be the perfect combination for producing some of the best wine grapes in the world. It is thought that the original vines were brought from neighbouring Peru, where missionaries and colonists had not found the climate to be entirely suitable. These early varieties included 'Mission', which is undoubtedly the oldest *V. vinifera* grape cultivated in the United States and South America. Nowadays, the Mission grape has far more historical than commercial significance, but, for well over a century, certainly until the 1850s, the Mission grape defined California viticulture. The genetic heritage of Mission is still uncertain, but it shares many biological and ampelographic characteristics with the old Spanish variety 'Monica' (or 'Criolla'), of which it may be a clone. Jesuit Missionaries probably transported the original vines from Spain to Mexico in the middle-1500s, where it was cultivated for nearly a century before migrating north. It seems to have spread south during the mid-16th century. During its travels, the grape may have undergone mutation, or hybridisation (which seems likely, because of the relative hardiness of the Mission variety, compared to other *V. vinifera* types) before Missionaries brought it to Texas and New Mexico in the 1620s. More than a century later, in 1769, Franciscan monk Junipero Serra first planted the Mission variety at Mission San Diego in California. Father Serra spread vineyards northwards, as he established eight other missions before his death in 1784. Early California vineyards were so linked with the missions, that this name became attached to the grape variety.

As European immigrant wine growers developed increasing influence in California, and more vineyards were planted in cooler coastal valleys, Mission quickly lost its 'favoured variety' status, and, from the 1900s, most of what remained was in the Central Valley and in the foothills around Los Angeles where it primarily survived for making brandy and 'Angelica', a sweet dessert 'wine' created by blending brandy with unfermented juice from the Mission grape. In 1888, there were 4000 acres of Mission grapes in Napa Valley alone. At the beginning of the 21st century, the current total in the whole of California is just over 1000 acres.

Chile's vineyards flourish in the warm, fertile valleys that are positioned between the arid, rocky, mountainous desert to the north and the icy, cold, Antarctic expanse in the south. A wide selection of global wine varieties have been planted, including Cabernet Sauvignon, Merlot and Chardonnay, and, interestingly, many grapes that were believed to be Merlot have recently been shown to be Carmenère, which is a scarcely planted variety from Bordeaux. This would make sense, because, during the second half of the 19th century, the French replaced the Spaniards in the winemaking business in Chile, and brought some excellent new grapevine stock with them. Fortunately, phylloxera, that wiped out grapevines throughout most of the world, particularly in Europe, in the late 19th century (see Chapter 10), did not affect Chile, and much of the original French grapevine stock is still growing today. In fact, Chilean grapevines were used later to replenish many French after the ravages. The story of Chilean wine has been written recently by Alvarado.[29]

Argentina is one of the world's largest wine-producing countries, and grapes were introduced from Chile during the mid-16th century. There are claims that it is the world's 4th largest wine producer, and certainly the largest in the Americas. The Climate of Argentina is similar to the Andes and it is this similarity that supports the Argentinean wine industry. Argentina is mainly an arid landscape that profits from irrigating waters off the mountains. Argentina's warmer inland region encourages vine growth down the entire length of the country. In the north, the vineyards lie at the same latitude as Morocco; and in the south, vineyards share latitude with New Zealand. One of the vital aspects to growing quality wine grapes here is altitude, with vineyards planted at 2000 and 3000 feet to take advantage of the cooler temperatures. Argentine wines, which have improved tremendously over the last few decades, are made from popular traditional grape cultivars, such as Chardonnay, Merlot, Cabernet Sauvignon as well as lesser known varieties like Tempranillo, Bonarda, Barbera, Torrontés and Malbec. Young[30] has provided a thorough appraisal of the Argentine wine industry.

Uruguay, where wineries are mainly run by 3rd or 4th generation European immigrants, is now an emerging force in the international wine trade, and there are small volumes of wine produced in Bolivia, Brazil, Paraguay, Peru and Venezuela. The reader interested in South American wines generally should consult two recent works on the subject.[31,32]

Most of the vast expanse of Canada is too cold for growing grapes, but wine is produced on a large scale on the Niagara Peninsula in southern Ontario, and the Okanagan Valley in southern British Columbia. Other, smaller wine-producing areas include the shores of Lake Erie in Ontario, and the southern Fraser River and southern Vancouver Island in British Columbia. There are also a few wineries in southern Quebec and Nova Scotia. It is now pretty evident that the 'grapes' that Leif Erikson's party discovered in part of Newfoundland were, almost certainly, blueberries or cranberries. Some 500 years later, in 1535, Jacques Cartier found the real thing on an island in the Saint Lawrence River, and he named the island Île de Bacchus (known today as Île d'Orléans). Early settlers imported wine from Europe, an expensive business, although we know that the Catholic clergy made sacramental wine from indigenous, wild grapes. A wine industry, as such, did not emerge in Canada until the arrival of Italian and German immigrants during the 19th century. Like the USA, Canada went through the tribulations of Prohibition, which did not end until 1930 (except for Prince Edward Island, where the laws were in force until 1948!).

Until the 1970s, most Canadian wine was made from North American grape varieties like Catawba and Concord. Wine producers believed that nothing else would grow there successfully. However, the turning point in Canadian winemaking came with the widespread planting of some of the classic European grape (*V. vinifera*) varieties, such as Chardonnay, Riesling, Cabernet Franc and Merlot. These varieties, used in many quality wines throughout the world, became the basis of the modern Canadian wine industry. Over the last few decades, the number of Canadian wineries has risen dramatically, from about 20 in 1970 to around 220 in 2004.

'Ice wine', a sweet, dessert wine made from frozen grapes, is now an icon of the Canadian wine industry, since Canada has the only climate in the world where ice wine can be made every year. Canada is now the world's leading producer of the drink. Grapes (Riesling and Vidal are the best varieties, although others are used) are left on the vine well into the winter months, and the 'ice wine harvest' is done entirely by hand, and commences once the vineyard temperature drops below -10 to $-13°C$, and the grapes have naturally frozen on the vines. As the frozen grapes are pressed under conditions of extreme cold, the water content of the juice remains within the grape tissue in the form of ice crystals, and only a small fraction of highly concentrated juice is expressed. The freezing and thawing of the grapes dehydrates the fruit, and concentrates the sugars, acids, and other compounds in the berries, thereby intensifying the flavours and adding complexity to the resultant wine. The juice obtained from ice wine grapes is about one-fifth the amount one would normally get if one pressed unfrozen grapes. This juice is then fermented very slowly for several months, until it stops quite naturally. Genuine ice wine must be naturally produced, no artificial freezing is allowed, and winemakers must comply with VQA (Vintners Quality Alliance) rules to ensure this. For several years European countries banned Canadian ice wine from their markets, stating high sugar content as their justification. The ban was lifted in May 2001.

The discovery of ice wine was apparently accidental, and arose when a producer in Franconia, Germany made virtue of necessity by pressing juice from frozen, apparently ruined, grapes in 1794. Legend has it that the owner of the vineyard was out of town when he should have been at home harvesting his grapes. Upon returning from his protracted break, he decided to pick and process the frost-damaged crop. He was amazed by the high sugar concentration of the juice, and the quality of the wine produced, which he called 'winter wine'. It was not until the mid-19th century that 'eiswein' was intentionally made by the winery, and ice wine remained 'Germany's secret' until 1962, when it was produced commercially in other parts of Europe. German eiswein is a *Qualitätswein* and falls under those rules.

In South Africa, the Dutch East India Company set up a base on the Cape of Good Hope in 1652. Three years later, the first governor, Jan van Riebeeck, planted a vineyard in his garden, with vines imported from France, Spain and Germany. In his diary entry for 2nd February 1659, van Riebeeck enthuses: 'Today, praise be to God, wine was made for the first time from Cape grapes'. Most of these early Dutch settlers had no wine tradition, and it was only after French Huguenots settled in the Cape (mainly in the Franschhoek Valley) during the 1680s, that winemaking began to prosper. There was, however, a general reluctance to plant vines on a large scale. All this changed when Simon van der Stel, an enthusiastic viticulturalist, became leader of the colony in 1679. Besides founding the town of Stellenbosch, van der Stel had time to plant over 100,000 vines on his estate at Constantia, and to produce very respectable wine. From the mid-18th century onwards, the dessert wines of Constantia were exported to Europe, and received great acclaim. The excellence of Constantia wines continues to this day.

During the 18th century, there was a resistance to Cape wines from European, and other markets, largely because of their indifferent quality. A major problem for the South Africans was a shortage of suitable oak for cooperage in which to age. The industry prospered during the first half of the 19th century, when the British occupation of the Cape, plus British 'differences' with France, created entry into massive new markets. By 1861, Britain had overcome her difficulties with the French, and winemaking in South Africa all but collapsed. To make matters worse, phylloxera struck around 1886, and many vineyards were decimated. Additional problems ensued from the Anglo-Boer War, and industry remained moribund until Charles Kohler formed an umbrella organisation, essentially a co-operative, in 1918. This was the Koöperatieve Wijnbouwers Vereniging van Zuid-Afrika Beperkt (KWV), and it was to bring stability to the industry, laying the foundation for today's success. One of many landmarks during the 20th century occurred in 1924, when Professor Abraham Perold cross-pollinated Pinot Noir with the more humble variety Hermitage (Cinsault). The result was South Africa's very own cultivar, Pinotage. Perold, a chemist by training, became the first professor of viticulture in the University of Stellenbosch. *Spirit of the Vine*, a comprehensive study of wine in South Africa, was published[33] in 1968 to celebrate the 50th anniversary of KWV, and the editor, D.J. Opperman, has drawn together contributions from a number of leading lights of the day. More recently, James Seely has written an excellent book on South African wine.[34]

Although winemakers, at present, do not want to be associated with gene technology, the day is not far off when we shall be drinking wines made with genetically manipulated organisms. Vines are difficult to engineer, and the biological processes that control grape quality are poorly understood (see Chapter 2). But, as outlined in Chapter 3, yeast is a different matter. The genes that control the metabolic pathways that produce the chemicals that contribute to the organoleptic quality of a wine have proved obligingly easy to manipulate. Much of the pioneering work in this field has been conducted by Professor Florian Bauer's team at the Institute of Wine Biotechnology, which was founded in 1995 at the University of Stellenbosch, the first such institution of its kind.

REFERENCES

1. V.I. Zapriagaeva, *Wild Growing Fruits in Tadzikistan* (in Russian with English summary), Nauka, Moscow, 1964.
2. R.S. Jackson, *Wine Science: Principles and Applications*, Academic Press, London, 2000.
3. A.C. Renfrew, *Sci. Am.*, October 1989, 106–114.
4. A.C. Stevenson, *J. Biogeogr.*, 1985, **12**, 293.
5. H.J. de Blij, *Wine: A Geographic Appreciation*, Rowman & Allanheld, Totowa, NJ, 1983.
6. P.E. McGovern, D.L. Glusker, L.J. Exner and M.M. Voigt, *Nature*, 1996, **381**, 480.
7. P.E. McGovern, *Archaeology*, 1998, **51**(4), 32.
8. V.R. Badler, P.E. McGovern and R.H. Michel, *MASCA Res. Papers Sci. Archaeol.*, 1990, **7**, 25.
9. T.G.H. James, The early history of wine in ancient Egypt, in P.E. McGovern *et al.* (eds), *The Origins and Ancient History of Wine*, Gordon & Breach, Amsterdam, 1996.
10. D. Zohary and P. Spiegel-Roy, *Science*, 1975, **187**, 319.
11. D. Zohary and M. Hopf, *Domestication of Plants in the Old World: The Origin and Spread of Cultivated Plants in West Asia, Europe and the Nile Valley*, 3rd edn Clarendon Press, Oxford, 2000.
12. P.E. McGovern, U. Hartung, V. Badler, D.L. Glusker and L.J. Exner, *Expedition*, 1997, **39**, 3.
13. P. Kaplony, *Die Inschriften der ägyptischen Frühzeit*, 3 vols, O. Harrassowitz, Wiesbaden, 1963–1964.
14. P.E. Newberry, *Beni Hasan I*, Egyptian Exploration Fund, London, 1893.
15. M.A. Powell, Wine and the vine in ancient Mesopotamia: The Cuneiforme evidence, in P.E. McGovern *et al.* (eds), *The Origins and Ancient History of Wine*, Gordon & Breach, Amsterdam, 1996.

16. J. Janssen, *Commodity Prices from the Ramesside Period: An Economic Study of the Village Necropolis Workmen at Thebes*, E.J. Brill, Leiden, 1975.
17. R. Palmer, Background to wine as an agricultural commodity, in R. Palmer (ed), *Wine in The Mycenean Palace Economy*, University of Liège, Aegaeum 10, Liège, 1994.
18. M.A. Murray, Viticulture and wine production, in P.T. Nicholson and I. Shaw (eds), *Ancient Egyptian Materials and Technology*, Cambridge University Press, Cambridge, 2000.
19. P. Warren, *Myrtos: An Early Bronze Age Settlement in Crete*, British School of Archaeology, supplement Vol. 7, Thames and Hudson, London, 1972.
20. R. Palmer, Wine and viticulture in the linear A and B texts of the Bronze Age Aegean, in P.E. McGovern *et al.* (eds), *The Origins and Ancient History of Wine*, Gordon & Breach, Amsterdam, 1996.
21. P. Blom, *The Wines of Austria*, Mitchell Beazley, London, 2006.
22. G. Rohály and G. Mészáros, *Terra Benedicta*, AKÓ Publishing, Budapest, 2004.
23. J. Halliday, *Wine Atlas of Australia and New Zealand*, Harper Collins, Sydney, 1998.
24. M. Allen, *Crush. The New Australian Wine Book*, Mitchell Beazley, London, 2000.
25. P. Iland and P. Gago, *Australian Wine: Styles and Tastes*, Patrick Iland Wine Publications, Adelaide, 2002.
26. N. Faith, *Australia's Liquid Gold*, Mitchell Beazley, London, 2003.
27. T. Pinney, *A History of Wine in America: From the Beginnings to Prohibition*, University of California Press, Berkeley, CA, 1989.
28. T. Pinney, *A History of Wine in America: From Prohibition to Present*, University of California Press, Berkeley, CA, 2005.
29. R. Alvarado, *Chilean Wine – The Heritage*, Wine Appreciation Guild, San Francisco, 2005.
30. A. Young, *Wine Routes of Argentina*, International Wine Academy, San Francisco, 1998.
31. M. Waldin, *Wines of South America*, Mitchell Beazley, London, 2003.

32. C. Fielden, *Wines of Argentine, Chile and Latin America,* Mitchell Beazley, London, 2004.
33. D.J. Opperman (ed), *Spirit of the Vine, Republic of South Africa,* Human & Rousseau, Cape Town, 1968.
34. J. Seely, *The Wines of South Africa,* Faber & Faber, London, 1997.

CHAPTER 2

The Vine

2.1 THE ORIGIN, TAXONOMY AND BIOGEOGRAPHY OF THE GRAPEVINE

Like many plants that have been established in cultivation for a prolonged period of time, the grapevine is difficult to accommodate easily into any orthodox botanical taxonomy. What is generally accepted is that grape-bearing vines for making wine belong to the genus *Vitis*, a member of the family Vitaceae. Several other names have been proposed for the grapevine family, including Ampelideae and Ampelidaceae, but the accepted name under the International Code for Botanical Nomenclature is Vitaceae. The Vitaceae comprises some 1000 species in 12–14 genera (depending on taxonomy), and most of these are tropical or sub-tropical climbing plants. Members of the genus *Vitis*, however, are predominantly distributed in temperate zones, occurring extensively only in the northern hemisphere. Other temperate members of the Vitaceae are to be found in the genus *Parthenocissus*, where Virginia creeper (*P. quinquefolia*) and Boston ivy (*P. tricuspidata*) will be familiar to many readers. Systematics within the genus *Vitis* has created considerable controversy over the last century or so, and there are many more names in the literature than there are accepted species! Most modern taxonomic studies would classify the modern domesticated vine as *Vitis vinifera* L. subsp. *vinifera*, with the wild Eurasian plant being *V. vinifera* subsp. *sylvestris* (C.C. Gmelin) Berger. Thus, the two plants are now regarded as sub-species rather than species in their own right, as some past schemes have proposed (*e.g. V. sylvestris* C.C. Gmelin). Some works split *V. vinifera* L. into two sections (varieties) rather than sub-species. The status of the sub-species (variety) *sylvestris*, however, is somewhat controversial, not least because some authorities maintain that there is no difference between *sylvestris* and *vinifera*. Negrul,[1] for example, who travelled widely and examined thousands of samples, concluded that the *Vitis sylvestris* of Gmelin is taxonomically equivalent to the cultivated forms of *V. vinifera*. In many regions, the distinction between the wild and the cultivated forms of *V. vinifera* is often obscured by a continuum of intermediate weedy forms, some of which are the result of hybridisation.

One of the first classifications of the genus *Vitis* was by the French ampelographer Planchon,[2] who regarded American and Asiatic species as sufficiently different and that they warranted placement in separate series. The main problem with Planchon's scheme, and with some subsequent schemes, is that only those species with practical uses in propagation are dealt with. In particular, species found to be useful as germplasm in the 'war' against phylloxera are given priority status, while other species are more or less ignored. The result is a somewhat blinkered and unbalanced view of the genus. Another factor contributing to the difficulty sometimes experienced in delimiting wild *Vitis* species is the fact that some of the geographical and ecological barriers, which would normally ensure that different species were reproductively isolated from each other, are incomplete and permit hybridisation. Such a phenomenon, which occurs in some parts of North America and eastern Asia, causes a great deal of confusion, with one species merging into another. This would not occur, of course, where two species were reproductively isolated from each other, by either incompatibility or hybrid sterility.

Fossil evidence suggests that archaic vines can be traced back to Tertiary sediments some 60 million years old, and that by the beginning of the Pleistocene (*ca.* 2 million years BP) two sub-genera had evolved, *Euvitis* and *Muscadiniae*, both of which were distributed over the mid-latitude regions of North America and Eurasia. Glaciation decimated muscadine grapes and eliminated them from all but one area extending around the Gulf of Mexico into southeastern USA, and as a result there are now only three extant species assigned to this sub-genus. In the above-mentioned region of the Americas, one muscadine species, *V. rotundifolia*, has been commonly employed to make sweet wines, locally known as 'scuppernong'. All other remaining species of *Vitis* are assigned to the section *Euvitis*. As we shall see, the two sub-genera are sufficiently distinct and that some authorities have elevated muscadine grapes to genus level. Biogeographically, wild *Euvitis* species in North America were concentrated to east of the Rocky Mountains, and of the thirty-plus species identified, *V. lambrusca* is by far the most important oenologically – its genes being traceable to most of the major grape cultivars in the USA. The current distribution of species in the *Euvitis* sub-genus includes North America, Central America, northern South America (the Andean chain in Colombia and Venezuela), Europe and Asia, whereas *Muscadiniae*, as indicated above, are restricted to northeastern Mexico and the southeastern USA. Across Eurasia, wherever wine drinking is a part of tradition, one species of *Euvitis*, *Vitis vinifera* L., dominates, suggesting that its superiority for making wine must have caused it to replace other competing species. This is a reasonable assumption because from central Asia eastwards to Japan and Southeast Asia – where drinking wine made from grapes is not ingrained into their culture and therefore the attributes of *V. vinifera* are not put at a premium – a number of different species can still be found. So, what are the oenologically important characters of *V. vinifera*? Firstly, the fruit achieves a fine balance between fermentable sugar content and acidity, and secondly, with human assistance, the plant has evolved into a vigorous, self-reproducing hermaphrodite, capable of producing a virtually limitless number of stable cultivars with subtle variations in taste, colour and aroma. Apart from being the result of some of its biological characteristics, the dominance of *V. vinifera* in Europe and western Asia may well be partly attributable to the physiography of parts of that area, which could have played an important role during past glacial and interglacial periods. It is fairly evident that the genus *Vitis* appears to have established its present geographical ranges by the end of the last major Quaternary glacial period (*ca.* 8000 BC), and it is thought that repeated glacial advances and retreats during the Quaternary, and the alignment of the major mountain ranges, dramatically affected the evolution and distribution of *Vitis* species, particularly *V. vinifera*. In the Americas, for example, the mountain ranges are to be found running predominantly north–south, whereas in Europe and western Asia, they are mostly aligned east–west.

Such arrangements would have allowed the displacement of North American and eastern Chinese taxa southwards in advance of the expanding icesheets, but prevented the same happening to species in Europe and western Asia. In the latter areas, upland masses such as the Himalayas, Caucasus, Alps and Pyrenees, would have provided effective barriers to any southward movement of *Vitis* in the face of advancing ice. Thus, we have the situation where there are around 30 species of *Vitis* in North and Central America, over 30 species in China, but only one, *V. vinifera*, in Eurasia. It is a sobering thought that the one surviving Eurasian taxon has, during the last 500 years, been broadcast all over the temperate regions of the planet by *Homo sapiens*, and is now responsible for almost all of the world's wine! The present distribution of the wild Eurasian grapevine is almost certainly a fraction of what it would have been some 50 million years ago, when there was a warmer climate. This ancient, more widespread distribution would have appertained until the advent of the most recent Quaternary Ice Age, which can be dated to some 2.5 million years ago. More latterly in the geological time scale, man has carried *V. vinifera* outside of its natural range, and the plant can now be found growing in the tropics as well as in its favoured mid-latitude haunts. Other *Vitis* species were undoubtedly present in Eurasia immediately prior to the most recent ice age, but only *V. vinifera* actually survived the glacial advances – thanks to the fact that it found shelter in certain areas that provided a suitable micro-climate (known as refugia). Evidence suggests that pockets of *V. vinifera*, apart from being able to merely survive the glacial advances, were vigorous enough to be able to expand the range of the plant (*i.e.* re-colonise) during interglacial times. In Europe, refugia occurred around the Mediterranean basin and south of the Black and Caspian seas, the areas that played an important role in the overall evolution of the grapevine. In some areas, the vine started to expand its range immediately after the ice retreated. This is certainly true in some regions of southern France, where, after displacement, *V. vinifera* was again apparently flourishing by 8000 BC, from which time onwards the climate in Europe started to warm up gradually. Partial domestication of the vine (as indicated by seed remains having properties intermediate between wild and domesticated forms) seems to have occurred around 5000–6000 BC in Europe. This is several thousand years before the agricultural package is

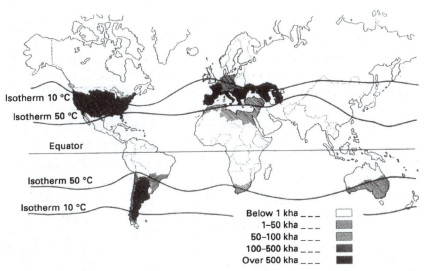

Isotherm 10 °C

Isotherm 50 °C

Equator

Isotherm 50 °C

Isotherm 10 °C

Below 1 kha _ _ _
1–50 kha _ _ _
50–100 kha _ _ _
100–500 kha _ _ _
Over 500 kha _ _ _

Figure 1   *The major worldwide plantings of* V. vinifera *(after Amerine and Joslyn[3])*
(Reproduced by kind permission of The Regents of the University of
California)

thought to have spread into western Europe. The distribution of
the main wine grape plantings in the world is shown in Figure 1.

One of the more intriguing contributions to *Vitis* taxonomy was
put forward by the aforementioned Negrul,[4] who divided grapevine cultivars into a series of 'proles' (probably best translated
as 'ecotypes'), according to their supposed origin. Negrul, whose
work was important in sorting out the ancestry of the Eurasian
grapevine, was a pupil of the great Russian botanist Nikolay
Ivanovich Vavilov (1887–1943), who conceived the theory that
the centre of origin of a cultivated plant would be found in the
region in which the wild relatives of that plant showed maximum
genetic diversity. Vavilov proposed 13 world centres of plant
origin, and became internationally renowned as a plant population
geneticist, before falling foul of Lysenko. Like many free-thinking
Russians of that era, he died in a concentration camp. It was
Vavilov who was the first to claim that the earliest 'wine culture' in
the world emerged in Transcaucasia, comprising modern Georgia,
Armenia and Azerbaijan. This region, stretching between the Black
Sea and the Caspian Sea, is dominated by the high, snow-covered
mountains of the Greater Caucasus, a natural barrier between
Europe and Asia. Much of Transcaucasia is a veritable treasure

trove of wild fruit and nut species, including pomegranate, plum and hazelnut, and these together with an amenable climate would have been highly conducive to early human occupation. The wild Eurasian grape still grows there in great profusion. Negrul had the opportunity to travel widely throughout Europe and Central Asia, and collected and described numerous varieties of *V. vinifera*. He subdivided the varieties according to geographic area, morphological characters and physiological reactions, and noted that local varieties possess many recessive mutant traits, such as smooth leaves, large branched grape clusters and medium-sized juicy fruit. He observed that a greater number of recessive traits were exhibited by the most ancient cultivars. Using classical methods of botanical taxonomy, Negrul proposed three distinct groups of grapevine cultivar: *Proles occidentalis*, *Proles orientalis* and *Proles pontica*. The oldest of these cultivars (having the greatest number of recessive traits), *P. orientalis*, arose close to the proposed centre of origin of viticulture in Transcaucasia, and then spread to the 'Fertile Crescent', Egypt and Balkans. Varieties showing fewer recessive traits, and found from Georgia to Balkans, were designated *P. pontica*, while cultivars found along the northern Mediterranean and in Central Europe (*P. occidentalis*) were considered to be of most recent origin, and hence possessed the least recessive characters. Grapevines that were introduced into the Middle East during the period 3000–2000 BC were probably *P. pontica*. The main areas of distribution within these three groups are *P. occidentalis* in France, Spain, Portugal and Germany; *P. orientalis* in Central Asia, Iran, Afghanistan, Armenia and Azerbaijan; and *P. pontica* in Asia Minor, Greece, Bulgaria, Romania, Hungary and Georgia. Along with these fairly well-marked geographical differences, each group has variations in leaf pubescence (hairiness), fruit cluster size, grape size, grape acidity and sugar content, seed size and tolerance to cold. *P. occidentalis* vines, for example, are more tolerant of cold climates than those in *P. orientalis*, and they also bear smaller grapes with lower sugar content and greater acidity. Vines placed in *P. pontica* have larger grapes containing fewer, smaller seeds than *P. occidentalis* and, being of low acidity and moderate sweetness, are ideal for vinification. Vines from *P. orientalis* generally bear the largest fruits, particularly the oval-berried table grapes. Some authorities regard *P. pontica* as being intermediate between *P. orientalis* and *P. occidentalis*. Red and white grape-bearing vines are about equally represented in all three proles. Several sub-classes (sub-proles) of the three main groups of cultivar were defined by Negrul as a result of extensive field investigations into natural populations of *V. vinifera*. Accordingly, he recognised two distinct 'sub-proles' within *P. pontica*, these being *georgica* and *balkanica*. He also concluded that cultivars from the region of the Caspian Sea were so different from the 'normal' *P. pontica*, that they must have arisen from a different wild form (he called it *V. sylvestris* var. *aberrans*, as opposed to *V. sylvestris* var. *typica*). These cultivars included the vines that were used for vinification before the advent of Islam, and were accorded the name *P. orientalis sub-proles caspica*! Muscat grapes belong to this group. During the period of Islamic influence in the Mediterranean (from AD 500–1100), some table grape cultivars arose from *sub-proles caspica*, which Negrul designated as *sub-proles antasiatica*. The best known of these is the Thompson Seedless (Sultanina) grape.

The earliest cultivars in western Europe were introduced by the Phoenicians and/or the Greeks, and the grapevines concerned almost certainly belonged to Negrul's *P. pontica* or *P. orientalis sub-proles caspica*. Such cultivars struggled to adapt to the cooler climates of northern Europe, and so cultivars with more tolerance to the cooler conditions were selected from native populations. The latter represent Negrul's *P. occidentalis*, to which most of the highly celebrated varieties of Germany and France belong. Most of the above is attributable to a translation and summary of Negrul's classification, which has been provided by Levadoux.[5]

2.1.1 Prehistoric Evidence for *Vitis*

For evidence of *Vitis* in prehistoric times, we are reliant on fossil seed morphology and palynology. Neither of the markers is perfect, because seed samples are apt to exhibit some degree of morphological variation, while *Vitis* pollen differs in size according to whether it is fertile (produced by male and bisexual flowers) or sterile (produced by female flowers). With caution, therefore, it is possible to indicate the relative frequency of dioecious (wild) and monoecious (cultivated) vines at prehistoric sites, and tentatively suggest when domestication might have arisen. Based on seed morphology, two main groups of fossilised grape have been recognised from Tertiary deposits in northern Europe: the *V. ludwigii*-type and the *V. teutonica*-type. Seeds of the latter, which resemble those of *Euvitis* grapes, have been found in samples dating back to the Eocene (40–55 Million Years BP), while seeds of the *V. ludwigii*-type, which are more akin to those of muscadine grapes, have been found in Europe since the Pliocene period (2–10 million years BP). It should be stressed that such information is based on a relatively small number of specimens but, even allowing for that fact, evidence strongly indicates that the genus *Vitis* was widely distributed in the northern hemisphere by the end of the Tertiary period. Some treatises have made use of the prints of 'grapevine leaves' found in Tertiary rocks as evidence for prehistoric vines. A few of these fossil prints have been assigned to archaic genera, such as *Cissites* and *Paleovitis*, while others have been attributed to fossil species of *Vitis*, such as *V. sezannensis*. Modern opinion regards many of these fossilised leaf prints as being doubtfully from members of the family Vitaceae, and therefore of little use as taxonomic markers, and that only fossilised prints of seeds are acceptable as evidence of the prehistoric presence of the grape family.

The presence of *V. ludwigii*-type seeds in Tertiary sediments in northern Europe insinuates that muscadine grapes may well have been distributed throughout some of the northern hemisphere prior to the ice ages, but that they became extinct there sometime during the Quaternary period. It has been proposed that muscadine vines and *Vitis* vines may have diverged sometime during the Tertiary period, and that the former can be regarded as intermediate between *Vitis* vines, which favour temperate conditions, and *Ampelocissus* species, which prefer tropical climes.

Presently, most grape fossil remains come from Europe, which probably reflects the availability of suitable sedimentary deposits, and/or the distribution of palaeobotanical interest, rather than the actual ancient distribution of *Vitis* spp.

The grapevine (*V. vinifera*) can be considered as one of the classical fruits of the Old World, together with the olive, the fig and the date palm. All of these fruit tree species were an integral part of the cohort of plants around which agriculture and horticulture were developed in the Near East and the Mediterranean basin. It is of interest to note that the domestication of all of these species was accompanied by a change from sexual reproduction by seed, which yields variable progeny, to the vegetative propagation of cuttings, which are identical clones of their parent plants. By adopting the latter mode of reproduction, desirable vine traits could be selected for and be proliferated. In this respect, it should be noted that in the wild vine, red is the genetically dominant colour in the grape until a white mutant arises. This albino mutant then has to be selected and maintained by propagation in order to cultivate vines with white grapes. There is a major difference in the rate of new species innovation in open-field agriculture (*e.g.* cereal growing) and horticulture (*e.g.* viticulture), and this can be attributed to the time it takes to introduce new cultivars. In cereal crops, for instance, new varieties may be developed and introduced within a relatively short time span (say, 10 years), simply because the innovation (*e.g.* disease resistance) is genotypic. Cultivars of fruit crops on the other hand change very slowly, if at all, and any changes are usually attributable to horticultural practice rather than genetics. As a consequence, most vineyards are planted with traditional grapevines that have been perpetuated for centuries by vegetative propagation. Commercially, grapevines are rarely asked to perform for more than 40 years, even though carefully tended samples can bear fruit for a very long time. The most renowned, long-lived example is the Great Vine at Hampton Court Palace, London, which was planted in 1769. Since the early Bronze Age, grapes have provided staple food and drink in these areas, the plants with large, sweet, juicy fruits being used as table grapes, while those with smaller, juicy, but more acidic fruits were selected for wine-making. Vines yielding small, seedless berries were favoured for raisin production. Irrespective of its exact origins, the grapevine can be considered as one of the oldest cultivated plants for which there are hardly any extant ancestors. This is tempered by the fact that many wild vines are situated in areas of the world in which their very existence is being compromised by human activity and, as a consequence, wild forms are disappearing from the planet at a disturbing rate. Mankind will have to act quickly if he wishes to preserve these wild populations, which carry the genes that have contributed immensely to our modern cultivated vines.

It has been suggested that the ancestral forms of *Vitis* were sun-loving plants, and were bushy rather than climbing in habit. It is then proposed that when forests expanded during the Eocene period, these plants were forced to aim for a higher stratum in the forest canopy in order to satisfy their requirement for sunlight. Thus, they adopted a climbing habit by the differentiation of some of their leaves, or floral clusters, into tendrils. It is interesting to note that, as long as other factors are favourable, the modern wild vine of Eurasia inhabits only regions with relatively intact woodland.

The vines of Europe are certainly derived, at least in part, by a selection from sub-species *sylvestris*, which is native in a large part of central and southeastern Europe. Confusion of the native sub-species with naturalised plants of sub-species *vinifera* has made difficult the precise determination of the original geographical limits, and some wild vines are probably recent hybrids between the two sub-species (Some authors, moreover, have suggested that other species of *Vitis* from the eastern Mediterranean region, now extinct in the wild state, have contributed to the cultivated vine, but the evidence for this theory is slender). The situation has been further complicated in the 19th century by the introduction to Europe of many species of *Vitis* from America. These are more or less resistant to attacks of 'phylloxera', a parasitic aphid that has been damaging European vines from 1867 onwards (see Chapter 10). American vines are now used almost exclusively as stocks – the scions grafted on these are either cultivars of *V. vinifera* subsp. *vinifera*, or hybrids between it and American species, or of purely American species or hybrids. These American vines are locally naturalised, especially around neglected or abandoned vineyards. The species planted on a large scale in Europe include *V. aestivalis*, *V. berlandieri*, *V. cordifolia*, *V. labrusca*, *V. rotundifolia*, *V. rupestris* and *V. vulpina*.

2.1.2   Differences between Muscadine Grapes and *Euvitis* spp.

One of the most obvious morphological differences between the two sub-genera is that the woody stems of muscadine grapes have an entire, non-shredding bark covered with prominent perforations (lenticels), whereas *Euvitis* grapevines have a bark that shreds profusely and inconspicuous lenticels. Internally, the pith of *Euvitis* stems is interrupted at intervals (the nodes) by a diaphragm of woody tissue, something that does not occur in the stems of muscadine grapes. Other easily visible morphological differences are the branched tendrils of *Euvitis* vines (tendrils of muscadine vines are unbranched), which also have more elongated flower clusters, and fruits that adhere to the stalk at maturity. Muscadine vines bear small floral clusters, and their berries are liberated individually from the cluster as soon as they reach maturity. There are small differences in seed morphology between the two sub-genera, and they also differ in chromosome composition, the diploid number of chromosomes (2n) being 38 in *Euvitis* and 40 in *Muscadiniae*. Because of this difference in diploidy, any crosses made between the two sub-genera show imprecise pairing and separation of chromosomes during meiosis, causing the resultant progeny to exhibit poor fertility.

2.1.2.1 Differences between Wild and Domesticated Vitis: *What Happened During Domestication?*. In some ways, the wild grapevine proved to be an ideal candidate for being cultivated because it required relatively little modification before it ultimately became domesticated. Its mineral and water requirements are not excessive, allowing it to grow on impoverished sites (*e.g.* hill sides) bearing soil that is unsuitable for other crops. Its tendrils gave it the ability to climb on other plants, and so it would have required relatively little attention if grown with other crops, or in a scrub or woodland environment. In addition, the wild vine possessed considerable regenerative capabilities, and so was indifferent to the intense pruning regimes, which converted a liane into a shorter, shrubbier, more manageable plant. The shorter stature of the cultivated vine reduced the necessity for structural support, and probably reduced water stress, thus enabling it to survive in warmer climes. The regenerative powers and general hardiness of the grapevine allowed the plant to survive in cooler conditions than it would have liked, and so permitted the spread of viticulture into central Europe. The variation of seed morphology in wild and domesticated vines is such that they can be effectively distinguished on that criterion. The most obvious difference is that wild vine seeds have a prominent beak, and are more elongated than those of *V. vinifera* subsp. *vinifera*, which lack a beak and are, by definition, more rounded. The degree of 'roundness' of a seed can be expressed by its 'seed index', which is simply seed width/seed length. In wild grape seeds the index is, on average 0.64, while in the more rounded, cultivated seed the index is approximately 0.55. Why this would have happened is unclear because there is no obvious biological advantage for the domesticated vine in having more rounded seeds. Pollen of wild and domesticated vines is also distinguishable, and in *V. vinifera* there are differences between fertile pollen (produced by male and bisexual flowers) and sterile pollen (produced by female flowers).

As already indicated, a major difference between wild and domesticated grapevines is the sexuality of their flowers and the size of their fruit. Wild vines are dioecious, that is, there are separate male and female plants bearing the appropriate sex organs. As a result of domestication, *V. vinifera* has become hermaphrodite (monoecious), with functional anthers ($\male$) and a pistil ($\female$) on the same flower. Such a condition permits self-fertilisation by wind (or, even by gravity), which has obvious biological benefits for the plant itself, and for the viticulturalist aiming to maintain plants with a standard fruit size. In addition, nearly every hermaphrodite flower produces fruit of a consistently larger size than would be found in wild *Vitis*. If the ancient, dioecious habit had persisted into the cultivated form, then viticulturalists would have been forced to grow male and female vines with only the latter bearing fruit. So, in effect, genetic events that gave rise to a biological advantage for the vine (*i.e.* hermaphroditism), have also been a boon for the grower. It is almost certain that biologically useful mutations in wild *Vitis* would have occurred very slowly in nature, until man began to encourage selection of desired (domestic) traits in the plant, from where change was somewhat accelerated. Early attempts at cultivation would have involved selecting fruit-bearing ($\female$) vines from their woodland habitat, thus separating them from their sources of pollen ($\male$ vines). Favouritism would almost certainly have been shown towards vines with large fruits, and/or those with enhanced flavour and aroma characters. By selectively propagating such plants away from male vines, unisexuality would have been progressively discouraged. Nowadays, nearly all cultivars bear bisexual flowers and set fruit after self-pollination. One of the unheralded consequences of the shift from being a cross-pollinating plant to a self-pollinating one is that the nectarines in the monoecious flowers are poorly developed and non-functional, so that they are no longer attractive to pollinating insects. As is the case with most other fruit trees, viticulture is based on the 'fixation' and maintenance of vegetative clones, and vines are propagated vegetatively by rooting winter-dormant twigs or by grafting. Although wild vines are essentially dioecious, male plants sometimes possess non-functional pistils and some female plants may contain non-functional anthers. Such a phenomenon hints that the 'archaetypal *Vitis*' may have been bisexual (monoecious), but that they reverted to being dioecious at some stage during evolution. Indeed, primitive forms of *Vitis* from the Tertiary have been shown to have been hermaphrodite, just like contemporaneous, cultivated *V. vinifera* subsp. *vinifera*. This means that at either end of its long existence on planet Earth, the grapevine has been capable of self-fertilisation. Somewhere in between, for reasons not yet fully apparent but probably related to the harsh environment during the last Ice Age, the wild grape became dioecious throughout its range. The segregated male and female plants each had an anther and a pistil, but a mutation suppressed the development of one, or the other, on different plants so that, functionally, flowers on an individual plant can be considered to be either male or female. In what was to become the male plant, a dominant mutation of one of the nuclear chromosomes (designated Su^F) suppressed the development of the female reproductive apparatus. In the nascent female, a recessive mutation (designated Su^m) prevented the development of the male apparatus. The upshot of all this is that, in the wild, male vines seldom yield any grapes and female plants produce fruit that is highly variable. As we find it today, the Eurasian wild grape is hardly likely to be a source of good wine. It is small and many-seeded, has relatively low sugar content, is of high acidity and has a tough skin. Colourwise, they are usually black or dark red, and very rarely white.

It would appear that the modifications that occurred to transform the wild vine into a domesticated form were effected at an early stage in history. Certainly, by the time that we have the first depictions of vines from ancient Egypt, it would appear that the wild vine had been transformed into its domestic counterpart. There are no known illustrations of non-grape-bearing, male vines from ancient Egypt, even though a fairly high percentage of plants would have been fruitless in the days when the sexes were separated (dioecy). Let us envisage how the transformation from wild (dioecious) to domesticated (monoecious) grapevine might have occurred. Unfruitful male vines would have been grubbed up, thus depriving the fruitful female vines of their source of pollen. This would eventually result in the female vines ceasing to bear fruit. Meanwhile, there would have been a few hermaphrodite vines among the wild population, bearing both pollen and female organs (ultimately fruit). Their fruit size would have been small, certainly smaller than that from female vines, but better than nothing, and so they would have been persevered with. Because pollen was obtainable from these hermaphrodite vines, male, non-fruiting plants could be dispensed with and as a result of selection by man, fruit yields on hermaphrodite plants were improved (in size and, presumably, taste). Dioecious vines grown from seed are notoriously heterozygous and unlikely to bear much resemblance to their parent plants. Young vines from self-pollinated herma-phrodite (monoecious) plants are very weak because a number of harmful recessive genes are expressed, and without the interven-tion of man, natural selection would lead to the elimination of hermaphroditism within a population. As it has turned out, man has been responsible for the hermaphrodite condition being synonymous with cultivated *Vitis*. It is interesting to note that by back-crossing, viticulturalists can obtain the primitive dioecious condition in *Vitis* plants within a few generations.

2.2   THE GRAPE AND MATURATION PROCESSES

As the source of the basic raw material, the grape requires special consideration, especially since its condition at harvest is probably the overriding factor in determining wine quality. As the old adage goes, 'good wine is made in the vineyard'. The mature grape is the result of complex physiological and biochemical phenomena, which themselves are intimately linked to environmental condi-tions. Botanically, the grape, being a multi-seeded fleshy fruit, is classified as a berry, and when compared to other types of fruit, it does not lend itself easily to detailed study. This is partly due to the fact that berry growth and development are the result of a long and complex reproductive cycle, and partly because berries are aggregated into a cluster. The study of grape maturation is greatly complicated because of the variability of individual berries within a bunch. Having said that, it is possible to use a number of biochemical markers to assess fruit maturation. Probably the most important outward manifestation of berry maturation is *véraison*, which marks the beginning of berry ripening. In red cultivars, this point is marked by a change in grape colour from green to red.

2.2.1 Berry Structure

The grape, being a true berry, is a simple fruit with a pulpy pericarp, and it is this fleshy region that is of major significance to the winemaker. In the simplest of terms, the grape berry contains three major tissue types: flesh, skin and seed. All three tissues vary considerably in composition, and therefore contribute differently to overall wine composition. The skin typically represents around 15% of the berry weight and is the principal source of aromatic compounds and flavour precursors. It also contains flavonoid phenolic compounds (including flavonols, anthocyanins and large polymeric flavonoids called tannins). Phenolic compounds in skins represent some 30% of the total berry phenolics. The seeds, which represent *ca.* 4% of berry weight, contain both non-flavonoid and flavonoids phenolics, including relatively large amounts of tannin. Seed phenolics approximate to 60% of those compounds found in the berry. Seeds also contain significant levels of nitrogenous compounds, minerals and oils (primarily oleic and linoleic acids). The pulp accounts for about 78% of the berry weight, and its primary constituents are hexose sugars (notably glucose and fructose), organic acids (mainly tartaric and malic), mineral cations (especially K^+), nitrogenous compounds (soluble proteins, ammonia and amino acids), pectic substances (cell wall structural material composed of galacturonic acid polymers) and non-flavonoid phenolics (primarily benzoic and cinnamic acid derivatives). Phenolic compounds in the pulp represent around 10% of the total phenolic content of berries. Division of the grape berry into its constituent regions has engendered a number of different systems, but most authorities would agree that the pericarp is divisible into three main regions: the exocarp (skin), the mesocarp (flesh) and the endocarp. The latter is, by definition, the tissue that surrounds the seeds and is scarcely distinguishable from the rest of the pulp. Again, most authorities would agree that the 'skin' of the grape should include the epidermis and its waxy covering, and the hypodermis in which the cell walls are thickened with cellulose (called collenchyma cells). The cells in the hypodermis are separable into two regions, with the outer cells being rectangular in shape and the inner ones being polygonal. The cells in this dermal area contain a high percentage of the substances responsible for pigmentation, flavour and aroma. The mesoderm is largely composed of large, unthickened, polygonal cells, and this tissue forms the bulk of the berry flesh. On the outside, the cuticle is secreted by the epidermis and is responsible for the 'bloom' of grapes. It is composed of overlapping platelets of wax, which look fascinating under the electron microscope. Wax thickness is relatively constant throughout berry development and is usually in the order of 100 μg wax per cm^2 berry surface. Platelets are hydrophobic and help to prevent water loss from the fruit. Chemically, epicuticular wax consists two-third of the hard wax (mainly oleanolic acid) and one-third of soft wax – a complex mixture of organics including fatty alcohols such as lauryl, stearyl and cetyl.[6]

The grape is richly supplied with vascular tissue, which after it enters the fruit via the pedicel, branches out to supply the developing seeds (via an ovular network), the flesh (via a central network) and the skin by a peripheral system of veins, which have been likened to a mesh of 'chicken wire'.

From an oenologist's point of view, the grape berry is an independent biochemical factory because, in addition to the primary metabolites essential for plant survival, it has the ability to synthesise all the other secondary components, such as flavour and aroma compounds, that go towards defining a particular wine. There is tremendous potential for variability in ripening berries within a cluster and therefore within a vineyard, and it is often difficult to determine when a vineyard, with a large discrepancy in berry maturity, is at its optimum stage for harvest. One of the main aims of modern viticulture, and certainly one of its greatest challenges, is to be able to produce a uniformly ripe crop.

The profile of a wine can be manipulated by simply changing berry size. As a rule, wines made from smaller berries will contain a higher proportion of skin- and seed-derived compounds. In addition, the number of seeds in a berry can considerably affect a wine. The 'normal' or 'perfect' number is four per berry, but there is often fewer than that. Environmental and nutritional conditions at bloom often affect the success of fertilisation and the subsequent number of seeds in a berry.

If we consider the biological reason for the production of the grape berry, that is that the vine is trying to reproduce itself, then all of the observed changes in grape maturation make sense. During what we have called the Stage I, the vine seed is immature and requires protection from what are essentially its dispersal agents (birds and other animals). The plant protects its young seeds by making them as unappealing as possible to likely predators, that is by synthesising organic acids, tannins and pyrazines, which taste pretty disgusting. Such compounds are protective agents, as far as the vine is concerned. Having produced a viable seed, Stage III – the second period of berry growth – is all about making the seed covering (the juicy berry) as appealing as possible to birds and mammals so that the seeds will be dispersed. It is as simple as that.

As can be seen from Figure 2, the initial branch from the vine cane is called the peduncle, which itself is furcated into rachi (singular rachis). Each rachis bears numerous pedicels on which the berries, themselves, are borne. The berry is supplied via the pedicel with a vascular system composed of xylem and phloem elements. Xylem is responsible for transporting water, minerals, growth regulators and nutrients from the root system, while the phloem transports photosynthate from the leaves. Berry formation, of course, is preceded by flowering and then sexual reproduction, and botanically the arrangement of the flower head in *V. vinifera* is called a panicle. Each berry is lavishly supplied with vascular tissue, and the way in which berries are spatially arranged into a cluster is determined by the length of the pedicels. If they are long and thin, then the cluster will be loose, while if pedicels are short, bunches will be more compact.

Under temperate conditions, flowering in the mature vine is usually a three-step process, the first of which is the formation of uncommitted primordia sometimes referred to as *anlagen* (singular *anlage*). *Anlagen* are formed by the apices of latent buds on shoots of the current season, and will develop as either inflorescence primordia or tendril primordia, at which point they enter dormancy. Flowers are formed from inflorescence primordia in the spring of the following season – at the onset of 'bud burst'. Using scanning electron microscopy, the origin of inflorescence primordia in latent buds and the subsequent development of the inflorescence was studied extensively and reported on by Srinivasan and

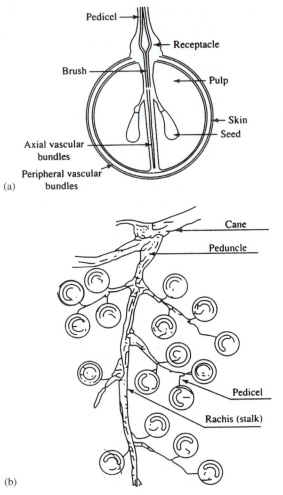

Figure 2   *The grape: (a) sectioned berry at maturity (b) composition of grape cluster*
(Reproduced by kind permission of John Wiley & Sons Ltd.)

Mullins.[7] These workers discerned 11 developmental stages, which were all related to changes in the shape of organs, or to the addition of new structures. In this scheme, the latent bud apex is considered to be purely vegetative (until the emergence of the first *anlage*), and buds containing only leaf primordia are classified as Stage 0. The earliest manifestation of reproduction in the vine, the *anlage*, is designated as Stage 1 and the first *anlagen* are formed around mid-summer in latent buds at the base of the cane on newly ripened wood. After this initial manifestation, *anlagen* start to appear in latent buds further towards the growing tip, their formation coinciding with a change in stem colour from green to brown (called *aoûtement*).

In Srinivasan and Mullins's scheme, the formation of *anlagen* comprises Stages 1–3, the formation of inflorescence primordia occurs through Stages 4–7 and the formation of flowers results from Stages 8–11. Within each flower, the sepals, petals (calyptra) and sexual apparatus (stamens and pistils) develop one after another (Stages 9, 10 and 11, respectively). Stage 11, therefore, represents the fully formed grape flower just before anthesis (the opening of the flower ready for pollination). Depending on the cultivar, the latent bud apex produced from three to eight leaf primordia (Stage 0) before the first *anlage* is initiated; Shiraz, for example, produces five.

As we have noted, in nature most *Vitis* species are dioecious but nearly all important commercial cultivars are hermaphrodite, their flowers containing functional stamens and pistils. According to Negi and Olmo,[8] the sexuality of such a flower is determined by three alleles: Su^+, Su^F and Su^m, and depending on matters of dominance, flowers become functioning males or females. The primitive hermaphrodite state is Su^+Su^+, and a dominant mutation, Su^F, suppresses ovary development to give functional male flowers. Conversely, in the presence of the recessive Su^m, sterile pollen is produced and the flower is effectively female. The dominance relationships of these three alleles are, $Su^F > Su^+ > Su^m$.

The exact means of pollination in *V. vinifera* is imprecisely known, and our knowledge is beset by anomaly. Most authorities agree that the vine is primarily anemophilous (wind pollinated), even though the flower structure does not fundamentally imply this method. For example, at the base of the ovary is a whorl of nectaries called the 'disc', and this is thought to be responsible for the odour of the vine flower. Production of nectar, however, has not been unequivocally proven, and in any case, nectaries are not compatible with anemophily. Bud pollination, involving dehiscence of the anthers before anthesis, is fairly common, and it is also thought that some insects may play a role in pollination. Similar questions exist in the debate about whether *V. vinifera* is cross-pollinated or self-pollinated, but most evidence points to the latter mechanism as being the norm. The dioecious progenitors of the

European vine would, of course, have been cross-pollinated. Where cross-pollination is necessary with *V. vinifera* plants, this is effected by either hand or machine.

Whatever the case, pollination is normally followed by fertilisation, which signals the onset of berry formation. Grapevine embryology has not been the subject of extensive study, and information about the embryo, endosperm and seed formation is fragmentary, to say the least. When the seed is formed, its outer layers are formed from the integuments that supported the egg cell (ovule). The hardening of the grape seed is due to lignification of part of the outer integument. The inner integument remains un-lignified and closely aligned to the seed's food reserve (endosperm). Grape seeds are very resistant to decay, and therefore of considerable archaeological and palaeobotanical interest. Poor levels of fertilisation can lead to the formation of rudimentary (aborted) seeds, a condition called stenospermocarpic seedlessness, while a lack of fertilisation leads to totally seedless berries, a condition known as parthenocarpic seedlessness. In the latter situation, there is a degeneration of the embryo sac, and pollination alone is a trigger for fruit development. With stenospermocarpy, both pollination and fertilisation are triggers for fruit development, but the embryo and endosperm fail to develop. As a consequence, the outer integument around the ovule contains little or no lignin, and the aborted ovules appear as small, white specks in the flesh of the fruit. Seedlessness can be a varietal character, and is much in demand for table grapes (*e.g.* Thompson Seedless) or for making raisins (Corinthe).

In general, not all of the flowers in a cluster are successfully pollinated, and non-pollinated stigmas lead to non-fertilised ovaries, and these eventually give rise to small green berries. It has been shown that the fertilisation/berry-setting ratio decreases as the number of flowers in the cluster increases. After berry-set, a variable proportion of ostensibly fertilised young berries cease to grow and fall from the vine. This is called 'shatter' (*Coulure* in French) and is caused by the formation of an abscission layer at the base of the pedicel. Under temperate growth conditions, shatter is linked to the carbohydrate status of the vine, but in more northerly vineyards climate plays a part in the phenomenon and shatter is worst when conditions are cold and wet. There is also a varietal variation in propensity to shatter, with Chardonnay, for example, being

much more likely to shatter than Riesling. Some *coulure* is beneficial as a vine might have difficulty in ripening a full, heavy crop,
resulting in loss of overall grape quality. Heavy shatter will result
in a very small grape crop.

2.2.2 Developmental Stages of the Grape

The transition from ovary to ripe grape can be conveniently
divided into three phases, and irrespective of the way in which
the fruit is set, the growth of the berry follows a double-sigmoid
pattern (Figure 3), which essentially represents two successive sig-
moidal growth periods separated by a lull in growth. The physiology behind this pattern has not been fully elucidated, but it is most
likely that fruit growth is in no small way controlled by plant
hormones emanating from the seeds.

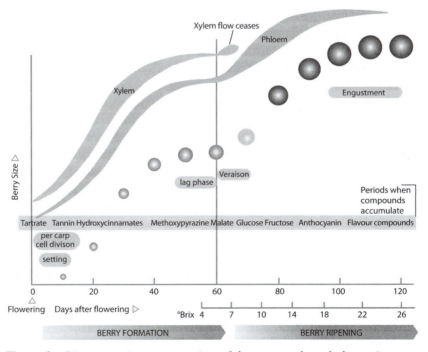

Figure 3 *Diagrammatic representation of berry growth and the main events
accompanying it*
(Reproduced by kind permission of Dr James Kennedy, University of
Oregon)

2.2.2.1   Stage I.   This stage can last up to 9 weeks after anthesis, depending upon cultivar and growth conditions. It consists of a period of rapid vegetative berry growth, which increase in both size and mass. There is some seed growth, but very little increase in the size of the embryo. Initially, cells in the pericarp divide rapidly, and this cytokinesis is followed by a period of cell enlargement. The rapid cell divisions that occur during the first weeks of Stage I produce a certain number of cells that will, in essence, determine the eventual size of the berry. The intensity of growth seems to be correlated with the presence of seeds in the pericarp tissue. It is known that grape seeds are a lucrative source of cytokinins and gibberellins, which diffuse into the surrounding tissue, but whether these phytohormones actually affect berry growth is a matter of debate. On the 'anti' side, it should be noted that berries of seedless cultivars (whether seedlessness is due to parthenocarpy or steno-spermocarpy) also exhibit a double-sigmoid growth pattern. Inter-estingly, it has been shown[9] that exogenously supplied gibberellin could stimulate the development of the seedless cultivar Delaware, and as a result of this work, application of this phytohormone has become a common viticultural practice. Chlorophyll is the domi-nant pigment during Stage I, and the berries appear hard and green with a high respiratory activity. The grape rapidly accumulates organic acids during this phase, and this is measureable as 'titrable acidity' (TA).

Several solutes accumulate during this phase, all of which reach maximum levels around véraison. The most prevalent are tartaric and malic acids. Tartaric acid levels are highest towards the outside of the developing berry, whereas malic acid is concentrated more in the flesh. Most tartaric acid accumulates during the start of Stage I, unlike malic acid, which peaks its accumulation just prior to véraison. Also accumulating during this first phase of growth are the hydroxycinnamic acids, which are distributed throughout the flesh and skin. Tannins, including monomeric catechins, are also synthesised during this first growth phase. They are present in skin and seeds, but are almost absent in flesh. Some 'varietal' aroma compounds, such as methoxypyrazines, are detectable dur-ing Stage I and persist right through the way to harvest.

2.2.2.2   Stage II.   This stage lasts for at least a week, but can be as long as six depending upon cultivar and growing conditions. It is often referred to as the 'lag' phase of growth because there is little or no growth in the pericarp tissue. There is, however, considerable growth and development of the embryo, and the seeds themselves mature. As a consequence, it is noticeable that this lag is less prominent in seedless varieties. Rates of respiration and photo-synthesis are depressed, and there is a diminution in the chloro-phyll content of the berry, which remains as a hard, green structure throughout. Phytohormone synthesis is reduced, and there is a build-up of the growth inhibitor, abscissic acid (ABA). It is during this phase that TA reaches its maximum in the grape, and it is the length of Stage II that will largely determine whether a cultivar is early- or late-ripening.

2.2.2.3 Stage III. This stage is essentially a second phase of berry growth, and corresponds to berry maturation since growth is mostly by cell enlargement rather than by cytokinesis. It can last from 5 to 8 weeks. The onset of this phase is manifested by the softening of berry tissue, and in pigmented varieties, by a change of berry colour. As mentioned earlier, this stage, by definition, is called *véraison* (some authorities maintain that véraison signals the end of Stage II). In white cultivars, the colour change is from green to a lighter shade of green (*i.e.* a 'whitening' of the berry). Respi-ration rate increases, and the berry reaches its maximum dimen-sions, there being a direct positive relationship between seed number and berry size, a feature first reported by Müller-Thurgau in 1898.[10] Overall, the berry approximately doubles in size between the end of Stage II and the harvest. As one would expect, seeded varieties almost invariably yield larger berries than their non-seeded sisters. A number of complicated biochemical transforma-tions occur during this period, including a significant increase in hexose sugar level and a concomitant decrease in organic acids, especially malic acid. Levels of potassium, amino acids (notably proline and arginine) and phenolics also increase in the berry. During véraison, potassium from the soil is translocated into the fruit, where it forms potassium tartrate. The level of potassium in grape musts varies from 600 to over 2500 mg L^{-1}, the latter levels being found in certain red cultivars. The vitamin, ascorbic acid, is present in ripe grapes and can be found in juice at levels up to 50 mg L^{-1}, but this grape-derived material is largely destroyed during vinification. Anthocyanins, which are usually restricted to the dermal layers of the berry, are also synthesised during the second growth period.

The increase in hexose sugar content of the berry in Stage III is due to a massive influx of sucrose, which is immediately hydrolysed to its two components, glucose and fructose. Sucrose influx commences at véraison, and the eventual concentration of the two hexoses is dependent on how long the berry is allowed to remain on the vine. Conversely, many of the other solutes that accumulated during Stage I, such as malate, remain until harvest but their concentration is reduced. This is often due to simple dilution caused by water ingress into the berry, but not always. Malic acid levels, for example, are actively reduced, especially in warmer growing areas. Tannins also decline on a per-berry basis during Stage III, but this appears to be due to oxidation as they become attached to the seed coat. Some volatile aroma compounds formed during Stage I decline on a per-berry basis during this second growth phase. Most notable, perhaps, are the methoxypyridines that can impart 'vegetal' character to some wines, such as those made with Cabernet Sauvignon and Sauvignon Blanc. Decline in these compounds is thought to be associated with sunlight.

It is thought that the resumption of growth in Stage III is attributable to an increase in turgor pressure and some increase in elasticity of pericarp cell walls. The driving force for cell expansion is due to a greater propensity for water uptake, and this is due to an increased water potential difference between the vascular tissue in the region of the pedicel and the pericarp cells. Maturing grapes lose only small amounts of water through transpiration. Not only do they have a low frequency of stomata, but those that are present become blocked by cuticular wax deposits. The importance of the roles played by the two components of the vascular strands alters during Stage III. Available evidence suggests that xylem remains functional up to véraison, and then becomes less active or even defunct. Conversely, phloem, which has a reduced function early in berry development, becomes the primary source of ingress of materials into the berry after véraison. Increases in berry volume, which are primarily due to water uptake, are normally closely associated with increased sugar content after véraison, but in some varieties, most notably Syrah, the increase in sugar levels during the latter stages of ripening is not accompanied by an increase in berry volume but by berry shrinkage. This shrinkage seems to be caused by the loss of water from the fruit by transpiration.

When the berry grows, mesocarp cells are capable of expanding at a much faster rate than skin cells, and expansion of the berry seems to be partly governed by the capacity of the epidermal cells to expand. In the vineyard, this disparity in elasticity may lead to fissures appearing in the fruit (berry crack). If skins are removed from berries, it is noticeable that they can expand twice as rapidly as intact samples. The transition from Stage II to Stage III represents the beginning of berry ripening, and most of the physiological changes occur rapidly, usually within 24–48 h.

It is now pertinent to summarise the main changes that occur in the status of carbohydrates, organic acids, nitrogen compounds and phenolics during the growth and ripening of the grape, and to overview their relationships. Because of their fundamental importance, plant growth regulators are also considered in this section. In simple terms, the concentration of sugars increases, while that of organic acids decreases during the ripening period. Since the concentration of sugars at berry maturity (the end of Stage III) greatly exceeds that of organic acids at the end of Stage I, it is unlikely that there is a wholesale biochemical conversion of acids to sugars within the berry. In fact, the enhanced sugar status is mainly due to more vine photosynthate reaching the berry. This is not so much caused by an elevated rate of photosynthesis, but more by a change in the pattern of translocation within the vine; hence, more sucrose is directed to the fruit. Hand in hand with this is an increase in invertase activity in the berry, such that glucose and fructose become the main fruit sugars. Invertase activity starts to increase markedly at the onset of véraison. Glucose and fructose are stored primarily in the vacuoles of mesocarp cells, although small quantities of both sugars may be found in the exocarp.

Decrease in berry acidity on ripening is primarily due to a decrease in malic acid content of the berry. When berries are immature (Stages I and II), malate is actively synthesised from sucrose, a mechanism that ceases to operate after véraison. In addition, there is increased activity in the pathway that converts malate to glucose. This conversion is an important means of controlling the intracellular pH of berry cells when the energy needs of the grape are reduced (*e.g.* at night). The conversion of excess malate to glucose is called gluconeogenesis, and it is known that after véraison the berry shows an increase in the level of enzymes associated with this pathway, such as glucose-6-phosphatase, fructose 1,6-diphosphate and malate dehydrogenase. Coincident with the presence of these enzymes is the production of enhanced levels of ABA. Although gluconeogenesis is elevated during véraison, the actual amount of malic acid converted to glucose only represents around 5% of that present, and the extent to which the mechanism contributes to the increase in berry hexose sugar content is difficult to assess.

In the immature grape, the ammonium ion accounts for more than 50% of the total nitrogen content, the other major sources of the element being amino acids and proteins. During ripening, the nitrogen content of the berry increases, especially that of amino acids, such that in ripe fruit over 50% (can be as high as 90%) of berry nitrogen is in the form of these compounds. As already indicated, it is the levels of arginine and proline that increase most markedly in most cultivars, but it is glutamic acid that generally constitutes around 50% of the nitrogenous material translocated into the grape berry. Free amino acid concentrations can increase five-fold during maturation and, as would be expected from increased levels of amino acid synthesis, ammonium ion content gradually decreases after véraison. The amino acid profile in berries varies considerably with cultivar and growing conditions, but alanine, α-aminobutyric acid, arginine, glutamic acid, proline and threonine are often conspicuous. As a rule, arginine is usually the predominant species in mature grapes, and it is unsurprising that this amino acid plays an important role in grape berry nitrogen metabolism, being 'on hand' for conversion into growth regulators, such as the polyamines, spermine, spermidine and putrescine, which play a role in berry set. Arginine also has a close biosynthetic relationship with other amino acids, such as ornithine, proline, glutamic acid and aspartic acid. The latter is extremely important because it acts as a 'reserve' for oxaloacetate. The tripeptide glutathione is the major soluble compound in the grape with a free sulfydryl group, and this compound plays a role later on in the prevention of must browning. Perceptible levels of protein synthesis occur during berry maturation, the soluble fraction largely remaining in the juice and the insoluble fraction remaining bound to cell walls. It is interesting to note that, at harvest, around

50% of the nitrogenous content of the non-woody parts of the vine are stored in the berries, and that when such grapes are crushed, the expressed juice would normally be expected to contain only around 20% of the total berry nitrogen – the remainder being retained in the seeds and skins.

Gibberellin and cytokinin activity is significant during Stage I but decreases as ripening approaches, whereas the level of the growth inhibitor ABA increases. This is the general pattern observed during ripening – a decrease in phytohormone level and an increase in growth inhibitor level. Indeed, *in vivo* exogenous additions of ABA can accelerate hexose sugar accumulation, and it has been suggested that this compound may be a trigger for fruit ripening. Addition of phytohormones to vines, however, produces varying results. In general, berry growth is not affected by exogenous auxin application, the exception being 4-chlorophenoxyacetic acid that enhances the growth of seedless grapes. Gibberellic acid exogenously supplied to seedless cultivars induces an immediate increase in berry growth, again suggesting that seeds may well be the main source of these hormones in nature, whereas seeded varieties respond less markedly.

Ethylene (ethene) also acts as a plant hormone, since it affects the growth of many plants. It is produced by most parts of a plant, but especially where cells are dividing rapidly; thus, it is particularly important during fruit ripening. Ethene acts at the genetic level, activating regulatory genes called 'ethene-responsive elements' (EREs). Some plants respond to the gas at minute levels in air (1 ppm), and some fruits (*e.g.* banana) show a dramatic ripening response to ethylene, characterised by a large, transient evolution of respiratory carbon dioxide. Such fruits are called 'climacteric'. Grapes do not exhibit any such response to ethylene at the onset of ripening, a fact which classifies them as 'non-climacteric' fruits. This means that there is no guaranteed enhancement of grape ripening in response to exogenous ethylene supply. It would appear that ethylene does play a role in grape ripening because exogenous application of ethephon (2-chloroethylphosphoric acid), an ethylene-releasing compound, elicits enhanced colour (anthocyanin) formation in coloured cultivars. All in all, it would seem that the total control of grape maturation is the responsibility of several hormonal substances (auxins, cytokinins, gibberellins and ethylene), and that their activity is eventually modified by a number of naturally occurring inhibitors, which include ABA and certain phenolics.

For more complete discussions of grape berry development and the ripening process, please see the review by Coombe,[11] the book chapter by Kanellis and Roubelakis-Angelakis[12] and the articles by Bisson[13] and Kennedy.[14]

2.2.3 Flavour and Aroma Compounds in the Mature Grape

The grape berry contributes several groups of compounds that can contribute to the final flavour and aroma of a wine. Three of the basic wine's tastes – sweetness, sourness and bitterness – are accounted for by major grape components, namely sugars, acids and polyphenols, but if these were the only contributors to flavour, then wine would be relatively boring. Fortunately, there is a myriad of other components, often volatile and usually present in minute quantity, which contribute to the wide variety of wine styles that we see today. Much of the original, grape-derived sugar content, of course, is modified during alcoholic fermentation, and the grape organic acids may be metabolised during secondary fermentation. This means that the volatile flavour components tend to become the most important grape-derived contributors to flavour and aroma. Overall, the most significant grapevine compounds contributing to finished wine flavours are organic acids, proanthocyanins (tannins), terpenoids (monoterpenoids, sesquiterpenoids and C_{13}-isoprenoids) and various precursors of aromatic aldehydes, esters and thiols.

Three main types of volatile compounds are found in wines according to their origin: (1) primary aroma compounds, which are those present in the grape and persist through vinification to end up in the finished wine; (2) secondary aromas, which are generated primarily by yeast activity during fermentation; and (3) tertiary aromas, which are generated during maturation, in either wood or bottle. Huge lists of compounds that have been identified in wines have been compiled over the years, but the reader should be aware that based on analytical techniques post-1980 lists are likely to be most reliable. Over 400 such compounds have now been reported and modern methods now permit the scientist to detect the source of many compounds. As a result, it is now evident that some primary flavours and aromas may also be generated during fermentation and ageing.

Some highly flavour-active compounds, such as esters, ketones, lactones and fusel alcohols, are not predominantly grape-derived, but are the products of fermentation, and so will not be dealt with here. We shall consider only the primary volatile compounds, the terpenes and the phenolics, and some representatives of an interesting group of nitrogen-containing heterocyclics, the pyrazines. It is such compounds that are primarily responsible for the 'grapey' or 'varietal' aromas of certain wines. It should be noted, however, that, qualitatively and quantitatively, the largest fraction of wine volatiles comes from secondary aromas and the contribution of secondary/tertiary aromas makes a wine markedly different in flavour and aroma to that of the grapes/musts from which it originated.

In both red and white cultivars, most of the volatile grape-derived flavour components are produced during ripening. Some important flavour and aroma compounds are also synthesised as precursors during Stage III (often as glycosides), and these will be released on maturation of the wine. The period required to release free, volatile compounds is called 'engustment'. Most volatile aroma compounds are synthesised during the later phases of berry development, and are largely confined to dermal tissue. The accumulation of these aroma compounds, of which several hundreds have been identified in ripe grapes, does not seem to be closely correlated with sugar concentration. Few of them have been extensively studied, purely because they are present in such low concentration and have a tendency to degenerate during attempted isolation.

Terpenes and their derivatives (called terpenoids[*]) occur widely as primary constituents of grapes and many other plants, and are metabolites of mevalonic acid (3,5-dihydroxy-3-methyl-pentanoic acid), which is a key compound in biochemistry. It is a precursor of the biosynthetic pathway known as the HMG-CoA reductase pathway, which produces terpenes (and steroids). This is also known as the mevalonate pathway. The pathway commences when acetyl CoA is converted to acetoacetyl CoA via the activity of thiolase enzyme. Acetyl CoA then reacts with acetoacetyl CoA to form 3-hydroxy-3-methylglutaryl

---

[*]Terpenoids arise when terpenes are chemically modified, such as by oxidation or rearrangement of the carbon skeleton.

CoA (HMG-CoA), which is then reduced to mevalonate by NADPH.

Plant-derived volatile terpenoids are mainly stored as non-volatile, water-soluble glycoside derivatives (sugar conjugates) in exocarp cell vacuoles, although some may be present as free volatiles. Some plants sequester large quantities of lipophilic, terpenoid volatiles in specialised anatomical structures, such as glandular hairs, or glands as found in the skin of citrus fruits, or resin ducts of *Pinus* spp. Grapes lack such structures, so instead they are required to trap these compounds biochemically. Accordingly, they are stored as water-soluble conjugates (usually as glycosides) in cell vacuoles. Other important volatiles, such as the cysteinylated precursors of aromatic thiols, may be stored as amino acid conjugates. These compounds have to be volatilised later in order to enable the flavour to be perceived. Endogenous glycosidases and peptidases play a vital role in the timing of release of these 'natural fruit flavours' in wine, and physical crushing, and the events during fermentation can introduce grape and yeast enzymes to these conjugated substrates. Winemakers may also add exogenous enzymes to the fermentation to stimulate the volatilisation of conjugated flavour compounds (see page 199).

Terpenes (the name 'terpene' being derived from 'turpentine') encompass of a wide range of substances, which are regarded as being derived from a basic structure of a linear chain of five carbon atoms, namely isoprene.

Terpenes are derived biosynthetically from units of isoprene, which, as Figure 4 shows, has a basic molecular formula of C_5H_8. The basic molecular formula of terpenes is multiples of it, $(C_5H_8)_n$, where 'n' is the number of linked isoprene units. This represents the 'isoprene' or 'C5' rule. Isoprene units may be linked together 'head-to-head' to give linear chains or they may be arranged in rings. Isoprene itself does not undergo the building process, but is capable of aggregation when in an activated form, such as isopentyl pyrophosphate (IPP, or isopentyl diphosphate, Figure 5) or

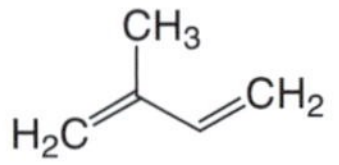

Figure 4   *Isoprene*

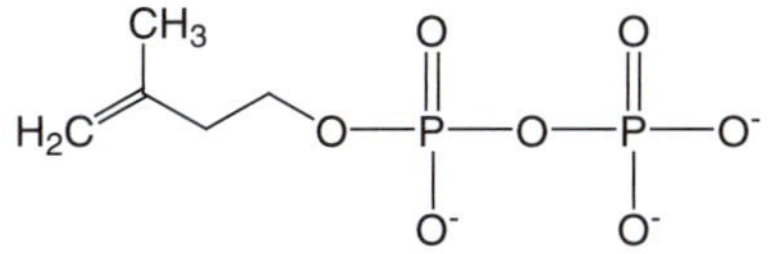

Figure 5 *Isopentyl pyrophosphate*

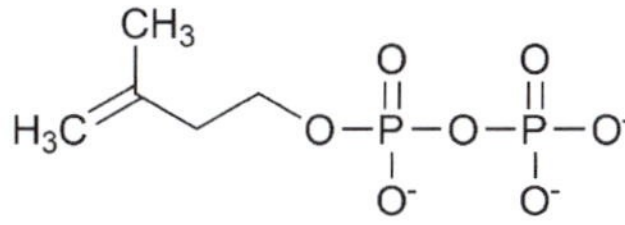

Figure 6 *Dimethyallyl pyrophosphate*

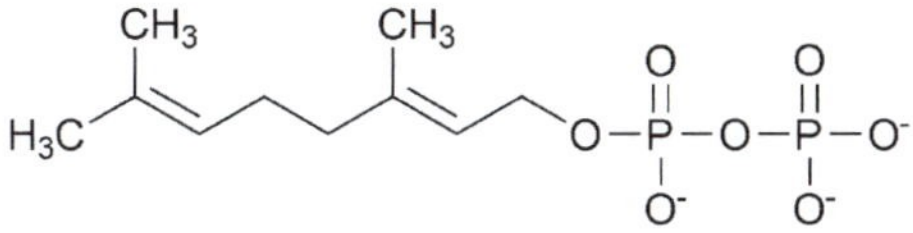

Figure 7 *Geranyl pyrophosphate*

dimethylallyl pyrophosphate (DMAPP, or dimethylallyl diphosphate; Figure 6). IPP is formed from acetyl-CoA via the intermediacy of mevalonic acid in the HMG-CoA reductase pathway. IPP can then be enzymatically isomerised to DMAPP.

As chains of isoprene units are built up, the resulting terpenes are classified sequentially according to their size, and so we get hemiterpenes, monoterpenes, sesquiterpenes, diterpenes, sesterterpenes, triterpenes and tetraterpenes. Hemiterpenes consist of a single isoprene unit, and isoprene is the sole example, although there are oxygenated derivatives, such as isovaleric acid, which are hemiterpenoid. Monoterpenes consist of two isoprene units (sesquiterpenes have three), and may be linear (acyclic) or contain rings (cyclic). Acyclic monoterpenes are formed when IPP and DMAPP combine to form geranyl pyrophosphate (Figure 7), which when the pyrophosphate group is eliminated, yields monoterpenes such as myrcene (Figure 8).

If the pyrophosphate groups are hydrolysed, geraniol results, and additional rearrangements and oxidations lead to compounds such as linalool (Figure 9).

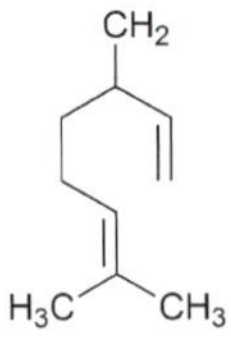

Figure 8   *Myrcene*

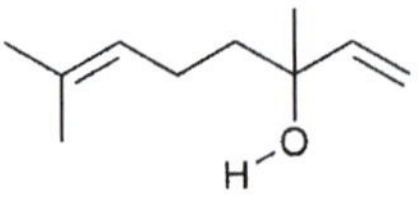

Figure 9   *Linalool*

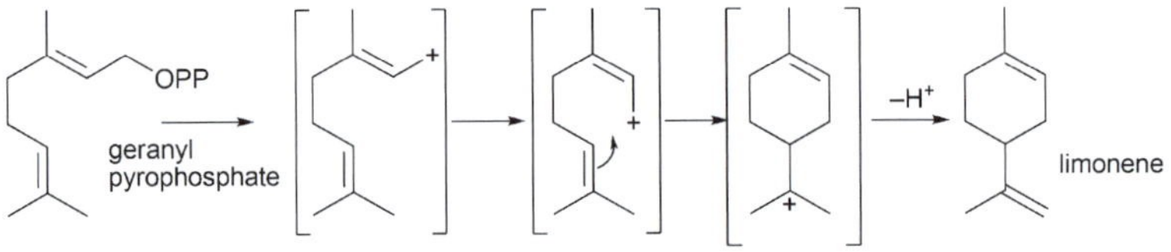

Figure 10   *Formation of limonene*

   The most common cyclic form of monoterpenes is a six-membered ring, which is formed by cyclisation of geranyl pyrophosphate (Figure 10), and if this molecule undergoes two separate cyclisations, then bicyclic monoterpenes result. Bicyclic forms are not known in grapes.
   The main compounds in grapes are the simple hydrocarbons, such as limonene and myrcene, and a number of hydrocarbon derivatives, both with linear and monocyclic structures. These derivatives can be alcohols, also known as terpenols (*e.g.* linalool and geraniol), aldehydes (*e.g.* lialal and geranial), acids (*e.g.* linalic and geranic) and even esters (*e.g.* linalyl acetate). To date, some 44 different terpenoids have been identified from grapes, six of which may be regarded, oenologically, as really significant. Some terpenoids, notably terpenols, occur in a free, volatile form, but many of them are in bound (especially glycosidic) form, and these molecules are too large and too water soluble to have odours, but when they are hydrolysed odouriferous compounds are released. The distribution and amount of grape monoterpenoids vary

according to time and temperature. Free forms decrease in concentration as temperature rises, while if grapes are allowed to remain on the plant and over-ripen, bound forms start to accumulate. Free and bonded forms of terpenols accumulate in ripening grapes from véraison onwards, and some works suggests that they accumulate continuously through to harvest. It is generally agreed, however, that these compounds reach peak levels just prior to the hexose sugar peak being attained. Vineyard conditions, such as temperature and water relations, may be responsible for the contradictions in results.

Certain volatile terpenes play an important role in the varietal character of Muscat grapes and of some non-Muscat cultivars, such as Riesling, Gewürztraminer and Müller-Thurgau. In Muscats, terpenols such as linalool are largely responsible for the fruit and wine aroma, together with some hydroxylated derivatives called polyols. In Muscat grapes, linalool can be detected about a fortnight after ripening commences. Other volatile terpenoids, such as geraniol and terpineol, are detectable about 1 month after véraison. In Riesling grapes, linalool and linolenic acid are prevalent and deliver the characteristic aroma of must and subsequent wine.

Carotenoids, examples of which are found in grapes, are tetraterpenes (*i.e.* 40 carbon atoms), and on oxidative degradation yield derivatives with 9, 10, 11 or 13 carbon atoms. Among these compounds, norisoprenoid derivatives with 13 carbon atoms, the 'C_{13}-norisoprenoids', have useful aromatic properties. First identified from tobacco, these compounds were found in grapes during the 1970s. Chemically, they may be divided into two groups: megastigmane and non-megastigmane. The megastigmane skeleton is characterised by a benzene ring substituted on carbons 1, 5 and 6, and an aliphatic chain of four carbon atoms attached to carbon 6. Megastigmanes are oxygenated C_{13}-norisoprenoids, with skeletons oxygenated at carbon 7 (damascone series) or carbon 9 (ionone series). Among these compounds, β-damascenone has a very low olfactory perception threshold in water (2 ng L^{-1}) and a relatively low one in dilute ethanolic solution (45 ng L^{-1}). In red wine, its recognition threshold is much higher (*ca.* 5000 ng L^{-1}). It is probably present in all grape varieties but was first identified in Riesling and Scheurebe grapes.[15] It has since been found in Muscat grapes. β-Damascenone has a complex, floral smell, often referred to as 'tropical fruit' or 'stewed apple'. β-Ionone has a perception threshold of 7 ng L^{-1} in water, 800 ng L^{-1} in dilute ethanolic solution and 1.5 µg L^{-1} in wine. It has a characteristic aroma of 'violets', and has been identified from several white grape cultivars as well as Muscat grapes. Other oxygenated C_{13}-norisoprenoids identified in wine are 3-oxo-α-ionol, 3-hydroxy-β-damascone and β-damascone. The perception thresholds of these substances are much higher than those of the compounds mentioned so far.

Non-megastigmane C_{13}-norisoprenoid derivatives have also been characterised in grapes, perhaps the most significant being 1,1, 6-trimethyl-1,2-dihydronaphthalene (TDN), which has a distinctive 'kerosene-like' odour and was first reported by Simpson in 1978.[16] TDN plays a major role in the 'petroleum' smell characteristic of some old Riesling wines, and may appear in levels up to 200 µg L^{-1} after bottle ageing (its perception threshold in wine is 20 µg L^{-1}). It is thought that TDA is generally absent from grapes. Also in the same family are vitispirane and actinidol, which have a 'camphor-like' odour.

Some of these non-megastigmane C_{13}-norisoprenoids are synthesised from grape-derived megastigmanes by biochemical modification in the acid medium provided by wine. Vitispirane, for example, with its camphorated smell is formed during ageing in bottle, and TDN is also thought to be derived from megastigmanes under acid conditions. Also at the pH of wine, some non-volatile, oxygenated C_{13}-norisoprenoids may be converted to β-damascenone, but most C_{13}-norisoprenoids are present in grapes in the form of non-volatile precursors (glucosides and carotenoids). A compound such as β-damascenone may, of course, be classified as a ketone. Examples of the main families of C_{13}-norisoprenoid derivatives found in grapes are shown in Figure 11.

One or two grape-derived nitrogen-containing flavour compounds are important in oenology, the most notable being the methoxypyrazines, methyl anthranilate and *o*-amino acetophenone. These compounds, which are present in small amounts, have only been investigated relatively recently and this has only been made possible by improved analytical techniques. Perhaps the most remarkable are the 2-methoxy pyrazines, which are flavour-active at minute concentrations and were the subject of an extensive study by Lacey *et al.*, who reported their results in 1991.[17] This group, from CSIRO, Australia, analysed juices and wines made

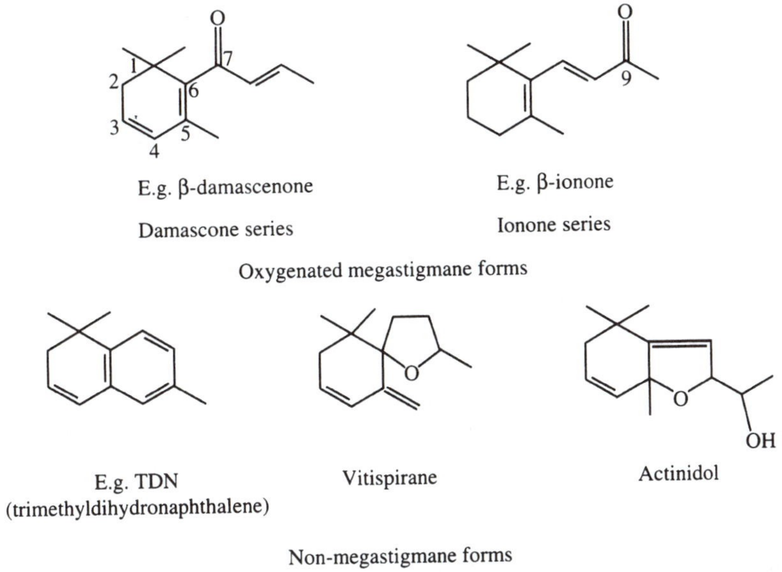

E.g. β-damascenone

Damascone series

E.g. β-ionone

Ionone series

Oxygenated megastigmane forms

E.g. TDN
(trimethyldihydronaphthalene)

Vitispirane

Actinidol

Non-megastigmane forms

Figure 11 *Examples of the main families of C_{13}-norisoprenoid derivatives found in grapes*

from the Sauvignon Blanc grape: some 22 wines from Australia, New Zealand and France, and 16 juices from different Australian regions. Three major 2-methoxy pyrazines were identified and quantified by gas chromatography/mass spectrometry, with 2-methoxy-3-isobutylpyrazine being present in all juice and wine samples at levels varying from 0.6–38.1 ng L^{-1} and 0.6–78.5 ng L^{-1}, respectively. This was the predominant methoxypyrazine in all samples. Next most abundant was 2-methoxy-3-isopropylpyrazine, which was found in 11 wines (0.9–5.6 ng L^{-1}) and almost all juice samples (0.2–6.8 ng L^{-1}). Finally, small quantities of 2-methoxy-3-*sec*-butylpyrazine (typically 0.1–1.0 ng L^{-1}) were detected in a few wine and juice samples. It was evident that fruit grown in cooler climes (*e.g.* New Zealand) contained greater levels of methoxypyrazines than that grown in hot conditions (Australia). It was also shown that methoxypyrazine levels were relatively high at véraison, but decreased markedly during the ripening period. 2-Methoxypyrazines have now been identified in a number of grape cultivars, including Sauvignon Blanc, Cabernet Sauvignon, Cabernet Franc, Semillon, Gewürztraminer, Chardonnay, Riesling, Pinot Noir and Merlot, but they are only usually present above the recognition threshold in the first three named above and occasionally in Merlot. The predominant species, 2-methoxy-3-(2-methylpropyl)pyrazine, is often referred to as the 'Cabernet pyrazine', and is responsible for imparting the aroma variously described as 'vegetative', 'green', 'bell-pepper' and 'capsicum' to some wines made with Cabernet Sauvignon grapes. The compound is detectable in water at *ca.* 2 ppt, although its threshold in wine is higher. It has been found at levels up to 50 ppt in some wines. In nature, these very pungent compounds, which are located in the skin of the grape, are used by the vine as a means of protecting the immature seed from predators and so as would be expected, the herbaceous methoxypyrazine aroma is most apparent when grapes are under ripening condition. 2-Methoxy-3-isobutylpyrazine was first identified in grapes by Bayonove *et al.*[18] A fourth member of the family, 2-methoxy-3-ethylpyrazine, has also been reported from grapes.

Methyl anthranilate (2-aminobenzoic acid methyl ester; $C_8H_9NO_2$) is the volatile compound mainly responsible for the distinctive 'foxy' aroma of some North American *Vitis* spp. It was long ago considered to be solely responsible for the phenomenon in *V. lambrusca* and *V. rotundifolia* grapes (Power and Chesnut, 1921),[19] but it is now known that other nitrogen-containing compounds are also involved, particularly ethyl anthranilate and *o*-amino-acetophenone (Figure 12). The variety Concord, derived from *V. lambrusca*, contains the highest levels of anthranilates thus far discovered. Methyl anthranilate is highly offensive to birds and is the active component of preparations such as Bird Shield™, which is used to protect commercial fruit crops such as cherries. Methyl anthranilate has been used by the food and drug industry for more than 40 years to flavour candies, sodas, gums and drugs, and is on the US Food & Drug Administration's GRAS list.

    o-Amino-acetophenone   Methyl anthranilate   Ethyl anthranilate

Figure 12   *Some compounds found in* V. lambrusca *and* V. rotundifolia *grapes and wines*

For a highly readable, succinct review of the role of volatiles in grape quality, the reader is directed to Lund and Bohlmann's recent paper.[20]

2.2.4 Grape-Derived Phenolic Substances

Compounds based on the phenol (hydroxybenzene) molecule play a very important role in winemaking, mainly as pigments and tannin agents, and several hundred have been identified in grapes and wine. They are generally referred to as 'phenolics', although in older literature they were often lumped together as 'tannins'. As with terpenes, the structures involved are mostly complicated and beyond the scope of this book, and so only the basic structure of important examples will be presented. Phenolics accumulate rapidly during berry maturation, and they are very important for grape (and wine) character because they include red pigments, astringent flavours and browning substrates. The primary classification of phenolics can be variable according to the basic criteria used, but from a viticultural viewpoint, two distinct groups are present in grapes: the non-flavonoids and the flavonoids with the former being mainly located in mesocarp cells, while the latter accumulate in the grape skin. Another convenient way of distinguishing them is into (1) monomeric compounds, (2) polymeric compounds, (3) combined phenolics and (4) red grape pigments. In this scheme the monomeric compounds can be divided into

(i) benzoic acid derivatives (C_6–C_1),
(ii) cinnamic acid derivatives (C_6–C_3), and
(iii) flavonoid derivatives (C_6–C_3–C_6).

Polymeric compounds are divisible into

(i) Condensed tannins, which are formed from flavonoids (*e.g.* catechin, epicatechin or leucoanthocyanidins) by C–C, or C–O bonds. They may contain 2–10 monomeric units.
(ii) Hydrolysable tannins, which are compounds constructed from gallic acid or ellagic acid residues together with a sugar molecule. These compounds are derived mostly from oak extractions.

Combined phenolics represent combinations of disparate molecules that are present in wine. They, themselves, are not formed in the grape, although their precursors may be. Examples are the combinations of cinnamic acid and tartaric acid. And, finally, red grape colourants which are based upon the flavylium ion.

At maturity, the grape skin contains derivatives of benzoic and cinnamic acids, flavonols and tannins, and these substances are distributed in the epidermal and sub-epidermal cells in both white and red grapes. In addition, the skin of red cultivars contains anthocyanins, which are mainly confined to the hypodermis and impart the red colour. In some growing years, the cells adjacent to the pulp can be coloured, and in *tenturier* varieties the pulp may be coloured. In general, phenolics accumulate very rapidly during berry maturation. A selection of basic wine phenolics is given in Figure 13.

The total phenol content of wine is less than that present in the fruit because traditional methods of destemming, crushing and fermenting usually give extraction rates of no more than 60%. Extraction efficiency is even less than this because phenolics are invariably added to the wine by microbial activity and ageing in wood. Processing protocol will largely determine the phenolic composition of a wine since different phenolic fractions are variously located in the berry. Flavonoids are derived primarily from the seeds, skins and stems of the grape, while anthocyanins

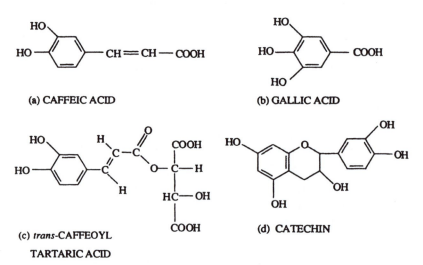

Figure 13   *Representative samples of some phenolics found in grapes and wines. Structures (a), (b) and (c) are non-flavonoids; (d) is a flavonoid*

and flavonols are extracted mainly from skins. Catechins and leucoanthocyanins reside mainly in seeds and stems. Increasing skin contact time and fermentation temperature, and the degree of berry disruption increase the flavonoid content of a wine, and this group of compounds usually accounts for 80–90% of the phenolic content of conventionally vinified red wines, and *ca.* 25% of the total in white wines vinified without skin contact. Owing to the broad chemical diversity of the phenolics found in grapes and wines, quantification of these important substances is not without difficulty. Accordingly, total wine phenolics are usually expressed in the arbitrary units of a phenolic standard that would produce the same analytical result. Gallic acid is the accepted 'standard', and wine phenolics are usually listed as 'gallic acid equivalents' (GAE).

The non-flavonoids are low-molecular-weight compounds, the phenolic acids, which in grape are often conjugated as esters or glycosides. These compounds reside in the vacuoles of berry cells, and winemaking procedures will determine how they are transferred to musts and wines. There are two basic groups and are based on the hydroxy derivatives of benzoic acid and cinnamic acid. In the former group the most important are protocatechinic acid (3,4-dihydroxybenzoic acid), syringic acid (3,4,5-trihydroxybenzoic acid), salicylic acid (2-hydroxybenzoic acid) and genistic acid (2,5-dihydroxybenzoic acid). In the cinnamic acid group, of greatest interest are caffeic acid (3,4-dihydroxycinnamic acid), *p*-coumaric acid (4-hydroxycinnamic acid), ferulic acid (3-methoxy-4-hydroxycinnamic acid) and sinapic acid (3,5-dimethoxy-4-hydroxycinnamic acid). In grapes, the benzoic acids are usually present as glycosides or esters, whereas the cinnamic acids can exist either in the free form or as esters. The most important esters are those formed with tartaric acid, and these are caftaric acid (caffeoyl tartaric acid) and coutaric acid (coumaroyl tartaric acid). Cinnamic acids can exist in two isomeric forms: *trans*, the most abundant in nature, and *cis*. Caftaric acid and coutaric acid are the major phenolics in unfermented grape juice, and the capacity of the berry to produce these acids persists throughout the ripening period. Non-flavonoid phenolics are subject to rapid oxidation and produce a faint yellow colour in ethanolic solution. They are also important as precursors of troublesome volatile phenols such as ethyl phenol, but otherwise they have little sensorial significance.

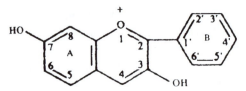

Figure 14   *Base structure of a typical flavonoid, showing rings and numbering system*

As we have said, the flavonoid phenols are located primarily in the skin, seed and stem of the grape, and they have a basic structural unit consisting of two hydroxybenzene rings (named A and B) joined by an oxygen-containing pyran ring (Figure 14).

There are a number of flavonoid sub-groups depending upon how the pyran ring is modified and changes in the state of oxidation of the core molecule that result from variation in hydrogen, hydroxyl and ketone groups associated with carbon atoms 2, 3 and 4. The three predominant groups found in grapes are flavonols, flavan-3-ols and anthocyanins, of which the latter are the most readily perceived. Much of the structure and colour of a wine is ultimately due to these flavonoids. Flavonoids may exist in the free, monomeric form, although this is somewhat unusual in nature, or they can be polymerised with other flavonoids to give dimers and then an array of larger, polymeric structures. They can also be esterified to other organic compounds, in which case they are termed acylated, or they may have sugars attached, thus becoming glycosylated (glycoside derivatives). Polymeric flavonoids make up the major fraction of all phenolics encountered during winemaking.

The most widespread flavonoids in *V. vinifera* berries are the flavonols, which are intense yellow pigments in the skins of red and white cultivars. To a slightly lesser extent we also find the flavanonols, which are pale yellow pigments. The major flavonols found in the grape are the 3-glucosides of quercetin, kaempferol and myricetin, and all three of these have a light requirement for their synthesis. All three pigments are found in red wine grapes, whereas white cultivars do not contain myricetin. The flavonols, of which quercetin is the most abundant, are all quite bitter and are believed to participate in co-pigmentation of anthocyanins, resulting in increased colour stability in red wines. These compounds reside in the grape skin, and the glycosidic forms are more

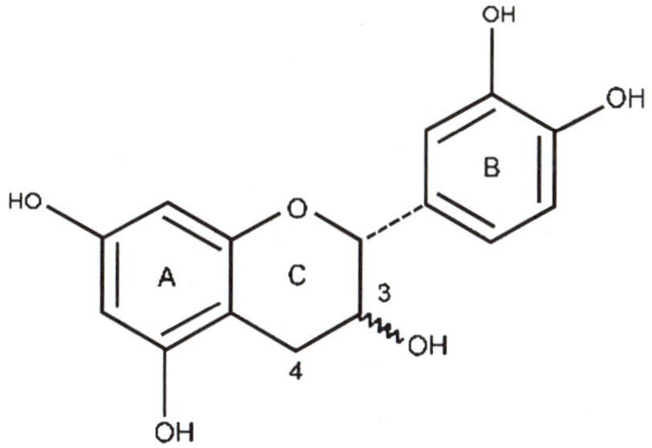

Figure 15   *The structures of (+)-catechin and (−)- epicatechin. (+)-Catechin has S-stereochemistry at carbon 3 (OH-group comes from the plane). (−)-Epicatechin has R-stereochemisty at carbon 3 (OH-group enters the plane) (after Thorngate and Noble[21])*

bitter than aglycones. Rhamnosyl derivatives appear to be the most common form of flavonols in the grape. The flavanonol most frequently reported from grapes, must, and wine is dihydroquercetin, which is also known as taxifolin.

In the context of winemaking, flavan-3-ols may be called 'catechins'. Catechins are primarily located in seeds and stems, although low levels are present in skins. They are the most important of the monomeric flavonoids, and their structure is based on (+)-catechin (Figure 15), of which four stereoisomers exist: (+)- and (−)-catechin and (+)- and (−)-epicatechin. Hydroxylation at the 5' position yields (+)-gallocatechin and its isomer (−)-epigallocatechin. Of these forms, (+)-catechin, (−)-epicatechin and (−)-epigallocatechin are the principal monomeric catechins in grapes. In white wines made with limited skin contact, catechins account for most of the flavonoid phenols and contribute significantly to the flavour profile. They act as precursors for browning in white wines and contribute to browning and bitterness in reds. Their concentration in white wines ranges from 10 to 50 mg L^{-1} and may reach 200 mg L^{-1} in reds. Also present in grapes are leucoanthocyanidins (flavan-3,4-diols), which differ from catechins in that there is a hydroxyl group at carbon position 4 instead of hydrogen. The introduction of this radical at carbon 4 results in three sites of asymmetry (at C2, C3 and C4), and so there are eight possible isomers. Upon heating in acid solution, leucoanthocyanidins, which, as their name suggests, are colourless, are converted to anthocyanidins, which are coloured.

Besides existing as monomers, flavan-3-ols and flavan-3,4-diols may exist in dimeric, oligomeric and even larger groupings. Polymerisation of catechin and leucoanthocyanidin flavonoids produces a group of compounds known as procyanidins. Their classification is based on the nature of the flavonoid monomers, bonding, esterification to other compounds or functional properties. Of the many groups of procyanidins, only the dimers and some of the trimers have been completely elucidated, even though these polymeric flavonoids make up the major fraction of total phenolics encountered in all stages of winemaking. Polymerisation, either oxidative or non-oxidative, yields tannins and condensed tannins, respectively, with the latter being the commonest functional class of procyanidins.

By definition, tannins are compounds capable of producing stable combinations with phyto-proteins and other plant-derived polymers, such as polysaccharides. Chemically, tannins are formed by the polymerisation of phenolic molecules, and they may exist in a variety of configurations. The resulting structures must be sufficiently large to form stable complexes with proteins, but not so gross that they become inactive. The molecular weight of tannins varies from 600 to 3500 kDa. Depending on the nature of their monomeric units, hydrolysable or 'gallic' tannins are distinguishable from condensed or 'catechin' tannins. Hydrolysable tannins, which encompass gallotannins and ellagitannins, liberate gallic acid and ellagic acid, respectively, on hydrolysis and also include glucose moieties. They are not normally synthesised by the grape.

Condensed tannins are largely grape-derived and are polymers of the catechins, their basic structural units being (+)-catechin and (−)-epicatechin. Catechins are constructed from two benzene rings that are joined by saturated, oxygenated heterocyclic ring. This is the phenyl-2-chromane nucleus (Figure 15). This structure has two asymmetrical carbon atoms (C2 and C3), with (+)-catechin and (−)-epicatechin being the most stable of the four isomers.

Heating these polymers in acid solution releases unstable carbocations that are converted into red/brown condensation products, mainly cyanidins. This is the reason why these compounds are referred to as 'procyanidins', a term that replaces the previously used 'leucocyanidin'. These molecules exhibit great structural diversity, something that results from the number of hydroxyl groups, their position on the aromatic rings, the stereochemistry of the asymmetrical carbon atoms in the pyran ring, as well as the number and type of bonds between the basic units. This structural diversity largely explains the wide variation in the properties of tannins. Basic catechin units should not be regarded as tannins because their molecular weight is too low. In the dimeric form, however, they have sufficient mass to be able to form stable combinations with proteins.

Dimeric procyanidins may be divided into two categories, which are identified by a letter and a number:

(i) Type-B procyanidins ($C_{30}H_{26}O_{12}$) are dimers resulting from the condensation of two catechin units linked by a C4–C8 or a C4–C6 interflavan bond (see Figure 16).

(ii) Type-A procyanidins ($C_{30}H_{24}O_{12}$) are dimers that, in addition to the C4–C8 or C4–C6 interflavan bond, also have an ether bond between the C5 or C7 carbons atoms of the terminal unit and the C2 carbon of the upper unit.

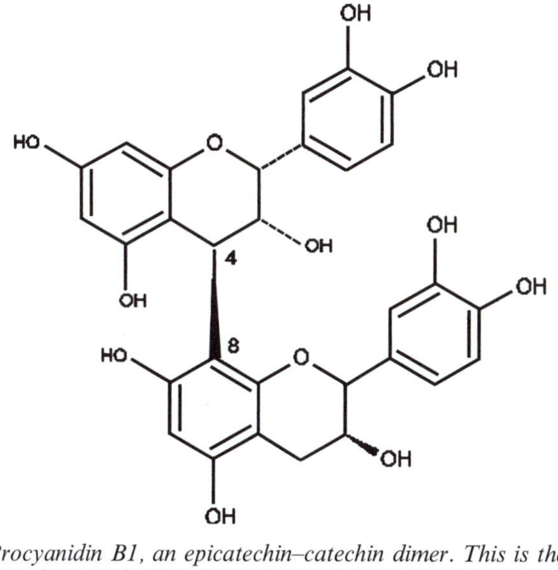

Figure 16 *Procyanidin B1, an epicatechin–catechin dimer. This is the most important dimer to be found in red and white grape skins (after Gawel[22])*

Trimeric procyanidins may also be divided into two categories:

(i) Type-C are trimers with two interflavan bonds corresponding to those of Type-B dimers.

(ii) Type-D are trimers with two interflavan bonds, one Type-A and another Type-B. Oligomeric procyanidins are polymers formed from 3–10 flavan units, linked by C4–C8 or C4–C6 bonds. An almost infinite number of isomers are possible. Condensed procyanidins contain more than 10 flavan units, which have a molecular weight in excess of 3000 kDa .

Anthocyanins are glycosides of the flavanoid anthocyanidins or, put another way, are glycosylated polyhydroxy and polymethoxy derivatives of the 2-phenylbenzopyrylium cation (*i.e.* the flavilyum cation). The most common sugar components of anthocyanins are the monosaccharides glucose, rhamnose, galactose, arabinose and xylose (in that order of frequency). These compounds were first studied thoroughly by the Nobel prize-winning German chemist, R. M. Willstätter before the First World War, while in the UK they were the subject of extensive early research by the pioneer plant biochemist Muriel Wheldale, who published a classic text in 1915.[23] Willstätter, who was investigating pigmentation in plants, found that many of the common pigments were based on molecules called anthocyanidins (in Greek *anthos*, flower; *cuanos*, dark blue), which together with a sugar moiety form the core of an anthocyanin molecule. The most important part of an anthocyanin molecule, the flavylium cation, is by definition an aglycone[†], and contains conjugated double bonds responsible for absorption of light around 500 nm, thus causing them to appear red to the human eye. The aglycones (anthocyanidins) are usually penta- $(3,5,7,3',4')$ or hexa- $(3,5,7,3',4',5')$ substituted, and are very unstable, but when they are linked to a sugar molecule, they become much more stable, more soluble and more chromogenic.

There are numerous reported anthocyanins but only 22 different anthocyanidins, of which five are to be found in red wines: cyanidin, peonidin, delphinidin, malvidin and petunidin. These

---

[†] Technically defined as the non-sugar compound remaining after replacement of the glycosyl group of a glycoside by a hydrogen atom.

aglycones differ in the number of hydroxyl and methoxyl groups in the B-ring of the flavylium cation (see Figure 14). The anthocyanin molecule exists in several different chemical forms (most of which are colourless), the equilibrium between them being pH-dependent. The appropriately low pH of wine helps to shift the equilibrium of the different forms of these anthocyanins towards the flavylium ion, which is the red form of the molecule. In nature, the most widespread anthocyanins are 3-glucosidic derivatives of antho-cyanidins, and in *V. vinifera* they are all of this type. Glucose is bound at its 1-position to the 3-position in the pyran ring (ring C). An additional glyceride link is available at the 5-hydroxyl position in the benzo ring, but this is not used in *V. vinifera* anthocyanins. In non-*V. vinifera* grapes, however, 3,5-diglycerides are very common. The 3-glucoside in *V. vinifera* may also be esterified at the C-6 of the glucose ring, with acetic acid, *p*-coumaric acid or caffeic acid, but such acylation of the sugar renders the molecule less soluble in water. Malvidin 3-glucoside is the major colouring compound in grapes, comprising up to 90% of the anthocyanin content in Grenache, for example. The properties of, and amount of each of the five types of *V. vinifera* anthocyanidin vary widely among cultivars/varieties and also according to cultural conditions. The proportion of each markedly influences grape colour and colour stability. Colour is determined by the number of hydroxyl groups in the B-ring of the anthocyanidin, and ranges from blue to orange through to red. Resistance to oxidation is greater when there are no *ortho*, or adjacent hydroxyl groups in this ring, as is the case with malvidin and peonidin. As we have said, it is malvidin, with two methoxy groups, and only one hydroxyl group in the B-phenyl ring, that is the most abundant anthocyanidin in red wines, and is largely responsible for their colour when young. Wine anthocyanins are affected by temperature, light, pH, oxygen, aldehydes, SO_2, some metallic ions and sugar degradation pro-ducts, and they are also prone to condensation and polymerisation reactions with other phenols. The chemistry of the anthocyanins and the changes that they undergo during fermentation and ageing is complex, but a useful outline has been given by Clarke and Bakker.[24]

Polyphenolics found in wines, including monomers of flavanols, flavanonols, anthocyanins and tannins, have been subjected to a variety of classifications, of which the size-based, quadripartite system proposed by Montedoro and Bertuccioli[25] is the one found most useful:

(i) Molecular weight below 300 kDa (monomers) or 500 (glycosylated monomers). These are responsible for colour (anthocyanins, red; procyanidins, colourless; flavonols, yellow) and bitter taste (flavanols and flavonols).
(ii) Molecular weight of 500–1500 kDa (2–5 monomer units). These are responsible for astringency, body and yellow colour (flavan tannins).
(iii) Molecular weight of 1500–5000 kDa (6–10 monomer units). These confer astringency and body, and may impart a yellow–red to yellow–brown colour (condensed flavan tannins).
(iv) Molecular weight above 5000 kDa (above 10 monomer units). These are highly condensed flavan tannins and are generally insoluble in water.

To give some idea of the phenolic profile likely to be encountered, the analysis of a Pinot Noir wine is documented in Table 1.

Plant phenolics are synthesised by several different routes, but two major pathways are involved: the shikimic acid pathway and the malonic acid pathway. The former participates in the biosynthesis of most plant phenolics, and this is certainly the case in grapes. The shikimic acid pathway (Figure 17) converts 'simple' carbohydrate precursors, phosphoenolpyruvic acid and erythrose-4-phosphate, derived from glycolysis and the pentose phosphate

Table 1   *Phenolic content of a Pinot Noir wine after vinification with manual punch-down (after Fischer et al.[26])*
(Reproduced by kind permission of Blackwell Publishing)

Phenolic compound	Content (mg L^{-1})
Gallic acid (3,4,5-trihydroxy-benzoic acid)	6.6
Catechin (*trans*-flavan-3-ol)	21.8
Epi-catechin (*cis*-flavan-3-ol)	13.0
Procyanidins	36.4
Caftaric acid (caffeoyl-tartaric acid)	8.8
Quercetin-3-glycoside	<2.0
Quercetin (flavonol)	0.4
Malvidin-3-glycoside	52.8
Monomeric anthocyanins	61.5
Polymeric anthocyanins	0.0

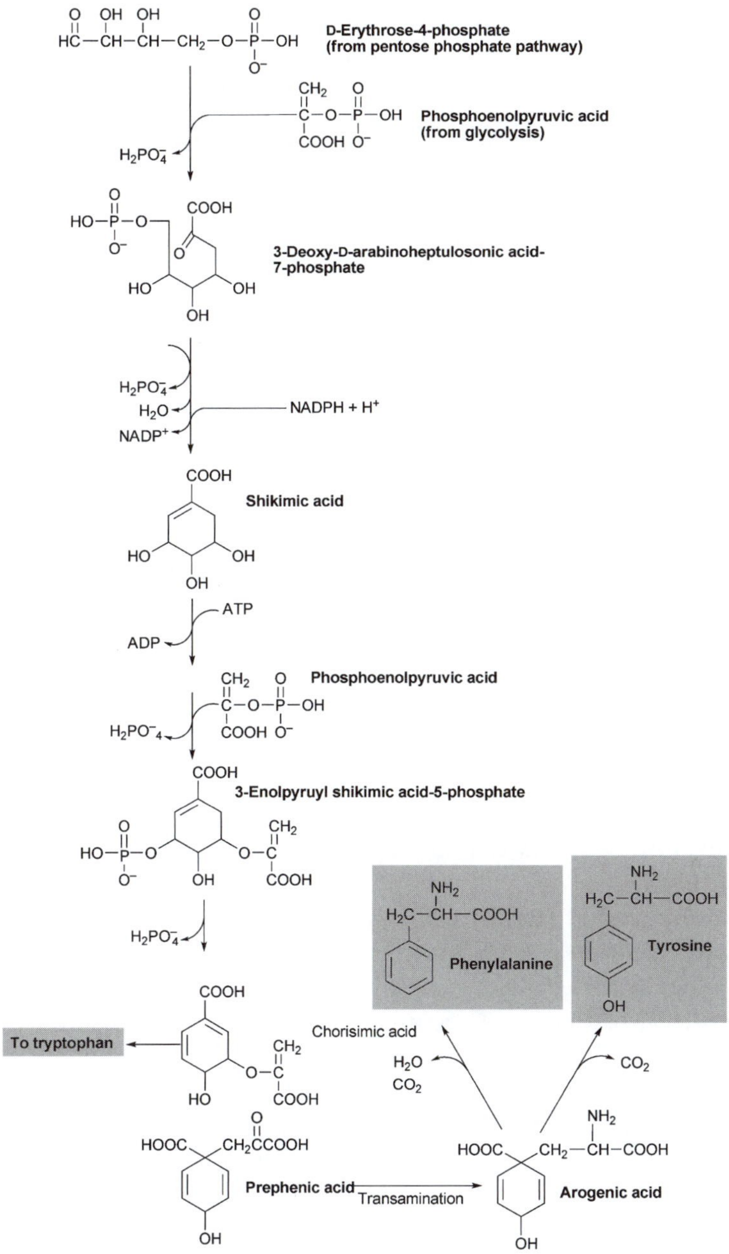

Figure 17 *The shikimic acid pathway*

pathway, respectively, to aromatic amino acids.[27] The pathway is present in green plants, fungi and bacteria, but is not found in animals, which therefore have no means of synthesising phenylalanins, tyrosine and tryptophan. Incidentally, the well-known, broad-spectrum herbicide glyphosate (available commercially as 'Roundup') kills plants by blocking a step in the shikimic acid pathway.

The condensation of phosphoenolpyruvate and erythrose-4-phosphate is catalysed by 3-deoxy-D-arabino-heptulosonate-7-phosphate (DAHP) synthase. Through a series of reactions, shikimic acid is produced, which is then converted to chorismate, the final precursor for the synthesis of phenylalanine, tyrosine and tryptophan. Phenylalanine and tyrosine then go on to produce phenolics (Figure 17).

REFERENCES

1. A.M. Negrul, *Atti. Acad. Ital. Vine e Vino.*, 1960, **12**, 113.
2. J.E. Planchon, *Monographia Phanerogamerum.*, 1887, **5**, 305.
3. M.A. Amerine and M.A. Joslyn, *Table Wines: The Technology of Their Production*, 2nd edn, University of California Press, Berkeley, 1970.
4. A.M. Negrul, *Ampelographia SSSR, Moscow*, 1946, **1**, 159 (in Russian).
5. L. Levadoux, *Annales de la' Amélioration des Plantes*, Ser. B, 1956, **1**, 59.
6. F. Radler and D.H.S. Horn, *Aust. J. Chem.*, 1965, **18**, 1059.
7. C. Srinivasan and M.G. Mullins, *Ann. Bot.*, 1976, **38**, 1079.
8. S.S. Negi and H.P. Olmo, *Vitis*, 1971, **9**, 265.
9. H. Ito, Y. Motomura, Y. Konno and T. Hatayama, *Tohoku J. Agric. Res.*, 1969, **20**, 1.
10. H. Müller-Thurgau, *Landw. Jahrb. Schweiz.*, 1898, **12**, 135.
11. B.G. Coombe, *Am. J. Enol. Vitic.*, 1992, **43**, 101.
12. A.K. Kanellis and K.A. Roubelakis-Angelakis, Grape, in G.B. Seymour, J.E. Taylor and G.A. Tucker, (eds), *Biochemistry of Fruit Ripening*, Chapman & Hall, London, 1993, 189–234.
13. L. Bisson, *Practical Winery and Vineyard*, July/August, 2001, 32–43.
14. J. Kennedy, *Practical Winery and Vineyard*, July/August, 2002, 14–18.

15. P. Schrier, F. Drawert and A. Junker, *Chem. Mikrobiol. Technol. Lebensm.*, 1976, **4**, 154.
16. R.F. Simpson, *Chemistry and Industry (London)*, 1978, **1**, 37.
17. M.J. Lacey, M.S. Allen, R.L.N. Harris and W.V. Brown, *Am. J. Enol. Vitic.*, 1991, **42**, 103.
18. C. Bayonove, R. Cordonnier and P. Dubois, *C. R. Acad. Sci. Paris D.*, 1975, **281**, 75.
19. F.B. Power and M.K. Chesnut, *J. Am. Chem. Soc.*, 1921, **43**, 1741.
20. S.T. Lund and J. Bohlmann, *Science*, 2006, **311**, 804.
21. J.H. Thorngate and A.C. Noble, *J. Sci. Food Agric.*, 1995, **67**, 531.
22. R. Gawel, *Aust. J. Grape Wine Res.*, 1998, **4**, 74.
23. M. Wheldale, *The Anthocyanin Pigments of Plants*, Cambridge University Press, Cambridge, 1915.
24. R.J. Clarke and J. Bakker, *Wine Flavour Chemistry*, Blackwell, Oxford, 2004.
25. G. Montedoro and M. Bertuccioli, The flavours of wines, vermouth and fortified wines, in I.D. Morton and A.J. Macleod (eds), *Food Flavours, Part B*, Elsevier, Amsterdam, 1986.
26. U. Fischer, M. Strasser and K. Gutzler, *Int'l. J. Food Sci. Technol.*, 2000, **35**, 81.
27. K.M. Herrmann and L.M. Weaver, *Annu. Rev. Plant Physiol. Plant Mol. Biol.*, 1999, **50**, 473.

CHAPTER 3

The Yeast and Fermentation

3.1 THE YEAST

The role of yeasts in the transformation of sound grape juice into wine was not elucidated until the second half of the 19th century, when, after much fundamental research by a number of scientists, Louis Pasteur finally gave credence to the notion that ethanolic fermentation was brought about by a small, living organism, not merely by a series of inanimate chemical reactions. The synthesis of many of Pasteur's ideas was presented in two classic works: *Etudes sur le vin*, published in 1866[1] and *Etudes sur la bière*,[2] published 10 years later. Pasteur actually commenced his researches into the fermentation of sugar in the mid-1850s, and, between 1855 and 1875, he established unequivocally: (1) the role of yeast in ethanolic fermentation, (2) fermentation as a physiological phenomenon and (3) differences between the aerobic and anaerobic utilisation of sugar by yeast; indeed, Pasteur invented the terms 'aerobic' and 'anaerobic'. Organisms that we now know to have been yeasts, had, in fact been observed some 200 years earlier, when Antonj van Leewenhoek, son of a cloth merchant from Delft, and amateur lensmaker, reported his microscopical findings to the Royal Society of London during the years 1674–83. Using his own, hand-built microscope, van Leewenhoek discerned and illustrated his famous 'animalcules' in samples of beer, but, at that time, the doctrine of 'spontaneous generation' (abiogenesis) held sway, and it was inappropriate to link small organisms, such as these, to fermentation. The significant events that occurred in the two centuries that elapsed between van Leewenhoek's original observations and Pasteur's publications, have been well documented (for example, see Harden[3] and Hornsey[4]), and involve such luminaries as Lavoisier, Schwann, Gay-Lussac, Liebig and Berzelius. Once the 'vitalistic' role of yeast had been generally accepted, the way was cleared for the study of many more small organisms, and, after another hundred years, the science of microbiology was well under way. Much of the foundation work concerned the yeast and its ability to effect ethanolic fermentation, and this organism, in addition to its industrial significance, has proved to be a useful tool in the related fields of biochemistry, molecular biology and genetics. The significance of yeast and the role played by ethanolic fermentation during the formative years of the sciences of biochemistry and microbiology have been admirably reviewed by Barnett.[5]

3.1.1 Taxonomy

Yeasts were the first microbes to be extensively studied scientifically, mainly because of their relatively large cell size, and inherent industrial importance. As microscopic fungi, yeasts are the least complicated of the eukaryotes, and the very name 'yeast' encompasses a wide range of unicellular ascomycete and basidiomycete genera, which are known to undergo some form of sexual reproduction during their life cycle, which culminates in the formation of ascospores and basidiospores, respectively. Also included, are a number of forms that reproduce only by asexual means, and these are known as 'imperfect yeasts', and are classified in the Deuteromycetes. Over the years, many of the latter have been shown to be non-sexual stages of the former two types. In a major, somewhat controversial, taxonomic work, Lodder[6] defined yeasts as 'those fungi whose predominant growth form is unicellular', a statement that implies, as is the case, that some species can produce transient filamentous (hyphal) outgrowths.

Emil Hansen's first classification of yeasts, which appeared around the dawn of the 20th century, only made a distinction between those that produced sexual spores, and those that did not (referred to as sporogenous and asporogenous, respectively). We now know that Hansen's sporogenous yeasts belong to the phyla Ascomycota and Basidiomycota, while the asporogenous forms are classified as Fungi Imperfecti. Yeasts have been traditionally classified by means of conventional cultural techniques that determine a range of morphological, biochemical, and physical properties. Criteria and methods for conducting such tests have been reviewed by Kreger-van Rij,[7] and include: cell shape and size; sporulation; fermentation and assimilation of different sugars; assimilation of a nitrogen source; growth-factor requirement and resistance to cycloheximide. Workload for such tests is demanding, and, in some cases, results for final identification may not be available for a fortnight. Progress in the development of simpler, more rapid methods was made during the late 1970s and the 1980s, and several easy-to-use diagnostic kits have been made available commercially. Such kits, however, are mainly applicable to a limited range of yeasts of industrial and medical significance; API 20C, and API 50CH are examples.

As modern fungal taxonomy would have it, classic ethanolic fermentations are carried out by strains of the yeast, *Saccharomyces cerevisiae*, regardless of whether the end product is ale, lager, wine, or a distilled beverage. Over the years, there have been a variety of specific synonyms given to certain industrial strains of this fungus that have been used to produce specific alcoholic beverages, and some of these names have elevated the organism concerned to species status. Examples are *Saccharomyces carlsbergensis/Saccharomyces uvarum* for the bottom-fermenting yeast used in lager beer production, and *Saccharomyces ellipsoideus* for the wine yeast. Such names have now largely disappeared from the literature, and are only used by microbiologists in the specialised fields in which they have been relevant historically.

In their monumental study, Barnett *et al.*[8] report 678 recognised yeast species for which cultures, and the results of standard physiological tests were available. The same work provides a register of almost 4000 taxonomic names that have been used at various times, since Hansen's work, together with their provenances and synonyms. The register is important, because some yeast names have changed so regularly that it is difficult for non-systematists to know which species is being referred to in different publications. Like other plants, the naming of yeasts is now governed by the International Code for Botanical Nomenclature, and the accepted name for the major wine yeast is: *S. cerevisiae* Meyen *ex* E.C.Hansen var. *ellipsoideus* (E.C.Hansen) Dekker. Barnett *et al.*,[8] use the following scheme for classifying this organism:

KINGDOM: Fungi
PHYLUM: Ascomycota
CLASS: Hemiascomycetes
ORDER: Saccharomycetales
FAMILY: Saccharomycetaceae
GENUS: *Saccharomyces*
SPECIES: *S. cerevisiae*

In a novel approach to the taxonomy of *S. cerevisiae*, Naumov[9] proposed six cultivars:

(i) 'cerevisiae', for brewer's top-fermenting yeasts, and baker's and distiller's strains;

(ii) 'ellipsoideus', for strains from primary wine fermentations;

(iii) 'cheresanus', for strains from secondary wine fermentations (*e.g.* those forming a film on sherry during oxidation of ethanol);

(iv) 'oviformis', for wine yeasts that do not ferment galactose, and that are resistant to high concentrations of ethanol and sulfites;

(v) 'diastaticus', for strains able to metabolise dextrins (*e.g.* soluble starch);

(vi) 'Logos', for strains able to ferment melibiose.

3.1.2 Cell Structure

S. cerevisiae cells are generally ellipsoidal in shape, ranging from 5 to 10 μm at the large diameter, and 1–7 μm at the small diameter, the cells increasing in size with age. Mean cell volumes are 29 μm^3 for a haploid cell, and 55 μm^3 for a diploid cell. Yeast cells show most of the structural and functional features of higher eukaryotic cells, and have been used as a useful model for eukaryotic cell biology. The main features of a 'typical' cell are illustrated in Figure 1, while the molecular composition of a freeze-dried wine yeast is given in Table 1.

The macromolecular composition of a similar dried organism would be as follows: moisture, 2–5%; protein, 42–46%; carbohydrate, 30–37%; nucleic acid, 6–8%; lipids, 4–5%; minerals, 7–8%.

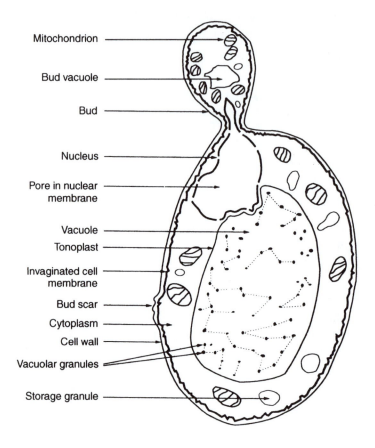

Mitochondrion
Bud vacuole
Bud
Nucleus
Pore in nuclear membrane
Vacuole
Tonoplast
Invaginated cell membrane
Bud scar
Cytoplasm
Cell wall
Vacuolar granules
Storage granule

Figure 1   *Diagrammatic representation of a section of a budding yeast cell as seen under the electron microscope (after Hornsey[10])*

From the data presented in Table 1, Rosen gave this particular yeast the approximate molecular formula of $C_{4.02}H_{6.5}O_{2.11}N_{0.43}P_{0.03}$.

The carbohydrate fraction of yeast includes structural components, such as the cell wall (majority), and compounds concerned with food storage, and/or stress resistance, such as glycogen and trehalose. Quite large differences may be observed in the concentrations of individual minerals during the various stages of the cell cycle, and, of course, the nucleic acid complement of cells will vary according to their ploidy. Some 75–80% of the actively growing yeast cell consists of water, much of which forms the basis of the cytoplasm (cytosol). The cell is surrounded by a rigid envelope, the

Table 1 *Molecular composition of dried wine yeast (after Rosen[11])*
(Reproduced by kind permission of Reed Elsevier)

Component	Percentage of dry weight
Carbon	48.2
Hydrogen	6.5
Oxygen	33.8
Nitrogen	6.0
Phosphorus	1.0
Magnesium	0.1
Calcium	0.04
Potassium	2.1
Sulfur	0.01
Iron	0.005

cell wall, which comprises some 20–25% of the wet weight of the cell, and which determines the shape of the cell, provides a barrier against the high osmotic pressure in the cytoplasm and generally maintains the integrity of the cell. The wall overlies the cytoplasmic membrane (plasmalemma), and is separated from it by a discrete gap, called the periplasmic space. The cell wall, periplasmic space and plasmalemma are often referred to as the 'cell envelope', and constitute some 15–25% of the total volume of the cell. The yeast cell wall is a complex extracellular organelle, capable of dynamic change in response to changes in external conditions and to different stages of the life cycle. Rigid the yeast cell wall might be; inert it certainly is not. The wall can be 150–300-nm thick, according to strain and growth conditions, and is metabolically active, containing enzymes capable of facilitating the transfer of macromolecules into the cytosol. Under the electron microscope, the yeast cell membrane appears 7–10 nm in thickness, and the periplasmic space 35–45-Å thick.

As a rule, plant cell walls are constructed of different polysaccharide moieties to which a variety of proteins are attached, and consist of two basic components: a malleable gel-like substance, and a fibrous matrix, which provides great strength. In most fungi, the fibrous component is chitin, a linear polymer of β-1,4 linked *N*-acetyl glucosamine (GlcNAc) residues. Cells of *S. cerevisiae* contain relatively little chitin (*ca.* 1% dry weight), most of it being concentrated at the bud neck, where it plays a role in cytokinesis (in this case, budding). In yeasts, the basic fibrous network is based upon 1,3-β-glucan, which accounts for some 40% of the cell wall dry weight, and is mainly responsible for the unusually high mechanical stability of the wall. 1,3-β-Glucan is synthesised at the cell surface, and consists of linear chains, with an average of 1500 glucose units. When viewed under the electron microscope, it appears to exist in two forms; fibrous and amorphous. The former is responsible for the shape and rigidity of the wall, while the latter contributes elasticity, and acts as an anchor for the outer layer of mannoproteins (glycoproteins). The fibrous glucan component is insoluble in water, alkali and acetic acid, whereas the amorphous fraction is insoluble in water and acid, but soluble in alkali. Two other polysaccharides, 1,6-β-glucan and chitin are, linked to 1,3-β-glucan, and have functions other than preserving cellular shape and rigidity. 1,6-β-Glucan, which constitutes approximately 10% of the cell wall dry weight, consists of polymers with an average length of 350 glucose units, and plays a critical role in the cell wall architecture by linking cell wall components together, and by anchoring mannoproteins. 1,3-β-Glucan is a sparsely branched polymer, whereas 1,6-β-glucan is much branched. Most of the mannosylated proteins are located externally to the 1,3-β-glucan network, resulting in a discrete outer 'mannan' layer, which may comprise 25–30% of the cell wall. The degree of glycosylation in the mannoproteins varies, one group, for example, consisting of 90% mannose and 10% peptide, and their molecular weight varies accordingly (between 20 and 450 kDa). Some mannoproteins are purely structural, and are anchored to the wall, while others manifest themselves as extracellular enzymes, such as invertase.

Small quantities of lipid and inorganic phosphate are also incorporated into the wall matrix. The reported lipid content in the wall of *S. cerevisiae* varies from 2 to 15%, but, in truth much of this may arise by contamination through membrane lipid. In a healthy yeast cell, the cell wall surface carries a net negative charge, which is attributable to phosphate chains located on the outer mannoprotein layer. The charge can be qualitatively demonstrated by alcian blue staining, and quantified by ion-exchange, or electrophoretic (zeta-potential) methods. The cell surface is also hydrophobic, a property conferred upon it by lipids, again located in the outer wall complex. As we have said, a number of important enzymes, such as invertase, have been reported as being associated with the cell wall, although some of these are actually located in the periplasmic space.

To partly illustrate the importance of the cell wall, consider, for a moment, the problems and stresses confronting yeast cells during the course of a wine fermentation. Initially, they find themselves in a nutrient-rich environment of low pH, but high sugar content. The latter, in particular, presents a considerable osmotic challenge. By the end of the fermentation, the cells find themselves in an environment that is more osmotically friendly, but contains around 12–15% ethanol, a level that would be toxic to many organisms. In addition, some of the nutrients may now start to become depleted, and, depending on how well the fermentation can be attemperated, there may well have been considerable temperature fluctuations. The yeast cell wall stands between all of these environmental changes and the cytoplasm and its organelles within the cell membrane. The importance of the wall to the yeast may be gauged by the fact that some 20% of *S. cerevisiae* genes control functions that are in some way related to cell wall biogenesis.

Most of the lipids in the cell are structural components of the plasmalemma, and the intracellular membrane-bound organelles. When viewed under the electron microscope, the cell membrane has a series of invaginations on its outer surface, which become most prominent during the stationary phase of growth. In frozen-etched surface view these invaginations can be seen as elongated folds, some 4 μm long. Also visible are clusters of hexagonally arranged particles, around 180Å in diameter. The role of these invaginations has not yet been fully elucidated. The membrane consists of lipids and proteins in approximately equal proportions, the proteins being functional rather than structural (see the 'fluid mosaic model' of Singer and Nicholson[12]). Membrane proteins are involved in the regulation of solute transport, and include enzymes that mediate cell wall synthesis, as well as constituting the ATPase responsible for maintaining the plasma membrane proton-motive force. Intriguingly, the enzymes for synthesising membrane lipids are located elsewhere, it being known, for instance, that the initial steps of sterol synthesis occur in the mitochondria (as do some enzymes associated with sterol degradation). The major lipid components of the membrane are phospholipids and sterols, and it is these compounds that are responsible for its hydrophobicity. The principal phospholipids are phosphotidylinositol, phosphotidylserine, phosphotidylcholine and phosphotidylethanolamine, while the main sterol is ergosterol. The last named can, if necessary, be added to the early stages of a fermentation as a substitute for the oxygen requirement during the synthesis of membrane components. Zymosterol is also a relatively abundant sterol in the yeast cell membrane. Sterols are orientated in perpendicular fashion within the membrane, between the chains of other lipids, and modulate its fluidity.

The periplasmic space (periplasm) is not an organelle, as such, but is more than a mere cavity between wall and membrane. Through the periplasm there is a dynamic migration of cellular components passing to the cell from the external medium (must) mixing with those moving out of the cell as a consequence of metabolism (*e.g.* ethanol). It appears that a number of enzymes are 'stored' here, most significantly invertase and acid phosphatase. The protein (mannoprotein) concentration in the periplasm is very high, much higher than would be needed if the protein was for purely enzymatic purposes, and it has been suggested that the resultant gel-like consistency provides a 'protective' layer for the plasmalemma.

The yeast cytoplasm (cytosol) is an acidic (pH 5.25) colloidal fluid, mainly containing ions and low/intermediate molecular weight organic compounds and soluble macromolecules (*e.g.* enzyme proteins and glycogen). The main cytosolic enzymes are those involved in glycolysis, fatty acid synthesis and protein synthesis (some). The internal stability, and structural organisation of the cell is guaranteed by the cytoskeleton, which is composed of microtubules (tubulin) and microfilaments (actin). These are dynamic structures that fulfil their function by regulated assembly and disassembly of protein subunits. Thus, α- and β-tubulin monomers polymerise as heterodimers to give microtubules, while globular monomers of G-actin polymerise into double-stranded microfilaments of F-actin. Microtubules and microfilaments are important in several cell reproductive processes, including mitosis, meiosis, septum formation and organelle motility. The cytoskeleton is invisible using conventional microscopy, but can be demonstrated by specialised electron microscope techniques. In the 1980s, it was shown that yeast had genes for actin and tubulin, and actin and tubulin proteins behaved like their mammalian counterparts. The main protein in the actin cytoskeleton is 86% identical to mammalian actins. It has been shown that the actin cytoskeleton consisted of two filament-based structures: the actin cortical patch and the actin cables. The actin cortical patches show a polarised distribution that changes during the cell cycle: first they appear at the incipient bud site, suggesting a role in bud emergence; soon thereafter they are found within the growing bud, indicating a role in bud growth; and late in the cell cycle they reorganise into two rings in the neck, where they are believed to be involved in septation and cytokinesis. Electron microscopy has shown actin cortical patches to be invaginations of the plasmalemma around which actin filaments and actin-associated proteins are organised. It has also been shown that subsets of actin cortical patches can move of speeds of up to 1 μm s^{-1}. The actin cables, which consist of bundled actin filaments, were observed to run along the long axis of budding cells. Such information fits in well with the understanding that actin is involved in: polarised cell growth; dynamic reorganisation of the cell cortex; membrane trafficking at the cell cortex and organelle segregation at cell division (*e.g.* actin filaments direct cell growth to the emergent bud). Yeast has a single actin gene; ACT 1. The actin cytoskeleton disassembles in response to osmotic stress, and is induced to re-assemble only after osmotic balance has been restored.

Yeast mitochondria have been the subject of much intensive research, and the literature describing the structure and function of these organelles is vast (Guérin[13]). As in other eukaryotes, they contain their own ribosomes (70S), which are different from cytoplasmic (80S) ribosomes. Some mitochondria are distributed freely in the cytoplasm, while others are located on fragments of endoplasmic reticulum (ER) within the cytoplasm (called 'rough ER'). At certain phases of growth, yeast mitochondria can comprise around 12% of the total cell volume. These organelles contain autonomous self-replicating DNA and systems for protein synthesis. Mitochondrial DNA (mtDNA) was not included in the yeast genome project, but material from the same haploid strain used in that work has since been sequenced.[14] Weighing 86 kb, mtDNA is much less substantial than the smallest yeast chromosome, and represents only some 0.5% of the yeast genome. The mitochondrial genome codes for just 5% of all mitochondrial proteins (in fact, it codes for only about 25 identified proteins, and seven hypothetical proteins) the remainder being coded for by nuclear genes, some 300–400 proteins having to be imported. The mitochondrial genome is enriched, approximately ten-fold, by genes that code for transfer RNA. The significance of mitochondria is amply illustrated by the fact that some 8–10% of nuclear genes are involved in their biogenesis.

Classically, the mitochondrion is known as the site of the respiratory chain leading to ATP formation. The electron transport chain generates ATP as per the chemiosmotic theory of Mitchell,[15] which confirms that the components of the electron transport chain are located in the inner membrane of the mitochondrion, such that protons are translocated from the inner to the outer side as electrons pass down the chain. Several other enzyme systems are located within the mitochondria, in the inter-membrane space (which contains adenylate kinase) and on the outer membrane, where there are some enzymes associated with phospholipid synthesis. Unlike the mitochondria of higher eukaryotes, those in the yeast lack the facility for β-oxidation. As a consequence, the sites of fatty acid degradation in *S. cerevisiae* are the peroxisomes, small membrane-bound bodies that also exhibit catalase and glycollate cycle activity. Peroxisomes proliferate in the cytosol when oleic acid is present in the growth medium, but are repressed during growth on glucose-based media. Under the light microscope, the most obvious features of yeast cells are the vacuoles, whose size and number vary during the growth cycle. Under certain conditions of growth, there is one large vacuole present in the cell; under others they are much more dynamic. The vacuolar membrane (tonoplast) is similar in structure to the plasmalemma, but is more elastic, presumably to accommodate frequent vacuole shrinkage and expansion. Vacuoles have two main functions: firstly, they serve as stores of nutrients; secondly, they provide a site for the breakdown of certain macromolecules, particularly proteins, thus releasing intermediates for other metabolic pathways. Accordingly, amino acids are major soluble constituents of vacuoles, as are proteinases (such as carboxypeptidases and aminopeptidases). Vacuoles also serve as a food store for inorganic phosphorus, where the element exists as a linear polymer of polyphosphate linked by high-energy bonds.

The vacuoles represent only a fraction of a complex, internal system of membranes within the cytoplasm of the cell. There are other membrane-bound organelles, as well, such as the ER, the Golgi apparatus (dictyosomes) and numerous vesicles, both secretory and endocytic. The ER connects the plasmalemma with the nuclear membrane, and, in effect, divides the cytoplasm of the cell into two main areas; that which is enclosed within the membrane system, and that which is exterior to it (the cytoplasmic matrix).

The Golgi complex consists of a series of stacked membranes and associated vesicles, which, like the ER, is highly dynamic. Both organelles form part of a secretory pathway, by which proteins are sorted and trafficked back and forth to the plasmalemma, or vacuole. Dictyosomes are implicated in cell wall synthesis, and, at certain stages of the cell cycle, they are difficult to distinguish from ER fragments.

After the vacuole, the second largest organelle is the nucleus, which is typically double unit membrane-bound, and littered with nuclear pores (now called nuclear pore complexes, NPCs), that control the passage of molecules in and out of the nucleus. The yeast nucleus is some 0.1–2.0 μm in diameter, while the pores, which are dynamic entities, are in the order of 0.1 μm in diameter. Nuclear pores facilitate the exchange of low molecular weight proteins between nucleus and cytosol. The nuclear membrane, which, as we have stated, is connected to the ER, is unusual inasmuch as it does not disappear during cell division (mitosis), as it does in other eukaryotes. Within the nucleus there is a darker, more dense area, called the nucleolus, which disappears during mitosis, and reforms during interphase. Although it is seen as a distinct region within the nucleus, it is not strictly delimited from the rest of the nuclear sap by anything like a membrane. The nucleolus is rich in RNA and protein, and is the site of the synthesis of ribosomal RNA, and the synthesis and organisation of cytoplasmic (80S) ribosomes. High-resolution electron microscopy indicates that there are two distinct regions to the nucleolus; one particulate, one fibrillar.

There is relatively little nuclear DNA in yeasts, as compared to other eukaryotes (*ca.* 14,000 kb in a haploid strain), and this has, inevitably, made the study of their chromosomes, under the light microscope, a difficult process. Each chromosome contains a single molecule of DNA, which is associated with an equal mass of proteins. Collectively, the DNA with its associated proteins is called chromatin. Most of the protein consists of multiple copies of five kinds of histone. These are basic proteins bristling with positively charged arginine and lysine residues. Both arginine and lysine have a free amino group on their R group, which attracts protons (H^+), giving them a positive charge, which enables them to bind to the negatively charged phosphate groups of DNA. Chromatin also contains small amounts of a wide variety of non-histone proteins, most of which are transcription factors, and are not permanently associated with DNA.

The number of 'sets' of chromosomes is referred to as the 'ploidy', and most laboratory strains of *S. cerevisiae* usually have one set of chromosomes, that is, are haploid (*n*), or have two sets of chromosomes (diploid, 2*n*). Most industrial ('domesticated') yeasts, on the other hand, are polyploid, usually triploid (3*n*) or tetraploid (4*n*), although, in some strains, the number of copies of chromosomes is not necessarily a perfect multiple of the '*n*' number. If the latter occurs, the condition is known as 'aneuploidy', and the phenomenon allows for extra, or reduced, numbers of chromosomes to be present. Fundamentally, significant differences have been shown between specific chromosomes of laboratory haploid strains and the chromosomes of commercial strains. It is often difficult to ascertain the ploidy of industrial strains, but, in one study, the ploidy of wine strains was found to be 1.9*n*,[16] the results being obtained by using flow-cytometry to determine DNA content, rather than measurement of cellular DNA and comparison to a haploid reference strain. The copy number of each individual chromosome in polyploid (or aneuploid) strains is not necessarily identical. The recent technique of chromosomal fingerprinting, or karyotyping, has shown differences in chromosome size (called polymorphism) to be widespread. The *S. cerevisiae* nucleus is particularly suitable to karyotyping as it has 16 chromosomes[*] that range considerably in size from 230 kb (chromosome I) to *ca.* 1.5 Mb (chromosome IV).

S. cerevisiae was the first eukaryotic genome to be completely sequenced, work commencing in 1989, and ending with the publication of the sequence on 24th April 1996 by Goffeau *et al.*[17] This major project involved the collaboration of over 600 scientists, working in around 100 laboratories worldwide. The sequence, which was first 'published' on the World Wide Web, and is now easily sourced on the Internet, revealed some 6000, or so, protein-encoding genes. The first complete yeast chromosome sequence (and, indeed, the first chromosome from any organism) was from chromosome III (Oliver *et al.*, 1992[18]). When the sequencing was

---

[*]As recently as 1981, there were thought to be 17 chromosomes in *S. cerevisiae*.

commenced, there were some 37 genetic markers on the map of chromosome III, whereas the complete sequence revealed the presence of some 170 genes specifying proteins of ≥ 100 amino acids. The project involved the sequencing of some twelve million nucleotide bases, which encode about 6217 potential proteins (called 'open reading frames' – ORFs) that account for almost 70% of the total sequence. By the beginning of the new millennium, only about half of these proteins were 'known', that is were well characterised, biochemically or genetically. Of the remainder, some 30% had no recognised function, and are described as 'orphan genes'. The yeasts used in the genome project were closely related haploid strains derived from the progenitor strain S288C. Strain EM93, the main progenitor strain of S288C, which is estimated to constitute 88% of the genome of this strain, was originally isolated from a rotting fig in California in the late 1930s, and is, therefore, arguably of some relevance to winemaking strains of yeast. Some genes that have been found in industrial strains cannot be found in the genome sequence, and it is assumed that these have become redundant in the 'unreal' laboratory environment, and it is likely that there are many more apparently redundant genes that are necessary to cope with 'natural' environments, such as the skin of the grape. The answer will only be known when (if) the genome of a natural winemaking strain of yeast is sequenced. It seems to be generally accepted that polyploidy is advantageous for domesticated industrial strains (since such strains are, at the very least, diploid), even though there is still little concrete evidence to that effect. The most extensive study so far has been by Galitski *et al.*,[19] who suggest that there is a selective advantage in maintaining a number of copies of most chromosomes that are of utmost benefit to the cell. The lessons learned from the yeast genome project have been adequately summarised by Dujon.[20] The complete yeast genome can be searched and browsed at the Saccharomyces Gene Database (SGD) at Stanford University: http://genome-www. stanford.edu/Saccharomyces/. A list of useful addresses for yeast researchers has been provided by Brown.[21] As we have said, the DNA sequence of *S. cerevisiae* is relatively small (13.5 Mb); the genome of the ultimate prize, in terms of sequencing, *Homo sapiens*, is of the order of 3500 Mb!!

Several non-chromosomal genetic elements may be present in the yeast nucleus: (1) 2 μm DNA, a stably maintained circular DNA plasmid, which replicates once during the S-phase of the cell cycle; (2) double-stranded RNA and linear DNA, found in 'killer' strains. They harbour the genes for toxins, which are deleterious to non-killer strains; (3) retrotransposons, mobile fragments called Ty elements (originally designated Ty1, Ty2, Ty3, Ty4).

Sexual reproduction in *S. cerevisiae*, which involves mating, or conjugation, and culminates in the production of ascospores, is relevant only to haploid or diploid cells, not generally to polyploid industrial strains, and so will not be detailed here. The interested reader is recommended to Sprague.[22]

All strains of *Saccharomyces* undergo a unique, asexual, asymmetric form of cell division called 'budding', whereby one cell gives rise to two daughter cells that are genetically identical to their mother. Under suitable conditions, the daughter cells, themselves, become mother cells, and this sequence will continue until cell division is arrested. Cell division is not a never-ending process, and when nutrients become limiting cells become senescent and eventually die. The cell cycle of yeast has been much studied, not from the point of view of its industrial significance, but because of the role that the organism plays as a 'model' for eukaryotic cell research. It is now known that key elements of the eukaryotic cell cycle have been highly conserved during evolution, and so, what happens in yeast is peculiarly relevant to what happens in higher eukaryotes (man included). Add to this the fact that yeast has a small genome, is highly amenable to genetic manipulation, and the cell cycle is completed in quick time, and one can appreciate its usefulness in the laboratory. A comprehensive review of the yeast cell cycle has been penned by Wheals.[23]

To fully appreciate the biological events that occur during a wine fermentation it is necessary to briefly consider the cell cycle of *S. cerevisiae*, a representation of which is illustrated in Figure 2. In this organism, as in all eukaryotes, the mitotic cell cycle can be divided into four intervals, G1-, S-, G2- and M-phase. S-phase and M-phase are the periods when DNA replication and nuclear division (mitosis) occur, respectively, and overall control of cell division is achieved by regulating entry into these two phases. Interspersed between DNA replication and nuclear division are two 'gap' periods, G1 and G2, which are of variable length, and, during which, organelle production and normal cell processes, such as growth and development, take place. Phase G1 represents the

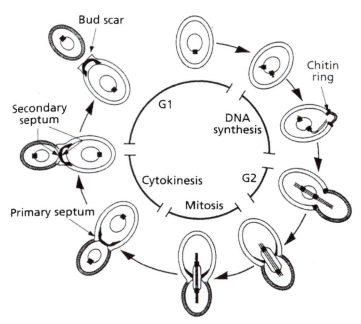

Figure 2   *The cell cycle of* Saccharomyces cerevisiae *(after Harold[24])*
(Reproduced by kind permission of the Society for General Microbiology)

period immediately after a daughter cell has separated from the
mother, and is immediately prior to the onset of another cell
division. As soon as the synthesis of DNA has been completed,
the cell, by definition, enters the G2 phase, where it remains until
mitosis commences. The major controlling event in the cell cycle,
termed 'START', occurs late in the G1 phase, and compels a cell to
undergo division; there is no going back.[25] After the initiation of
START, environmental factors, such as external stress and poor
nutrient levels, are no longer able to prevent cell division. Completion of START requires that a cell grows beyond a minimum
size, and commits that cell to a round of mitotic division, as
opposed to alternative developmental fates, such as conjugation,
or, for diploid cells, meiosis. Commitment to mitotic division sets
in train the pathways required for the initiation of DNA synthesis
and the transition to S-phase. If nutrients are in sufficient supply
(they are rarely a limiting factor in most stages of a wine fermentation), the S-phase is entered, and, in addition to embarking upon
DNA synthesis, an embryo bud is initiated. After the completion of

DNA synthesis, and entry into phase G2, extensive bud growth ensues, and nuclear events culminate in mitosis (M-phase). The final steps of the cell cycle are septation, which separates the cytoplasm of mother and daughter cells (cytokinesis), and total separation of the two cells, although complete separation is not a prerequisite for continued reproduction. Ineffective mother–daughter cell separation can result in chains of cells being formed, which is a characteristic phenotype of some individual strains. Total separation of the bud leaves a permanent mark on the mother cell, the bud scar and a temporary birth mark on the daughter cell. Some workers recognise a fifth phase to the cell cycle, G0, which represents cells in a state of 'suspended animation' (see below).

The genesis of a bud begins with the selection of a locus on the surface of the mother cell, where outgrowth will later take place. Placement of the bud is governed by a set of rules, which can be predicted from the genotype; for example: haploid 'α' and 'a' bud axially, as do homozygous diploids, but heterozygous diploids ('αa') bud in a bipolar manner. When the markers are activated, they induce a localised assembly of an annular structure that marks the new bud initial; this includes a ring of chitin, another of 10 nm neck filaments and several additional proteins. The daughter bud normally emerges within this ring. A newborn yeast cell grows and deposits cell wall polysaccharides uniformly all over its surface until it attains a critical size, whence it initiates a bud; thereafter, deposition of new wall material is confined to the growing bud. Expansion of the bud wall is patterned in space and in time: to begin with, growth takes place chiefly at the tip of the forming bud, then it becomes uniformly distributed and eventually expansion halts. Expansion of the wall is a secretory process, membrane-bound vesicles, which seem to originate in the Golgi apparatus, playing a prominent role. It is generally understood that these vesicles carry precursors and enzymes, such as mannosylated proteins, chitin synthase and β-glucan synthase, for wall biosynthesis. When the vesicles reach the plasmalemma, in the growth area, they fuse and undergo exocytosis. It appears that the prominent cables of actin microfilaments, that course through the mother cell cytoplasm and reach the bud, serve as 'tracks' upon which wall precursor vesicles move towards the site of exocytosis. The cables seem to steer the vesicles to specific fusion sites, which appear as prominent plaques on the plasmalemma that contain actin and actin-binding proteins. Figure 3 identifies processes known, or surmised, to contribute to bud enlargement, and represents a pictorial summary of the findings of a number of laboratories.

The nuclear events accompanying budding are now well known. Just prior to the formation of the bud primordium, the spindle-pole body (SPB), a centriole-equivalent situated on the nuclear membrane, undergoes duplication. The products separate and migrate, and this is followed by rotation of the nucleus and bringing the spindle-pole bodies in line with the axis of the emerging bud. The two spindle poles then nucleate development of the mitotic spindle, which guides separation of the duplicated chromosomes. Another set of microtubules sprouts from one spindle pole to reach deeper into the bud, and these are needed to guide the nucleus into the neck of the developing bud, and to guide other cytoplasmic organelles into the bud as well. Upon completion of mitosis, the maturing bud separates from its mother by ingrowth of the septum. Unlike the bulk of the cell wall, which is basically composed of β-glucan and mannoproteins, the septum is composed of chitin. Its

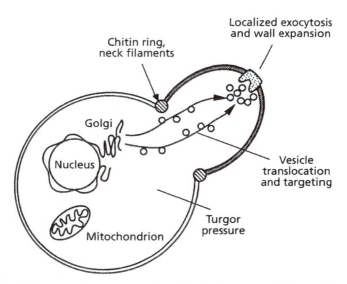

Figure 3   *Diagrammatic representation of bud enlargement in* S. cerevisiae *(after Harold[24])*
(Reproduced by kind permission of the Society for General Microbiology)

formation is a lengthy process, requiring two distinct chitin synthases, and actually begins prior to the emergence of a visible bud. Septation culminates with cytokinesis, and splitting of the septum to liberate the daughter cell. At the end of the cell cycle, after septation, daughter cells are invariably smaller than their mother cells. As a consequence a daughter cell has to increase its cell volume to a certain level (called the 'critical cell size') before it is capable of budding. As a result, the G1-phase of a daughter will be longer than that of its mother. As Figure 3 intimates, the growth (enlargement) of the bud is driven from within by the turgor pressure of the mother cell.

Because of the industrial importance of yeast, there have been many comprehensive treatises written over the years, but much of the basic biology of the species has been reviewed in admirable articles by Broach *et al.*[26] and Jones *et al.*[27]

3.2   FERMENTATION

Microorganisms have evolved different methods of regenerating NAD^+ from NADH. In *S. cerevisiae*, acetaldehyde serves as the terminal electron acceptor, whence it is reduced to ethanol. Pyruvate is first decarboxylated, by pyruvate decarboxylase, to yield CO_2 and acetaldehyde (a step requiring thiamine pyrophosphate). Acetaldehyde is then reduced to ethanol by alcohol dehydrogenase, thus regenerating NAD^+. There are four isoenzymes of alcohol dehydrogenase in yeast, designated ADHI, II, III and IV. ADHI, a constitutive enzyme is the species responsible for the conversion of acetaldehyde to ethanol, and its activity is dependent on the growth phase, stage of fermentation, temperature and yeast strain. ADHII and III are glucose-repressed, which means that they are not found during fermentative growth in high sugar media, but are expressed instead during aerobic growth, when ethanol acts as a carbon and energy source. ADHII is found in the cytoplasm, while ADHIII is located in the mitochondria. ADHII can compensate for the loss of ADHI at very low glucose concentrations, ADHIII cannot. ADHIV is present in the cell at miniscule levels, and its function is not fully understood. Most brewery fermentations employ relatively few species/strains of yeast, and the same can be said for the distillation industries, and for the yeasts used in a bakery. The situation regarding yeasts used in wine fermentations, however, is much more complicated, for, in total, over 200 species, representing some 28 genera, have been isolated from grapes, must and wine (see Kunkee and Goswell,[28] for example), and, amazingly, it is still imprecisely known which are the most 'important' yeasts in traditional wine fermentations. In addition, we still need to know a lot more about the kinetics and population dynamics of vinification fermentations generally. Grape juice represents an ideal *milieu* for ethanolic fermentation, and it is remarkable that, from the days of the 'discovery' of wine in ancient times, there had been relatively few innovations in must-handling technology until the 20th century; exposure of must to a suitably 'contaminated' environment was all that was needed to initiate a vinification fermentation. During the primary fermentation of wine, the two major grape sugars, D-glucose (dextrose) and D-fructose (laevulose), are converted to ethanol by the action of yeast. The by-products are carbon dioxide, aromas, flavours and heat; that is the processes are exothermic. During vinification, glucose is fermented more rapidly than fructose, because most wines yeasts are glucophiles rather than fructophiles (*i.e.* their cell membranes are more permeable to glucose), which usually means that residual sugar consists primarily of fructose.

In addition to being provided with a major source of carbon (*S. cerevisiae* can only grow on a limited range of carbohydrates, with glucose being the preferred carbon and energy source), and a nitrogen source, the yeast has day-to-day requirements for the macro-elements sulfur and phosphorus, and certain vitamins, such as biotin, pantothenic acid, inositol, thiamine, pyridoxine and nicotinic acid, if growth is to proceed normally. A range of metallic ions, such as, Na^+, K^+, Fe^{2+}, Zn^{2+}, Cu^{2+} and Mn^{2+}, must also be provided, together with a range of trace elements. Grape berries generally contain an ample quantity of these compounds, and a complex medium, such as grape must, invariably provide the yeast with the opportunities for maximal growth and metabolism. Vitamin deficiencies can arise in a must that has been prepared from nutrient-depleted grapes, the most common cause of this being due to *Botrytis cinerea*, or to some other mould infection. Thiamine deficiency may occur if grape juice, or grape juice concentrate has been stored for an extended period with lavish levels of sulfur dioxide.

The chemical composition of must is dependent upon the grape variety, time of harvest, and, of course with the climatic and edaphic factors (including fertiliser use) that affect vine physiology. On an average, wine grapes are harvested with a view to yielding a must with a fermentable sugar content of 180–250 g L^{-1}, depending upon the wine type intended. Table wines would normally demand a must sugar content of 200–220 g L^{-1}. Sugar-wise, most grape musts consist principally of roughly equal amount of fructose and glucose, with trace amounts of sucrose and a few pentose sugars. The two major hexose sugars have originated from sucrose, a product of photosynthesis, which is hydrolysed as it is translocated from the grape leaf to the berry. As a general rule, the later the date of harvest, the higher the glucose and fructose content of grapes.

The nitrogen content of grape musts is extremely variable, not only in overall level, but in the relative abundance of individual compounds, and is directly related to the nitrogen status of the soil, and to the fertilizer regime. In one of the more extensive studies, Monteiro and Bisson,[29] examined 29 different musts produced from three grape varieties (Chenin Blanc, French Colombard and Grenache), and the nitrogenous content (expressed as equivalents of ammonia), varied between 90 and 840 mg L^{-1}, with an average of 440 mg L^{-1}. The same workers calculated that to achieve a maximum yeast cell count (considered to be $1–2 \times 10^8$ cells mL^{-1}), an ammonia equivalent of 540 mg L^{-1} is required, and that for a fermentation to have any chance of running to completion, at least 140 mg L^{-1} is required (although, at this level, maximum cell concentrations would not be achieved). The yeast cannot use molecular nitrogen, or the nitrate ion, as the sole source of nitrogen, although the latter is invariably present in must, and the major utilisable nitrogen sources in a wine fermentation are ammonia (as NH_4^+), amino acids, polypeptides and proteins. It has been shown that some 80% of the utilisable nitrogen in grape musts is in the form of alanine, arginine, asparagine, aspartate, glutamate, glutamine, proline, serine, threonine and ammonia, with arginine and proline being the most abundant amino acids in grape juice. Other amino acids are either absent, or present in non-detectable amounts. The situation regarding proline is interesting because it is one of the most abundant amino acids in many musts, yet it is not utilised by the yeast under conditions of anaerobiosis. It is, however, assimilated at the start of fermentation, while conditions are still partially aerobic. This is due to the fact that molecular O_2 is required as the hydrogen acceptor for the first step in proline degradation (proline oxidase), which occurs in the mitochondria. In addition, proline permease, which allows proline to traverse the mitochondrial membrane, requires O_2 for its expression, and so proline is not even taken up when conditions are anaerobic. For reasons that will be explained later, the proline content of a finished wine can be even higher than that found in the original must. The actual amino acid profile of a must varies from one grape variety to another.

The requirement of yeast for phosphate is relatively constant, and varies from 1.5 to 1.6 µg per million cells. Most musts contain phosphate levels of between 200 and 500 µg mL^{-1}, which are, in theory, sufficient to cover the needs of the yeast population.[†]

There are two main ways by which the winemaker is able to start his fermentations, and these may be broadly categorised as 'spontaneous' and 'artificial'. Many wineries still take advantage of nature, and allow fermentation to occur spontaneously, the organisms for fermentation emanating from the grape, itself, or from the immediate winery environment (*e.g.* fermenting vats). Naturally occurring yeasts are able to start a fermentation if the must temperature is *ca.* 18°C. Some wineries transfer yeast harvested from the end of a previous fermentation (called 'pitching' in a brewery), in effect, a form of serial inoculation. Wineries using these methods consider that the diversity of species/strains of yeast, so imparted, is essential for the individual taste and aroma profiles of their products. Such processes may be considered to be artisanal, and the endogenous yeast source involved would undoubtedly represent a highly heterogeneous inoculum. In a non-sulfited and uninoculated must, grape surface yeasts begin to multiply within a few hours of its placement into a fermenting vat (*Kloeckera* and *Hanseniaspora* species predominating). There is also a tendency for some undesirable aerobic yeasts to develop (especially species of *Candida*, *Pichia* and *Hansenula*), if care is not taken, producing acetic acid and ethyl acetate from the grape must sugars. Fortunately, many of these 'contaminating' strains are relatively sensitive

---

[†] At the end of the yeast-growth phase of vinification, the average population density is *ca.* 2×10^8 cells mL^{-1}, meaning that a phosphate concentration of 300 µg mL^{-1} is required.

to sulfur dioxide. As has been indicated,[30] the unrestricted growth of some strains of indigenous yeast can result in serious sensory defects in the finished product. It is of interest that some micro-organisms indigenous to must, such as certain strains of the yeast *Zygosaccharomyces bisporus*, possess killer activity (see Chapter 9).

In fermentations that rely on spontaneous 'seeding' from the must, exactly how much of the inoculum originates from the grape, itself, is a matter of conjecture. The juice within the intact healthy grape is sterile, but the surface of the fruit is covered (albeit unevenly) with various microbes, most of which are in a slowly reproducing state (*i.e.* not in phase G0 of the cell cycle). The heterogeneous assemblage of microbes on the grape surface is known as the 'bloom'. Studies on the yeast composition of grape must have indicated that, on average, it will contain 10^3–10^4 colony-forming-units per millilitre, and that these organisms have all come from the surface of the grape. The dominant species on the grape skin have a relatively low ethanol tolerance, and are invariably apiculate yeasts. Some 50–75% of the yeast flora of musts can be attributed to species of *Kloeckera* and *Hanseniaspora*, with significant contributions from the genera *Candida*, *Hansenula* and *Kluyveromyces*. Because these organisms are not responsible for the primary wine fermentation, they are often referred to as 'wild yeasts'. It is significant that *S. cerevisiae*, which plays the dominant role in the rapid production of ethanol in wine fermentations, is either absent, or present only in low numbers on sound grapes, and in fresh must. In traditional wineries, therefore, it is evident that this dominant strain is being harboured somewhere in the process equipment, and that there is an 'accidental' inoculation of each batch of wine. The succession of principal yeast types in a 'typical' spontaneous wine fermentation is as follows:

Kloeckera spp. $\rightarrow$ *Hansenula* spp. $\rightarrow$ *Saccharomyces cerevisiae*
$\rightarrow$ *Saccharomyces bayanus*

Wineries that use spontaneous fermentations maintain that it is this very succession of organisms that contributes greatly to the formation of the desirable, flavoursome characteristics of their end products. In such a scheme, it is almost always *S. cerevisiae* that is the principal 'fermenting' yeast, the only other species of any significance, in this respect, being *S. bayanus*. According to the classification in Barnett *et al.*,[8] these two taxa are the sole representatives of the genus to be implicated in wine fermentations. These two species are ideally suited to their work in the winery, because they: (1) are ethanol tolerant, that is can still ferment during the latter stages of the fermentation, when sugar concentration is low and ethanol content is high; (2) can establish viable cell populations in solutions with high sugar level and low pH; (3) are strong and consistent fermenters, even at low temperatures; (4) ferment rapidly, and only stop when all the grape sugars have been depleted; (5) are more tolerant to SO_2 than other yeasts and bacteria) and (6) produce desirable 'wine-like' aromas and flavours.

The degree of disparity among taxonomists, regarding the classification of yeasts, can be gleaned from the fact that even modern oenological texts[31] continue to implicate other 'species' of *Saccharomyces* in vinification fermentations. To many, like Barnett, these are now regarded as strains of *S. cerevisiae*, or a related organism, or have been assigned to a completely different genus. To illustrate the point, Clarke and Bakker[31] mention the following as being of significance: *S. chevalieri*; *S. italicus*; *S. heterogenicus*; *S. bailii*; *S. uvarum*; *S. apiculatus* and *S. oviformis*. Of these, according to Barnett *et al.*,[8] *chevalieri*, *italicus*, *uvarum* and *oviformis* are now regarded as strains of *S. cerevisiae*; *heterogenicus* is now a strain of *S. bayanus*; *bailii* is now *Zygosaccharomyces bailii*, and *apiculatus* is classified as *Hanseniaspora apiculatus*.

In most modern wineries, certainly many of those in the New World, the must is inoculated with an aseptic culture of a selected strain of *S. cerevisiae*. Such a yeast is called a 'starter culture', and inoculation is usually carried out in tandem with an addition of sulfur dioxide, which limits the growth of organisms from the natural surface flora of the grape. The methodology is more usual during the production of white wines, although some red wine vinifications can arise via the use of a starter. Inoculation is certainly necessary for post-fermentation techniques, such as are encountered in the making of sparkling wines, as it is in wineries that pasteurise their juice before fermentation. Starter cultures are freeze-dried (lyophilised), and vacuum-packed and have to be rehydrated in warm, sterile water before they can be added to the must. In general, the inoculated yeast is introduced at a rate that will give a final level of 10^6 cells per millilitre must. The use of starter cultures has really only become popular during the last 40 odd years, a period that happens to coincide with the development and wide availability of aseptic, active, freeze-dried cultures, but it is not a new phenomenon, for the practice was used in France and Germany, for a while, during the early years of the 20th century. The first yeast starter cultures for wine were prepared by Herman Müller-Thurgau and Julius Wortmann in the 1890s at the Geisenheim Research Institute, Germany. Starter culture technology was re-introduced into some Californian wineries during the early 1960s, and is now standard practice in that region, as it is in most of the newer winemaking countries of the world, such as Australia, New Zealand, South Africa and South America. Traditionally, starter cultures were isolated from a fermentation, kept, refrigerated, as a stock culture on agar slopes and cell numbers were built up, by basic microbiological propagation methods, prior to inoculation into a must. With modern active dried yeasts available, such methodology is now largely obsolescent. The wide range of starter cultures now available permit the winemaker to select a strain to satisfy his/her requirements.

One of the most important features of a starter yeast, is that it should be able to establish a viable cell population in an environment of high acidity and high osmotic pressure, and be stable throughout the course of the fermentation. It must also, of course, be highly ethanol-tolerant, having to routinely withstand concentrations of up to 15% v/v, a level that is toxic for many microbes. A practical advantage of using a starter culture is that it shortens the lag phase of the yeast growth-cycle, and ensures a rapid build-up of actively fermenting cells. This, in turn, serves to competitively restrict the growth of 'wild' strains. Thus, it is far easier to predict events that are likely to occur during fermentation, because one is dealing with a yeast of known (*i.e.* proven) characteristics. One of the problems of using a cultivated yeast is that there is always a risk of mutation, and there is a need, therefore, to constantly re-evaluate and re-select the desired strain. It is of significance that the wider use of starter cultures has coincided with legislation restricting the levels of sulfur dioxide permitted in foodstuffs. With lower legal levels, some winemakers are no longer willing to risk spontaneous fermentations, and the increased likelihood of heavy growths of undesirable yeast strains, which can accompany that method of initiating fermentation. In those wineries employing starter culture inoculations, measures need to be taken to ensure that the inoculated yeast is the one operating throughout the fermentation, rather than a member of the indigenous flora. In white wines, this is not usually a severe problem, because the pre-fermentation separation techniques remove most of the particulate matter, but problems can arise during red winemaking. Most indigenous must yeasts, however, will succumb to a mixture of sulfite and inoculation with a selected strain of *S. cerevisiae*. With latter day concerns about the use of sulfur dioxide, a heavy inoculum of the chosen yeast strain is usually enough to discourage the growth of 'unwanted' organisms.

The debate about spontaneous fermentation versus the use of starters goes on, but many traditional wineries scorn the use of starter cultures, because they maintain that they only give rise to uniform, bland wines. It is also said that wines made by sponta-neous fermentation have a better 'mouthfeel', that is, that they are 'softer' and 'creamier'. There is no scientific basis for this, but the heterogeneous nature of the grape skin flora might well be respon-sible. The relative advantages and disadvantages of using starter cultures, as compared to the reliance on a natural indigenous yeast flora, have been exhaustively debated by a number of workers (Benda,[32] Bisson and Kunkee[33] and Subden[34]) but the benefits known to accrue from the use of starter strains are:

(i) an insured early start to the fermentation (*i.e.* shortening of the lag phase);
(ii) a diminished reliance on sulfur dioxide; and
(iii) a more rapid and more predictable fermentation, leading to enhanced utilisation of the winemaking plant. In a highly automated winery, the uncertainty of spontaneous seeding is too much of a risk.

In today's 'green' environment, spontaneous fermentations are certainly less interventionalist, and, of course, they cost nothing to initiate.

However vinification is actuated, the cells used to initiate a wine fermentation have sometimes been subjected to near-starvation conditions, and are probably in the stationary phase of growth. Such cells are quiescent, unbudded and often referred to as being in phase G0. If seeding of a wine fermentation has arisen from cells naturally present in the atmosphere of the winery, or from winemaking equipment, then they will almost certainly be in the G0-phase. Cells that have come naturally from the grape surface will have come from a nutrient-rich environment, and so will probably not be in G0. Introduction of yeast cells, by whatever means, into a fresh must sample results in a period of apparent inactivity; something that is an integral feature of the pattern of growth exhibited by every microbe when it is introduced into a fresh batch of growth medium. Figure 4 shows the pattern of a 'normal' wine fermentation. The increase in absorbance is attributable to yeast growth, while the decrease in sugar concentration, measured here in Brix, is the result of fermentation, and the concomitant release of ethanol and CO_2. From the practical winemaker's point of view, vinification may be divided into three distinct phases: an initial phase of yeast growth (multiplication); a middle phase of vigorous fermentative activity and a final phase in which fermentation declines and the inhibitory effects of ethanol, and other compounds, become apparent.

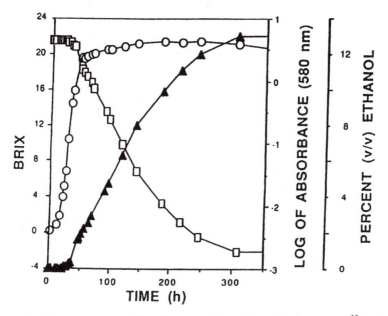

Figure 4 *Fermentation profile of grape juice (after Boulton et al.[35]). O = absorbance; ▲ = ethanol; □ = Brix*
(Reproduced by kind permission of Springer Science and Business Media)

The initial stage of the growth cycle is called the 'lag' phase, and represents a period when the introduced cells are adjusting to their new environment, in particular they are responding to the osmotic challenges presented by high sugar levels in must. Although there are no outward manifestations of metabolic activity, several important physiological and biochemical events are occurring within the cells; new enzymes, and enzyme-carrier systems are being synthesised to enable the yeast to utilise the wide variety of must constituents. The duration of the lag phase is dependent to some extent upon the origin and condition of the inoculated cells, but is also governed by factors such as must composition and temperature. In wineries that employ 'natural' methods of inoculating must, it is highly important that the lag phase of *S. cerevisiae*, and other wanted organisms, should be as short as possible, in order that the growth of less desirable microbes can be suppressed. As a general rule, the more metabolically active the inoculated cells (*i.e.* not in phase G0), the shorter the lag phase will be.

When cells are appropriately equipped, the lag period will give way to a short phase of the growth cycle in which the cells start to become metabolically active again, and embark upon cell division (budding). Thus, yeast multiplication commences as soon as the lag phase has ended, and, although there is some fermentative activity, most of the carbon and energy source consumed is converted into essential cell components. Under normal conditions of vinification, this short period of accelerated growth inextricably leads to a period of exponential growth, in which cellular activity and cell division (generation time) are at their zenith. This is called the 'logarithmic' ('log') phase of growth, and it is at this stage that the maximum production of ethanol is achieved. The observed high rates of cell multiplication are only possible if the original must contained sufficient dissolved oxygen to permit synthesis of the large quantity of cell membrane precursor material prerequisite for such an event. Normal grape handling during stemming and crushing invariably allows the must to absorb sufficient oxygen, and so, under most circumstances, a paucity of the element rarely becomes a limiting factor at the onset of fermentation. According to lore, winemakers would regularly stir their crushed grapes before fermentation commenced, a practice that would unintentionally introduce sufficient oxygen for the completion of fermentation. In the case of yeast starter cultures, it is important that they are aerated during resuscitation/propagation. After a short time, conditions in the fermenter become highly anaerobic, and the yeast converts fermentable sugars, principally glucose and fructose, to ethanol and CO_2 through the glycolytic, or Embden–Meyerhof–Parnas (EMP) pathway (Figure 5). This is the main energy-generating process for yeast, and can only proceed in the absence of oxygen. The high level of glycolysis generates copious quantities of CO_2 that forms a blanket over the fermentation, which, to the joy of the winemaker, prevents the ingress of oxygen. This blanket persists until the ferment subsides, from which point there is an increased likelihood of oxygenation occurring. Release of CO_2 bubbles at the base of the fermenting vat creates a turbulence that serves to disperse dissolved nutrients, and, if sufficiently vigorous, will dislodge settled yeasts cells.

Glycolysis also precludes any significant increase in yeast cell biomass, since most substrate is converted to ethanol and CO_2, and it is also evident that the accumulating level of ethanol does not adversely affect the EMP pathway, as much as it does the rate of vegetative yeast growth. The latter effect is not too pronounced now, because, as the concentration of ethanol rises, so the cells are entering their non-growing phase, and no longer actively taking up nitrogen sources. One of the known effects of ethanol on the growth of yeast cells is that it interferes with the uptake of amino acids and ammonia (as NH_4^+).

In the continued presence of oxygen, the yeast will undergo aerobic respiration *via* the oxidation and decarboxylation of pyruvate, and, through the Krebs' cycle, the ultimate production of CO_2 and water. The events relating to the Krebs' cycle also occur within the mitochondria. It was Pasteur who, in 1861, first described the deleterious effect of oxygen on wine fermentations. He noticed that when yeast was transferred from an anaerobic to an aerobic environment, growth was accelerated while uptake of sugar was diminished. This is known as the Pasteur effect, and is, in part, attributable to the difference in the Michaelis constant (K_M) for the accumulation of glucose under the two sets of conditions. It has been shown experimentally that, when yeast is grown anaerobically, the K_M value is 6.7 mM, while under aerobic conditions it is of the order of 17.4 mM.[36] These observations are linked to the affinity of glucose for its permease, which is decreased in the presence of oxygen. This reported enhancement of vegetative

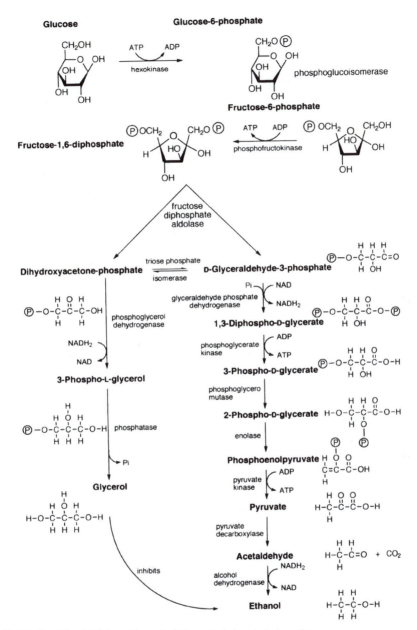

Figure 5   *The Embden–Meyerhof–Parnas (glycolytic) pathway*

growth (in effect, cell material) is simply explained by the fact that *S. cerevisiae* can produce more ATP from each molecule of glucose under aerobic conditions. In practice, because of the vast excess of fermentable sugar, and the fact that the yeast favours the fermentative mode of metabolism, aerobic respiration rarely reaches significant levels. In theory, industrial fermentations should proceed for as long as anaerobic conditions prevail, and there is a source of hexose sugar available in the growth medium, but this rarely occurs during the fermentation of must. Part of the reason for this is that, under anaerobic conditions, sterol and fatty acid synthesis cannot keep pace with the level of cell membrane synthesis required for cell replication (budding).

It has been known for many years (Crabtree, 1929[37]) that, at high glucose concentrations ($>0.4\%$), even under aerobic conditions, *S. cerevisiae* metabolism is fermentative rather than oxidative. This is associated with morphological and biochemical alterations in the mitochondria, which resemble those in cells grown under anaerobic conditions. Biochemically, the most significant differences are the deficiency of certain Krebs' cycle enzymes, and some respiratory chain components (notably cytochromes). This phenomenon is known as the Crabtree effect, the reverse-Pasteur effect or the glucose effect, and is an example of catabolite repression. In essence, all ethanolic fermentations are regulated by the Crabtree effect. It is the uptake of hexose sugars that is the rate-limiting step in glycolysis, and it has been shown that there is more than one type of glucose transport, depending primarily on substrate affinity. These mechanisms are referred to as 'low-affinity' and 'high-affinity' uptake. The regulation and expression of sugar uptake has been discussed by Kunkee and Bisson.[38]

Once yeast growth gets under way, musts are rapidly depleted of ammonia and amino acids, especially the former. Amino acids present in only trace amounts are removed from must very quickly indeed. Arginine, the main form of nitrogen storage in the grapevine, is the most abundant amino acid in many musts, and while it is metabolised in conjunction with other nitrogen sources, when these are present they seem to be taken up in preference (*i.e.* arginine does not appear to be a preferred nitrogen source). When other sources are depleted, arginine metabolism increases, but this does not prevent substantial amounts invariably ending up in the final wine. As we shall see, there are potential problems for the winemaker, when arginine is metabolised by some bacteria.

A lack of available nitrogen can lead to a 'stuck' fermentation, and this sometimes occurs in musts that have been over-clarified, thus reducing the protein content. One of the known adverse effects of nitrogen limitation is the irreversible inactivation of hexose sugar uptake, which is caused by a decrease in turnover rate of hexose transporter proteins. Conversely, an excess of nitrogen can cause excessive cell multiplication, but a reduced conversion rate of sugars to ethanol in individual cells. If this occurs, there will appear to be a stimulation of fermentation, attributable purely to the high concentration of cells.

Musts that are low in free amino nitrogen (FAN) are prone to the liberation of hydrogen sulfide (H_2S) during fermentation. The gas is released from sulfur-containing amino acids via the proteolytic activity of yeasts seeking an available nitrogen source. Yeasts strains vary greatly in their propensity to liberate H_2S. Where a 'sulfidic' fermentation is likely to arise, and needs to be prevented, it is common practice to add 200 mg L^{-1} diammonium phosphate (a convenient source of NH_4^+) at the start, and then another 100 mg L^{-1} during the fermentation should H_2S be detected. Such additions are only ameliorative during fermentation, not after it has subsided. Where H_2S persists after fermentation, the wine should be aerated in the presence of sulfur dioxide. If this fails, treatment with copper sulfate (which precipitates the sulfide as CuS) is necessary. If 'sulfidic' wine is left untreated, H_2S will form stable ethyl, or diethyl mercaptans, which, having a lower aroma threshold, are far more readily discernible to the drinker. These mercaptans impart an 'onion' or 'garlic' aroma. Volatile sulfur compounds are more likely to result from musts with a high level of suspended solids. H_2S is also produced by yeast autolysis after primary fermentation, since cell breakdown liberates sulfur-containing amino acids, which, if conditions are sufficiently anaerobic, lead to its formation. One way to reduce the likelihood of H_2S production by this method is to rack the wine as soon as possible after ethanolic fermentation. This is particularly appropriate for white wines that have been fermented in large tanks, where anaerobiosis is far more likely to prevail.

An active wine fermentation will support a yeast culture with cells in all four major stages of the cell cycle (one would not expect any cells to be in G0-phase). If any cells had been in phase G0 upon inoculation of the must, they will soon enter G1, a transition that is thought to be fuelled by a supply of endogenous glycogen within the metabolically inactive cell. The transition G0 → G1 is accompanied by renewed synthesis of membrane lipids, and, within about 2 hours, cells start to enter S-phase, and the first signs of budding. If the must has been inoculated with a starter culture, then around 90% of all cells will be showing signs of budding within 6–8 h. Bud initiation will take much longer in fermentations that have been left to inoculate naturally, a lag of 3–4 days being unexceptional. Either way, progression through the cell cycle is critical for the formation of sufficient yeast biomass to support vigorous fermentation.

The period of time taken for a mother cell to prepare, produce and liberate a daughter cell is called the 'generation time', and most of the strains of yeast that have found favour in wineries have a short period of replication under the conditions of their routine use. The Montrachet strain, for example, which is widely used in Californian wineries, has a generation time in grape juice of 1–3 hours at room temperature. Under conditions of industrial usage, most wine yeasts will multiply, through the log phase, until a population of around $1–2 \times 10^8$ cells mL^{-1} is attained. The overall yield of fermentable sugar to ethanol is around 50%, depending upon fermentation temperature. During active growth, cell division rate is limited by the growth of an individual cell. The initiation of a new round of division is regulated by signalling pathways comprised of networks of gene-mediated events that monitor cell size, age, nutritional status, the presence of mating pheromone and the completion of the previous round of cell division. These signalling pathways act together to form cell-cycle regulation points, which confirm that the previous steps have been completed without errors. Slow-growing cells have a relatively long generation time, and this additional time is spent in the G1 phase. In addition, daughter cells that have never produced a bud are smaller than their corresponding mother cells and so have longer G1 phases. This differential is known as mother–daughter asymmetry, and mother and daughter cells may be distinguished by both cell size and division rate. After the period of exponential growth, cells enter the phase of stationary growth, in which the number of viable cells in the fermentation appears to remain static. There is no stasis *per se*, but the number of cells being formed is being counteracted by the number that is becoming aged and non-viable. Most laboratory studies on yeast metabolism during the stationary phase of growth (which use defined media) have centred on its relationship to the gradual exhaustion of the carbon source. With yeast growing in grape must, entry into the stationary phase is not a simple matter of a response to carbon depletion. Indeed, it can occur when sugars are still present in high concentration (as much as 10% w/v). In a vinification fermentation, the onset of the stationary phase is rather more associated with the depletion of the nitrogen source, or some other macronutrient. Under such conditions, cells may still be metabolically very active, since there are plentiful reserves of ATP. The stationary phase of growth leads to the death, or decline phase, where the number of cells becoming moribund far exceeds the number of cells being liberated by budding, thus signalling a decline in overall viable cell number.

Cell ageing has been much studied in yeast, and, from time to time, the popular press highlights work from this field that may well be relevant to ageing and death in humans! The subject is, of course, of huge consequence to the fermentation industries. Ageing is a predetermined progressive transition of an individual cell from youth to old age that finally culminates in death. Yeast replicative ageing is a function of the number of divisions undertaken by an individual cell, and may be measured by enumerating the number of bud scars on the cell surface. An aged yeast cell is morphologically and physiologically distinct from younger cells, and this greatly affects fermentation performance. As a consequence of a fermentation, cell death may occur via one of two distinct pathways: necrosis and senescence. Necrosis may be defined as the accumulation of irreparable damage to intracellular components compromising cell integrity, leading to death and autolysis. Such damage occurs primarily as a result of exposure to excess stress, or repeated exposure to low-level stress. Necrosis may also occur following the deletion or disruption of specific essential genes due to a mutagenic event, or as a result of irreparable DNA damage. In contrast, senescence is the predetermined cessation of life as a result of the genetically controlled progression from youth to old age.

Even in the absence of lethal doses of stress or DNA damage during a wine fermentation, yeast will progress through a structured and defined lifespan, eventually reaching a senescent phase that will culminate in cell death. The lifespan of a fermenting yeast cell must not be thought of in purely chronological terms, but as the number of divisions it has undertaken. The number of daughters produced by a mother cell, termed the divisional age, indicates the relative age of the cell, while the maximum lifespan potential of a cell is referred to as the Hayflick limit. Yeast longevity is determined by genes and influenced by environmental factors. Each yeast cell is capable of dividing a number of times before reaching senescence, in which no cell division occurs, and death metabolism is initiated. This form of ageing is known as replicative senescence (and is a phenomenon shared by both yeasts and mammalian cells). Senescence is a consequence of termination of replication and is, therefore, intimately associated with cell division and the cell cycle.

In most industrial fermentations, cell division (budding) is not arrested primarily by nutrient deficiency, but by a lack of essential membrane lipids precursors. Should starvation conditions prevail, then yeast cells will stop budding, and enter the quiescent phase G0. Such shutting down of metabolism, and entry into this stationary phase of the cell cycle, is the cell's strategy for long-term survival (for months, or perhaps even years), and, because of it, wineries are able to purchase freeze-dried cultures!

It is when the yeast population reaches its maximum, and cells just begin to embark upon the stationary phase of growth, that fermentation rates reach their zenith. The velocity of the fermentation during this middle phase of vinification is almost entirely dependent upon the preceding phase of yeast growth. Exactly how much sugar is fermented, and when cells decide to enter the stationary phase appear to be determined by a combination of the fermentation temperature and the levels of nitrogenous matter present in the original must.

During the end-phase of fermentation, the inhibitory effects of the elevated level of ethanol begins to be felt by the yeast, and this, together with low levels of nutrients, results in a marked decrease in the rate of glycolysis, until, eventually, fermentation ceases. The final concentration of ethanol is dependent, not only upon the initial sugar concentration, but by the fermentation temperature (higher temperatures causing evaporative losses). In theory, there is a stoichiometric relationship between the amount of fermentable sugar initially present, and the amount of ethanol liberated by glycolysis, but such a relationship is rarely observed in practice. There are a number of reasons for this, one of which is the very way in which sugar levels are measured. Also, available must nitrogen plays a role, especially by determining how much carbon is converted into cellular components (*i.e.* yeast growth, rather than glycolysis). There is some evidence to show that the decline in fermentation rate is not necessarily associated with ethanol concentration, or loss of cell viability, but that it is simply the result of a permanent physiological alteration in the rate of glycolysis.

Under certain conditions, principally during the manufacture of sparkling wines, the build up of CO_2 can prevent the utilisation of the last traces of sugar. Such a phenomenon is rarely witnessed at the end of still wine fermentations, even though *in vitro* inhibition of fermentation by the gas has been observed at pressures as low as 0.3 atmosphere.

At the end of fermentation, amino acids are released into the wine by aged yeast cells. This release may be due to cell autolysis, or may result from ethanol toxicity. In respect of the latter, the phenomenon can be induced *in vitro* by adding alcohol to laboratory fermentations, thus enabling the role of ethanol in such leakage to be quantified.[39] The exact leakage mechanism, however, has yet to be elucidated. Leakage of nitrogenous compounds has two major implications for the winemaker, one beneficial, the other deleterious. Of benefit, is the role that these nitrogenous compounds play in the storage of wine *sur lie*, while the downside entails the fact that, under certain circumstances, amino acid breakdown culminates in the production of urea, a precursor of urethane, which is a suspected carcinogen.

In its simplest form, fermentation monitoring consists of daily checks of sugar level (via a hydrometer) and temperature, and the end of fermentation is the point at which all fermentable sugar has been converted to ethanol, something which may be signified by having two consecutive hydrometer readings with the same value. This can be confirmed chemically by assaying for residual sugar[‡]. The final stages of a fermentation should be closely monitored, including provisions for assessing odour and taste. When the density falls below 1.000, hydrometer readings become less precise, because the relationship between residual sugar and density is complex. In most circumstances, when fermentation is complete, wine density will vary between 0.991 and 0.996, according to ethanol content.

[‡] For many years, the reducing property of sugars was used to determine their concentration (*e.g.* Fehling's test), but such methods can encompass other substances as well.

When primary fermentation has finished, and there are no further processes (such as malo-lactic fermentation) to be undertaken, the wine should be chilled to around 4°C, to allow the gross lees to settle.

The amount of ethanol produced per unit of fermentable sugar during a wine fermentation is of considerable importance to the commercial oenologist. The theoretical conversion of 180 g sugar into 88 g CO_2 and 92 g ethanol would only be expected in the absence of yeast vegetative growth, and loss of ethanol by evaporation. Pasteur's original experiments indicated that he obtained an ethanol yield of 48.5% w/w (compared with the theoretical 51.1%), with 46.7% as CO_2, and the balance of 4.8% constituting glycerol, succinate and other products, including cell material.

The temperature at which a fermentation takes place must take into account, not only the production of ethanol, but the synthesis of desirable volatile compounds, and, in the case of red wines, the amount of extraction of phenolic compounds from grape skins. The control of fermentation temperature needs to be far more rigorous during the production of white wines, which are fermented at lower temperatures, than it does for their red counterparts. Generally speaking, white wine fermentations are carried out over the range 12–20°C, while reds demand 25–30°C, some French châteaux claiming temperatures as high as 32–34°C. At temperatures above these, there can be a considerable loss of desirable odour and flavour compounds, and unattractive ones, typically of 'burnt' or 'cooked' character, can be imparted. Jackson[40] maintains that winemakers in the New World vinify wine at the lower end of the 12–17°C range, so that the tropical fruit flavours (like pineapple) are generated. Conversely, he notes that the Europeans tend to ferment wines like chardonnays at 18–20°C. Red wines destined for 'early-drinking', particularly from Australia, tend to be fermented at lower temperatures (17–24°C). One general advantage of fermenting at lower temperatures is that there is less loss of ethanol by evaporation, and, in addition, volatile flavour components are more readily retained. Certainly, the higher fermentation temperatures associated with red wine production, while encouraging extraction of colour from grape skins, tend not to encourage the generation of fruity esters.

In the making of modern, floral, fruity white wines, fermentation normally takes place in large, closed (but vented), stainless steel vessels at between 15 and 20°C, although lower temperatures are occasionally used. Fermentation takes at least 2–3 weeks (maybe more), depending on the temperature. Lower temperatures enhance the production of esters, which are responsible for the fruity characters of these wines, while temperatures around 20°C favour the production of higher alcohols (fusel oils), such as *iso*-amyl and hexanol. Conversely, higher temperatures, in conjunction with low pH, give a considerably lower fusel oil content. Some fine white wines are still fermented in wooden casks, rather than metal containers, in order to impart desirable, full-bodied, characteristics. If wooden vessels are not available for fermentation, then such wines are fermented in stainless steel, and then aged in wood. Fermentation in wood is generally accepted as giving the best integration of fruit and wood. If fermented wine is aged in young, or poorly seasoned wood, than unacceptable 'sappy' notes can be imparted.

An unforeseen disadvantage in the use of some grades of stainless steel, used in the manufacture of wine tanks, manifested itself when it was discovered that some batches of inherently 'clean' wine could pick up sulfidic taints when they were transferred through certain, supposedly clean, vessels. Investigation showed that the tanks concerned had previously been used to store sulfidic wine, and subsequently scoured with caustic soda, and washed through with water. During storage of the sulfidic wine, manganese sulfide had been produced, and this was 'fixed' to the vessel walls by the caustic wash. Hydrogen sulfide was then released into the acid environment when the vessel was filled with wine. Rinsing such tanks with citric acid prevents this problem.

From a practical point of view, fermenting in wooden vats is much more labour-intensive than doing so in large stainless tanks. This is principally due to the amount of cleaning involved (which is invariably non-automated), and the fact that multiple monitoring is necessary.

The art of the oenologist is to ensure that yeast growth is uninterrupted and that the fermentation proceeds evenly to its completion, which, in the case of dry wines, means the complete removal of sugars. He/she has to retain the fruit characters of the grape, and encourage the formation of desirable flavour components resulting from yeast activity. Balanced against this, is the need to avoid the production of off-odours and off-flavours. One of the latent problems for any winemaker is that of a 'stuck' fermentation, whereby yeast activity is arrested before the desired level of dryness has been attained. The final stages of a fermentation are always the most difficult to accomplish, but, occasionally, a fermentation can stop much earlier, when there is plentiful sugar still present in the must. Such occurrences are potentially serious, because, with ample sugar still remaining, there is a likelihood of bacterial contamination, with all its dire consequences. All stuck fermentations are difficult to restart, but especially so when they occur in the final stages, when there is little remaining fermentable material (*e.g.* less than $10 \, g \, L^{-1}$), and malo-lactic fermentation has been initiated. Stuck fermentations are not quite so irreversible when there is more sugar still present (*ca.* $15 \, g \, L^{-1}$), and the ethanol content is less than 12%. Once the ethanol content has reached 13%, the chances of re-starting a fermentation are almost zero. The latter stages of a wine fermentation are also likely to present contamination problems for the oenologist because there is usually so little yeast activity that the growth of some bacteria is inevitable, especially if insufficient sulfite was originally added to the must. The two most likely causes of a stuck fermentation are excessive initial must sugar concentration, and excessive initial fermentation temperature, although there are other causes. Too low a temperature at the onset of fermentation can lead to an insufficient yeast population, while complete anaerobiosis can cause problems with the yeast population. Fermentability of a must can be influenced by the age of the vine, since old, less vigorous plants usually yield grapes whose juice has nutritional deficiencies and will not adequately support yeast growth, while, in white winemaking, grape crushing and extraction techniques considerably influence fermentability. Highly clarified grape juices, which have been processed and then held under anaerobic conditions, are often difficult to ferment to completion. Once extracted from the grape, there is a general need to protect juice from the adverse effects of oxygen, which principally causes browning and a loss of fruity aromas. The only circumstance whereby oxidised juice is desirable is in the manufacture of rancio wines.

The toxicity of ethanol, which is another possible cause of a stuck fermentation, can be exacerbated by several factors, including a late sugar addition to the must (a form of chaptalisation), and the build up of C_6, C_8 and C_{10} saturated fatty acids, most notably decanoic acid. During active fermentation, these acids are a result of yeast metabolism, and are normally catabolised and/or adsorbed by the yeast during fermentation end-phase. Indeed, one of the benefits of adding 'yeast hulls' to a fermentation is thought to lie in their ability to remove these acids from the wine. Conversely, certain sterols provide important protection against ethanol toxicity, there being several present in grape juice, including the unique oleanolic acid. Ingledew,[41] on the other hand, reported that yeast hulls stimulate yeast populations by providing a source of C_{16} and C_{18} unsaturated fatty acids, which act as oxygen substitutes under long-term fermentative conditions. One of the known cytotoxic properties of ethanol is that it inhibits the transport of carbohydrates into the yeast cell. Fortunately, this rarely occurs during the active phase of yeast growth, when ethanol levels are minimal.

Then there is the phenomenon of antagonism between strains of yeast, the most notable example being that of killer yeasts (K), which secrete proteinaceous toxins into the must (see Chapter 9), and can completely arrest a fermentation. Antagonism can also occur between yeasts and lactic acid bacteria, during red winemaking. Stuck fermentations are more likely to occur, and are more serious, if the pH of the must is too high.

Over the years, many theories have been forwarded to explain stuck fermentations; some plausible, others not, but it is interesting to note that some vineyards/wineries seem more prone than others, which seems to suggest an organic cause.

White wines are produced by the fermentation of grape juice, which means that extraction of juice, a limited amount of maceration and some degree of clarification, occur before the onset of fermentation (*i.e.* in the absence of ethanol). With red wines, juice extraction occurs during ethanolic fermentation, and most maceration takes place in the presence of ethanol. It is the absence of skin contact in ethanolic solution that distinguishes white wine making from red wine making, not the colour of the grape. White wines can be produced from red grapes, as long as they have white juice, and the winemaker ensures that the grapes are pressed under conditions that prevent the skin anthocyanins from getting into the must. In grapes generally, the desired varietal aromas, and aroma precursors, are situated in the grape skin, or in the layers of cells immediately underneath. It is an unfortunate fact that these regions of the skin will also contain most of the undesirable compounds in the grape, such as the biochemical consequences of any fungal disease, or bitterness associated with unripe grapes.

The ultimate taste of a dry white wine, made from a known grape variety, depends upon the pre-fermentation processes of harvesting, crushing, pressing (juice liberation) and clarification. Much of the art of making white wine involves knowing how to press the grapes, and how to clarify the juice, such that desired compounds are extracted and then adequately preserved. Juices with high levels of suspended material (*i.e.* turbidity), do not generally yield satisfactory white wines, and the winemaker would generally aim for a juice turbidity of between 100 and 250 nephelometric turbidity units (NTUs), which corresponds to *ca.* 0.3–0.5% of suspended particulate matter. Thus, as far as juice turbidity is concerned, the white wine maker will encounter problems if it is too low, and a different set of problems if it is too high. High levels of grape juice solids result in the formation of higher alcohols, such as *iso*-butanol, 2-methyl pentanol and 3-methyl pentanol, and the loss of the fruity ethyl esters. Some suspended solid material is necessary during fermentation in order to provide a site for budding yeast, and foci for the release of the major end products of fermentation (CO_2 and ethanol). The latter mechanism ensures that each yeast cell does not surround itself with toxic levels of its own 'waste products'. Where very bright juices are to be fermented, it is common practice to incorporate some inert biological material, such as cellulose fibres, to provide sites for yeast attachment. On this theme, 'expanded' cellulose fibres have been shown to be effective in preventing stuck fermentations. Their efficacy is due to oxygen being trapped within the fibrillar matrix, thus encouraging sterol synthesis during the latter stages of fermentation. Cellulose also traps toxins, which might otherwise interfere with yeast growth. Another means of avoiding protracted fermentations is to add preparations of fragmented yeast cells to the must. Such preparations, which are marketed as 'yeast ghosts', or 'yeast hulls', contain substantial amounts of micro- and macro-nutrients, membrane fragments and cell wall fragments, and are also thought to absorb some of the toxic fermentation end products.

Grape juice that is very low in solids (*i.e.* over-clarified, whether intentional, or not) often gives rise to an uneven fermentation and can lead to problems associated with protein instability. To counteract this, bentonite, or another stabiliser, may need to be added to the fermentation. Such an addition will also promote clarification of the wine after fermentation, lower levels being required than are necessary for post-fermentation additions. It is advisable to add the bentonite to the first half of the fermentation (it is usual to add while racking the juice into the fermenter), since later additions can cause the yeast to flocculate and so impede fermentation. Under certain extreme conditions, bentonite can even prematurely terminate a fermentation. When bentonite is added to the fermentation, it is necessary to do so in conjunction with a yeast nutrient addition, since bentonite can deplete must assimilable nitrogen due to electrostatic binding and adsorption. It should be realised that, while addition of bentonite to fermenting juice does aid clarification, and protein stability, it does not influence the ultimate overall stability of a wine. Laboratory trials are necessary to establish levels of bentonite required to achieve protein stability, particularly in view of the fact that the ethanol itself will denature and precipitate some of the proteinaceous material, thus causing 'over-fining'. Many winemakers use *ca.* 24 g hL^{-1} for additions during fermentation. Levels exceeding 48 g hL^{-1} may strip wine flavour and body, and impart an 'earthy' character. When wines of a differing nature are to be blended, post-fermentation, it is necessary to achieve protein stability via bentonite after blending has occurred, since any change in pH may serve to destabilise certain protein fractions. Fermentation with bentonite should always be seriously considered for wines that are not fermented in barrel, and/or do not have extensive contact with lees. Having said that, difficulties that have been encountered by using bentonite have led winemakers to examine a variety of other techniques for stabilising proteins, including the use of lees.

In red winemaking, maceration and the fractional extraction of the grape components occur during fermentation, rather than before it. Accordingly, the final taste, aroma and colour of a decent red wine are the result of how the course of fermentation proceeded, and to various operations during vatting. The red winemaker achieves the desired colour, aroma and tannin level over a period of time. If red wines are fermented out to low residual sugar levels in the presence of the skins (and, sometimes, the stems), then maceration will occur throughout. Most of the classic red wines are made in this way, and fermentation will normally occur at *ca.* 24–25°C for up to a fortnight. Such wines will have a deep colour, a high tannin content and will generally require a period of maturation to develop a full flavour. Conversely, lighter red wines can be made by running the juice off the skins, just after fermentation has commenced, and continuing the fermentation in the absence of the skins. During the manufacture of red wines, a shorter period of fermentation at an elevated temperature will favour the extraction of phenolic material, while a longer, cooler ferment will yield more aromatic components (especially esters). Generally speaking, it is agreed that 'time' is one of the best allies of the red winemaker.

The control of temperature during fermentation is an all-important facet of winemaking, white and red varieties requiring a different set of operating conditions. Vinification fermentations invariably take place at temperatures below the optimum for yeast growth and fermentation, and, with modern means of attempera-tion, rarely reach lethal levels (38–40°C for yeast, in the presence of alcohol). As we have said, ethanolic fermentation is an exothermic process, and, thus, while this is occurring at its highest velocity, a considerable amount of heat is evolved. To counteract this, a cooling system is usually required, and most modern fermentation tanks are equipped with some form of attemperation, either inside (cooling rings) or outside (cold jacket) the tank. Temperature control in wooden (oak) casks is more of a problem, because of the insulating properties of that material, and is pretty basic. It usually takes the form of keeping the casks in a temperature-controlled room, or cellar, where their high surface to volume ratio aids temperature control. One of the earliest means of cooling fermentation tanks in warm climates was to run a film of water over them, and allow it to evaporate slowly, an activity that presupposes an abundant (never-ending?) supply of running water!

3.3 YEAST STARTER CULTURES

As we have seen above, the task of vinification has been made much less problematical since the introduction of yeast starter cultures in the 1960s, dried active wine yeasts having been available since 1964. A number of companies specialise in such products, and to indicate the tremendous range available to the winemaker, both commercial and otherwise, I include a couple of the specifications from the many hundreds of dried yeast cultures now available. They give a feel of what the modern winemaker has to choose from. I use examples from the Lallemand catalogue, for no other reason than I happen to have been acquainted with their products. The company claim to be the world's largest producer of speciality yeasts for winemaking, and brand their products under the Lalvin® name.

3.3.1 Bourgovin

RC 212 *S. cerevisiae*

3.3.1.1 Origin. The Bourgovin RC 212 strain was selected from Burgundian fermentations by the Bureau Interprofessionnel des Vins de Bourgogne (BIVB). It was selected for its ability to ferment a traditional, heavy-style Pinot Noir Burgundy.

3.3.1.2 Oenological Properties and Applications. The RC 212 is a low-foaming moderate-speed fermenter with an optimum fermentation temperature ranging from 15 to 30°C (59–86°F). A very low producer of H_2S and SO_2, RC 212 shows a good alcohol tolerance (12–14% v/v). RC 212 is recommended for red varieties where full extraction is desired. Lighter red varieties also benefit from the improved extraction while colour stability is maintained throughout fermentation and ageing. Aromas of ripe berry and fruit are emphasised while respecting pepper and spicy notes.

3.3.2 EC-1118

S. bayanus.

3.3.2.1 Origin. The EC-1118 strain was isolated from Champagne fermentations. Due to its competitiveness and ability to ferment equally well over a wide temperature range, EC-1118 is one of the most widely used yeasts in the world.

3.3.2.2 Oenological Properties and Applications. The fermentation characteristics of EC-1118 include extremely low production of foam, volatile acid and H_2S, and this makes the strain an excellent choice. This strain ferments well over a very wide temperature range, from 7 to 35°C (45–95°F) and demonstrates high osmotic and alcohol tolerance. Good flocculation with compact lees and a relatively neutral flavour and aroma contribution are also properties of EC-1118. The strain is recommended for all types of vinification, including sparkling, and late harvest wines. It may also be used to restart stuck fermentations.

REFERENCES

1. L. Pasteur, *Etudes Sur Le Vin*, Victor Masson et Fils, Paris, 1866.
2. L. Pasteur, *Etudes Sur La Bière*, Gauthier Villars, Paris, 1876.
3. A. Harden, *Alcoholic Fermentation*, 4th edn, Longman, Green & Co, London, 1932.
4. I.S. Hornsey, *A History of Beer and Brewing*, Royal Society of Chemistry, Cambridge, 2003.
5. J.A. Barnett, *Microbiology*, 2003, **149**, 557.
6. J. Lodder (ed), *The Yeasts: A Taxonomic Study*, 2nd edn, North-Holland Publishing Co., Amsterdam, 1970.
7. N.J.W. Kreger-van Rij, *The Yeasts: A Taxonomic Study*, 3rd edn, Elsevier, Amsterdam, 1984.
8. J.A. Barnett, R.W. Payne and D. Farrow, *Yeasts: Characteristics and Identification*, 3rd edn, Cambridge University Press, Cambridge, 2000.
9. G.I. Naumov, *J. Ind. Microbiol.*, 1996, **17**, 295.
10. I.S. Hornsey, *Brewing*, Royal Society of Chemistry, Cambridge, 1999.
11. K. Rosen, Preparation of yeast for industrial use in the production of beverages, in C. Cantarelli and G. Lanzarini (eds), *Biotechnological Applications in Beverage Production*, Elsevier Science, New York, 1989.
12. S.J. Singer and G.L. Nicholson, *Science*, 1972, **175**, 720.
13. B. Guérin, Mitochondria, in A.H. Rose and J.S. Harrison (eds) *Yeasts*, 2nd edn, vol 4, Academic Press, London, 1991.
14. F. Foury, T. Roganti, N. Lecrenier and B. Purnelle, *FEBS Lett.*, 1998, **440**, 325.
15. P. Mitchell, *Science*, 1979, **206**, 1148.
16. A.C. Codon, T. Benitez and M. Korhola, *Appl. Microbiol. Technol.*, 1998, **49**, 154.
17. A. Goffeau, B.G. Barrell and H. Bussey, *Science*, 1996, **274**, 546.
18. S.G. Oliver, Q.J.M. van de Aart and M.L. Agostoni-Carbone *et al.*,, *Nature*, 1992, **357**, 38.

19. T. Galitski, A.L. Saldanha, C.A. Styles, E.S. Lander and G.R. Fink, *Science*, 1999, **285**, 251.
20. B. Dujon, *Trends Genet.*, 1996, **12**, 263.
21. A.J.P. Brown, Appendix III: Useful www. addresses for yeast researchers, in A.J.P. Brown and M.F. Tuite (eds), *Yeast Gene Analysis, Methods in Microbiology*, vol 26, Academic Press, London, 1998.
22. G.F. Sprague, Mating and Mating-Type Interconversion in *Saccharomyces cerevisiae* and *Schizosaccharomyces pombe*, in A.H. Rose, A.E. Wheals and J.S. Harrison (eds), *The Yeasts*, vol 6, 2nd edn, Academic Press, London, 1995.
23. A.E. Wheals, Biology of the cell cycle in yeasts, in A.H. Rose and J.S. Harrison (eds), *The Yeasts*, 2nd edn, vol 1, Academic Press, London, 1987.
24. F.M. Harold, *Microbiology*, 1995, **141**, 2765.
25. G. Sherlock and J. Rosamond, *J. Gen. Microbiol.*, 1993, **139**, 2531.
26. J.R. Broach, J.R. Pringle and E.W. Jones (eds), *The Molecular and Cellular Biology of the Yeast Saccharomyces*, vol 1, Cold Spring Harbor Laboratory Press, New York, 1991.
27. E.W. Jones, J.R. Pringle and J.R. Broach (eds), *The Molecular and Cellular Biology of the Yeast Saccharomyces*, Cold Spring Harbor Laboratory Press, New York, 1992.
28. R.E. Kunkee and R.W. Goswell, Table wines, in A.H. Rose (ed), *Economic Microbiology Alcoholic Beverages*, Academic Press, London, 1977.
29. F.F. Monteiro and L.F. Bisson, *Am. J. Enol. Vitic.*, 1991, **42**, 199.
30. W.R. Sponholz, Wine spoilage by microorganisms, in G.H. Fleet (ed), *Wine Microbiology & Biotechnology*, 1st edn Harwood Academic Publishers, Chur Switzerland, 1993.
31. R.J. Clarke and J. Bakker, *Wine Flavour Chemistry*, Blackwell, Oxford, 2004.
32. I. Benda, Wine and brandy, in G. Reed (ed), *Prescott and Dunn's Industrial Microbiology*, 4th edn AVI Publishing, Westport CT, 1982.
33. L.F. Bisson and R.E. Kunkee, Microbial interaction during wine production, in J.G. Zeikus and E.A. Johnson (eds), *Mixed Cultures in Biotechnology*, McGraw-Hill, New York, 1991.

34. R.E. Subden, *CRC Crit. Rev. Biotechnol.*, 1987, **5**, 49.
35. R.B. Boulton, V.L. Singleton, L.F. Bisson and R.E. Kunkee, *Principles and Practices of Winemaking*, Chapman and Hall, New York, 1996.
36. A. Kotyk and A. Kleinzeller, *Biochimica Biophysica Acta*, 1967, **135**, 106.
37. H.G. Crabtree, *Biochem. J.*, 1929, **23**, 536.
38. R.E. Kunkee and L.F. Bisson, Wine-making yeasts, in A.H. Rose and J.S. Harrison (eds), *The Yeasts, Volume 5 Yeast Technology*, 2nd edn Academic Press, London, 1993.
39. S.P. Salgueiro, I. Sá-Correia and J.M. Novais, *Appl. Environ. Microbiol.*, 1988, **54**, 903.
40. R.S. Jackson, *Wine Science: Principles and Applications*, Academic Press, London, 2000.
41. W.M. Ingledew, *Die Wein-Wissenschaft*, 1996, **51**, 141.

CHAPTER 4

Winemaking Processes

As many as 4000 varieties of *Vitis vinifera* have now been developed and are used in the production of wines. The diversity and quality of wine result from not only the type of grape used but also the distinctive qualities of soil, topography and climate. Over the years, a wide variety of vinification processes have evolved, and have been favoured for certain wine style requirements. Although the specifics of winemaking can vary from location to location and in the individual fermenting techniques employed, the basic steps involved in winemaking are similar for most wineries. It should be stressed, however, that modern science and technology has removed much of the mystique from winemaking, and we are now far closer to 'standard methodology' than we have ever been. In general, New World winemakers have been far more receptive to the new technology than those from the Old World. Change continues apace; the existence of a publication such as *Growing Quality Grapes to Winery Specifications* (Krstic *et al.*[1]), would have been unheard of 20 years ago.

4.1 RED WINE PRODUCTION

Of prime importance, of course, is choice of grape variety, which for a given winery will be the primary determining factor in deciding wine style and quality. Some red grape varieties are so disparate – for example Cabernet Sauvignon and Pinot Noir – that it is impossible for them to be converted into anything resembling comparable wine styles. The most significant variations in red grape varieties are in colour, flavour and tannin content. In general terms, the site of the vineyard, the choice of the cultivar and the mode of vine cultivation will pre-determine the potential for flavour and aroma components within the grape, while the seasonal growth conditions (*e.g.* a cool summer) will determine the practical concentrations of these components that will be available to the winemaker. In varietal wine styles, it is the responsibility of the winemaker to maximise extraction of the characteristics of the fruit. The vast array of flavours that exist within red grape cultivars has enabled a wide spectrum of distinctive red wine styles to evolve.

4.1.1   Time of Harvest

Much of what is said here is also applicable to white wine grapes. The condition of the grapes at harvest is crucial to wine quality, and the concept of 'optimal ripeness' is an important one. In many cases, the optimal time for picking lasts for only a very few days, and ripeness has to be assessed by rigorous sampling, which in a large vineyard can be an onerous task. Sampling should take into account the position of vine in the garden, the position of bunch on the vine and the position of berry in the bunch, all of which can influence sugar concentration and other ripening parameters. Samples are then returned to the laboratory for sugar and pH analysis. As Rankine[2] states quite forcefully, 'There is no way that a brief stroll through the vineyard with a hand-held refractometer, sampling a berry here and there, can provide a reliable assessment of the state of ripeness of the vineyard. To achieve this one must do a lot of walking and sampling.' In addition, it should be stressed that measurement of sugar content is simply a convenient, easily measured indicator of berry development and does not necessarily enlighten the winemaker about the status of other important grape components.

As is shown in Chapter 2, the development of the grape typically follows a double-sigmoid curve, in which three distinct phases are discernible: (1) The berry undergoes rapid cell division and growth and acid accumulates, but there is little or no sugar. This is called the 'green' stage. (2) The berry growth rate decreases markedly, and the berry 'rests' until véraison (typically around 60 days after flowering). (3) At véraison, the berry starts to soften and accumulate sugars, organic acids become degraded and colour begins to appear. The berry enlarges and flavour and aroma compounds build up. Once these developmental stages have been completed, the grape begins to lose water while still on the vine, this dehydration being of value for making late-picked and fortified wine styles. The two main grape sugars produce different levels of sweetness, fructose being sweeter than glucose. If sweetness was measured on a scale of 1–100 and fructose scored 100, then glucose would score 66 (sucrose would score 84). During ripening of the berry, glucose accumulates first, but by véraison the proportions of glucose and fructose are roughly equal. In very ripe berries, fructose levels will exceed those of glucose. Individual grape varieties have differing abilities to accumulate sugars – Chardonnay and Pinot Blanc, for example, being high-fructose varieties, while Zinfandel and Chenin Blanc are high-glucose varieties.

The major organic acids of ripe grapes are L(+)-tartaric and L(−)-malic, which together constitute some 90% of total acidity (usually around 5–8 g L^{-1}). Citric acid is the next most abundant, and there are minute amounts of a number of other acids. As far as contribution to 'acid' taste is concerned, tartaric and malic acids are roughly equivalent. High-tartaric acid varieties include Riesling and Palomino, while higher levels of malic acid are found in Pinot Noir and Chenin Blanc grapes.

Under-ripe grapes can have sugar levels below those necessary for giving wine a normal alcohol content (11–13%), and their enhanced acidity gives rise to a harsh palate. In addition, experience tells the winemaker that only grapes at, or close to, physiological ripeness will yield the desired varietal flavours and prevent the wine from tasting 'green' and 'thin'. Under certain circumstances, 'mother nature' determines that it is necessary to harvest immature berries, particularly in cool climates where there is a likelihood of an early frost, and in damp localities where disease pressure mounts as the season progresses. Conversely, grapes picked when over-ripe can lead to sub-standard wines. This is really a problem only in warm regions, and the berries will contain too much sugar and not enough acidity. The high pH, if left unadjusted, can lead to bacterial contamination in the wine, which, in addition, can often take on a 'stewed fruit' or 'jammy' flavour. As a rule of thumb, the faster the rate of ripening of grapes between the moment that they change colour and harvest, the better will be the colour, flavour, sugar and acidity. This can be illustrated by the fact that in cool climates, the best wines are made in warmer years.

The significance of harvesting time can be ascertained by referring to the Italian Piedmont region, where the native Nebbiolo and Barbera vines have long been grown. Many wineries there now tend to favour higher acidity, more pronounced grape flavour and less tannin than previously, which means that they prefer to pick their grapes early. Highly traditional makers, on the other hand, who seek higher levels of alcohol and tannin, will harvest at a later date (October to November). In some parts of Australia, it is possible to take advantage of the warm conditions by picking grapes over a wide range of ripeness levels such that a greater range of flavours can be captured by blending the resultant musts. This is seen as preferable for a variety such as Cabernet Sauvignon, where grapes harvested all at the same time tend to produce one-dimensional wines.

The pH of grapes is not necessarily a reliable indicator of the acidity of the final wine, simply because it can alter considerably during processing, especially during alcoholic fermentation, and after any malo-lactic fermentation. In summary, the most important parameters at harvest are the sugar level and fruit flavour of the grapes; pH can always be adjusted in the winery. Red wines are normally made from grapes with a ripeness measure from 18 to 25° Brix (10–14° Baumé), which should result in wines with an ABV in the range of 10–14%. Having said all this, it should be noted that, in some areas, the harvesting decision is still primarily governed by the prevailing weather conditions and/or the availability of labour.

4.1.2 Harvesting

Grapes can be picked either by hand or with the aid of machinery, there being advantages and disadvantages attached to each method. Hand-picking allows bunch selection (*i.e.* whether ripe, healthy or botrytised), and does not cause significant damage to the fruit, nor does it ever outpace the ability of the winery to process the yield! Mechanical harvesting is much more rapid and can be carried out at night, which is a considerable bonus in warm regions. It is generally cheaper than hand-picking (allowing for machinery costs) and delivers only grapes (no stems) to the winery. On the down side, there may be damage caused to grapes and vines and the collection of material other than grapes (known in the trade as

'MOG'). A harvester cannot differentiate between ripe and unripe bunches and, of course, there is always the possibility of disastrous circumstances in the case of breakdown. Harvesting method is often closely linked to grape variety and the style of wine intended. For instance, with Cabernet Sauvignon, mechanical means are usually satisfactory because the grapes are usually crushed before fermentation. With Pinot Noir, however, there will be an emphasis on hand-picking because there is a premium on undamaged grapes. Damaged Pinot Noir berries begin to oxidise almost immediately, and, once picked, grapes should be transferred quickly to the winery. Temperature at harvest has been touched upon, and it goes without saying that grapes should be as cool as possible at harvest in order to minimise oxidation and unwanted microbial growth. Some wineries make their first addition of SO_2 into the picking bin. Transfer from vineyard to winery should be as quick as possible, and it is rarely a problem in small European businesses. In the New World, however, where distances can be vast, it can take around 24 hours from harvesting to the first stages of winery processing.

Before proceeding to the activities inside the winery, it will be useful to ascertain exactly where the 'goodness' lies within the grape bunch, since the location of important compounds often determines the path taken by the vigneron. It has been calculated that only around one-third of the colour and tannin of the grape will be released and incorporated into wine, and that vastly different contributions are made by flesh, skins, seeds and stems. Black and white grape varieties contain roughly equal amounts of non-pigmented tannins, but the skins of black varieties contain around twice the total phenolic content of white varieties (due to pigmented anthocyanins). The flesh consists of large cells with conspicuous, sap-filled vacuoles, which yield most of the free-run juice after crushing and draining. It has the lowest pH (3.0–3.8) of all grape bunch components (*i.e.* a high free acid to potassium ratio), and less than 5% of all the phenolics of the berry. Varietal flavour compounds occur in both flesh and skin. Grape skins can contain high levels of sugar, sometimes to the tune of 80% of the amount found in flesh. In white grape varieties, some 10% of total phenolics are located in the skin, while in black varieties this may be as high as 65%. Much of the berry potassium content is to be found in the skin. The waxy bloom of skins contains fatty acids and sterols, which are thought to stimulate yeast growth. Stalks have a sugar content of less than 1%, and they contain between 0.5 and 3.5% polyphenols by weight – these being mainly catechins and leucoanthocyanins, which have a harsh, astringent taste. Similar tannins are found in seeds, and can be the major source of them in wine (up to 50% of the grape content of these compounds is located in the seed).

4.1.3   Crushing

The first piece of equipment to be used in most wineries is the crusher, an invention of the late 19th century. Crushing was originally performed by simply 'treading the grapes', a process often carried out at night and known as *pigeage* nowadays. This age-old technique still survives in some traditional wineries. The aim of crushing is to split the skins and release the juice ready for the onset of fermentation. Having successfully harvested the crop, the first decision facing the winemaker is whether to incorporate stems into the vinification, or whether to de-stem. Stalks contain a lot of soluble tannins, which can enhance phenolic levels and increase astringency in a potentially soft wine, and so a particular protocol is often characteristic of wine-producing region. Some Burgundian producers add whole bunches or even just the stalks to the fermenter in order to redress softness, a practice that has been adopted by some New World producers, who refer to it as 'stalks return'. In Bordeaux, on the other hand, crushing and de-stemming is widely carried out. Most crushers first pass the grape bunches through a series of rollers, which crushes them and then subjects them to beaters, which removes the stalks (the crushed berries falling through a slotted floor). Some modern machines de-stem first. The extent of the crush can be varied by means of adjustable rollers, as some makers require almost whole berries. It would be true to say that the majority of modern red wines are made from well-crushed and de-stemmed berries. Fully crushed and de-stemmed grapes will yield a wine with good colour but rather shallow taste and aroma profile. SO_2 is often added to crushed grapes in order to suppress the growth of the components of their natural micro-flora, and to bind with anthocyanin pigments to make them more soluble.

4.1.4 Fermentation

The essential elements of vinification fermentations are documented in Chapter 3, and, as intimated, there are many variations
in fermentation protocol for red wines. The following will describe
some of the processes peculiar to red wine fermentations, which are
generally regarded as being more of an art than those required for
white wine production. The reason for this is that the composition
and quality of a red wine generally depend on far more processing
variables than are applicable to a white wine. Let us assume that we
are going to make a dry red table wine from Cabernet Sauvignon
grapes, and that crushed and de-stemmed must has been introduced into a fermentation vessel, together with a dose of SO_2,
around 50 mg L^{-1}. Tartaric acid may also be added at this stage
(regulations permitting), if a pH adjustment is necessary (and if no
chaptalisation has occurred). Selection of fermentation vessel type
has a considerable effect on wine style, and is one of the first crucial
decisions for the oenologist. Red wine fermentations have to take
into account the extraction of pigments from the 'cap' of skins that
floats to the top of the ferment, and vessels should be designed with
this in mind. The initial choice is between 'open' or 'closed' vessels,
and each type has advantages and disadvantages. Open vessels,
considered more traditional, allow for oxygen contact during the
early stages of fermentation, and for manual (*i.e.* gentle) manipulation of the skin cap; they also encourage dissipation of heat
during fermentation. On the down side, they are only practical for
small volumes, they are not suitable for extended maceration and
they allow free escape of ethanol. Relatively free access to oxygen
during the later stages of fermentation is also usually regarded as
disadvantageous. Closed vessels are multi-purpose, and can be
used for white wine fermentation and storage. They can hold huge
volumes of must, are immune to the ingress of oxygen and facilitate
easier cooling and warming operations. To their detriment, closed
vessels have a tendency to allow fermentations to get too hot and
make cap management difficult.

The second major choice to be made is whether spontaneous
fermentation should be encouraged, or whether the must should be
inoculated. The 'pros' and 'cons' are discussed in Chapter 3. The
normal range over which red wine fermentations are carried out is
from 18 to 35 °C. At the lower end of this range, fermentations will

be slower, less ethanol will be lost through evaporation and the resultant wine will contain more fruit flavours. It is likely to be used for early drinking. Cabernet Sauvignon (and Pinot Noir) wines are normally vinified at more elevated temperatures.

During red wine fermentation, a mass of skin debris called the cap floats to the surface, carried there by bubbles of CO_2. If allowed to remain suspended, this cap will overheat and dry out, thus minimising extraction of colour (especially) and flavour components. A dried cap also becomes a repository for undesirable microbes such as acetic acid bacteria. Various methods exist for re-suspending cap material in the fermenting wine, the simplest being to physically 'punch' it back down into the liquid phase with a paddle, something that can only be carried out in open fermenters. Alternatively, wine can be pumped from the bottom of the tank and discharged over the top of the cap (called 'pumping-over'), thus irrigating and dispersing it. Various means of irrigation have been devised, the Ducellier system (also known as the 'autoferm-enter') probably being the most useful (Figure 1).

The apparatus consists of a tank with an upper and a lower compartment, and it enables actively fermenting juice to be regu-larly sprayed over the mass of grape skins under anaerobic condi-tions. This allows rapid extraction of colour, and can be operated on a semi-continuous basis. The lower chamber holds fermenting must in an atmosphere of CO_2, while the upper chamber holds fermenting juice, which has been forced upwards by the pressure of CO_2 generated during fermentation. When the upper chamber is full, a release valve opens and the juice descends rapidly through a central tube and sprays over the mass of skins in the lower section. The bottom chamber must be kept full if the system is to work properly.

Some wineries fit heading-down boards inside their tanks, which prevent the skins from floating to the top. Finally, there are now rotary fermenters, which contain a system of internal vanes and revolve slowly about their long axis, thus ensuring thorough con-tact between skins and wine. Notable examples are the Rototank and the Vinomatic fermenter. If the skin cap remains intact for too long, it has an insulating effect and creates temperature stratifica-tion in the fermenting tank. The temperature in the centre of the base of the layer of skins may be as much as 10°C higher than the wine below it.

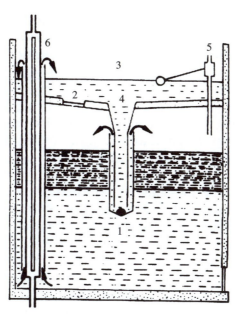

Figure 1 *Ducellier system: (1) fermenting must; (2) port for introducing must; (3) open juice fermentation space; (4) cooled fermenting juice; (5) valve for release of CO$_2$; (6) cooling column for fermenting upward-flowing juice (after Rankine[2])*
(Reproduced by kind permission of Bryce Rankine and Pan Macmillan Australia Ltd.)

Pressing the grape mass (pomace) occurs after the free-run wine has been removed from the fermentation vat, and takes place when the winemaker decrees that the required amounts of colour, flavour and tannin have been extracted. The timing can vary from 2 days to 3 weeks post-fermentation, according to wine style. Some wineries consistently leave the wine in prolonged contact with skins (and, sometimes, seeds and stalks) after fermentation has been completed, usually for a period of 2 or 3 weeks. This practice, which was at one time a characteristic of Bordeaux wines, is called 'extended maceration', and can often have a pronounced effect on the wine, increasing phenolic content and diminishing colour. There is also some evidence that wines produced in this way have a better ageing capability.

The archetypal wine press is the 'basket press', which is still used in smaller (and more traditional) wineries, and historically it has been *de rigueur* to use this kind of press as a drainer, prior to the application of pressure. The basket press consists of an upright, slatted wooden frame (often circular in cross-section), which holds the grapes, and through which a solid plate descends. The passage of the plate downwards through the cage forces the wine out through the numerous gaps in the side walls. Pressing by this method is regarded as *méthode ancienne*. The first major development of the 20th century was to turn the cage over on its side, allowing it to revolve and permit mechanisation (*e.g.* the Vaslin horizontal basket press). A more recent development is the pneumatic or 'bladder press', in which a rubber bag is inflated inside the press to apply powerful but gentle pressure. The inflated bag presses the must/grapes against the side of the cage. Because of the large surface area under pressure, material is pressed more gently, with little or no chance of breaking the pips or damaging the stalks. Even in some highly conservative châteaux, the pneumatic press is replacing the traditional basket press, mainly on account of its gentle treatment of grape material. In very large wineries, a continuous press may be employed, which extrudes the pressed skins (*marc*) continuously from one end of the apparatus as fresh must is introduced at the other. While this method of pressing is very efficient, it is very harsh and it is not unusual to hear 'the pips squeak'. 'Press wine' is high in colour and tannin, and a percentage of it may be blended with free-run wine to add tannins, character and longevity. Not all red wines go through a pressing stage. When using high-tannin red grape varieties, it is invariably necessary to fractionate the pressings and then blend it at a later stage.

With red wines, practicalities demand that they be almost completely fermented in tank (to ensure appropriate extraction of colour and tannin from the skins) before being transferred (drained) into a barrel for the completion of fermentation. Any full-bodied red wine can benefit from at least partial barrel fermentation, and the contribution of wood to the character of red wine is discussed in Chapter 7.

4.2   WHITE WINE PRODUCTION

Much of what has been said about red winemaking is relevant to white wine production, and so only the main points of variance will be discussed here. During the production of white wines, varietal flavour can often be the most important organoleptic aspect of the final product, and a good Riesling or Gewürtztraminer, for example, will exhibit grape-derived aromas above everything else. Hence, it is highly important to match grape variety with growing region, so that sugar production and varietal flavour are synchronised. As an example, Chardonnay grapes grown in very warm climes will often over-produce sugar at ripeness, and thus yield wines with too much ethanol. If such grapes were picked sooner at a lower sugar content, they would not contain sufficient varietal flavour material. Either way, there is a problem for the winemaker. Rather similar problems can be encountered with levels of acidity in some white grapes (*e.g.* Gewürtztraminer), which can decrease sharply as ripeness approaches. In such instances, it is best to pick the grapes at ripeness and adjust pH at the crusher (thus obtaining maximum varietal flavour), rather than harvesting them earlier when acidity is higher, but before the flavours have developed. Most white varieties, however, are more likely to suffer from too much acidity, especially if grown in the cooler areas of the planet. In general, white grapes are far more tolerant of cooler climates than reds, and they often have a much more elongated period of optimal ripeness (the 'harvest window'). White wine juice may be entirely fermented in barrel, or may be partially fermented in tank and then transferred to barrel. Barrel fermentation is usually practised for premium, full-bodied styles, such as Chardonnay and Semillon.

4.2.1 White Wine Styles

It is usual to recognise four classes of still white wine:

   (i) dry, floral and fruity;
  (ii) medium–dry, floral and fruity;
 (iii) dry, full-bodied;
 (iv) sweet table wines.

In the first category, we may include the classic Rieslings from Germany, which typically have terpenes as the major contributors to aroma and flavour. They have an ethanol content of 9–11% v/v, high acidity (<7.5 g L^{-1} as tartaric acid), low pH (>3.3) and less than 7.5 g L^{-1} of residual sugar. There will be no organoleptic evidence of malo-lactic fermentation. Wines in the second category are of a similar alcoholic strength, and derive from the classic styles of the Mosel. They have residual sugars in the range of 10–30 g L^{-1}, with a good acid balance. Dry, full-bodied whites, as exemplified by Chardonnay, have higher levels of ethanol (13–14% v/v), lower titrable acidity (6–7 g L^{-1} as tartaric acid) and higher pH values (<3.5) than the floral and fruity wines. Wines in this category will typically possess oak aromas and the signs of a malo-lactic fermentation. Sweet, white table wines will normally have a residual sugar of more than 30 g L^{-1}, an acidity in the range of 8–10 g L^{-1} (as tartaric acid) and a pH in the 3.3–3.7 range. Sweet wines fall into two main categories: (1) those involving *Botrytis cinerea* infection and (2) those produced as a result of other means of sugar concentration. Wines in the former category are the most complex, the classic examples being French Sauternes and the Trochenbeeren auslese of Germany.

Sauternes have an alcohol content of *ca.* 14% v/v, a residual sugar of 65–100 g L^{-1} and a distinct 'oakiness'. The German offerings are less alcoholic (9–12% v/v) and higher in residual sugar (120–150 g L^{-1}). The use of *Botrytis* infection means that the varietal characters of the grape become lost, and almost any variety can be used to similar effect, the only constraint being susceptibility to infection by the fungus. Traditionally, Muscadelle, Riesling, Sauvignon Blanc and Semillon were used. Non-botrydised sweet white wines are made in three main ways: (1) from grapes partially dried on the vine or on mats; (2) by early termination of the fermentation; or (3) by late addition (back-blending) of concentrated grape juice. Such wines are often made with Muscat-flavoured grape varieties.

4.2.2 Harvesting

For most styles of white wine, grapes must be harvested and transported very gently in order to minimise damage, and subsequent release of phenolic compounds that might intrude upon delicate varietal flavour compounds. It is also important to aim for almost total exclusion of oxygen immediately after picking (and throughout processing), a technique often referred to as 'protective handling'. Protective handling involves the use of low temperatures, inert gases and antioxidants (such as SO_2), which obviously incurs some expense, but grapes and juices treated in this way give rise to wines that retain their primary fruit flavours and colour intensity.

4.2.3   Crushing

The first stage of a controlled white wine vinification is to carefully crush the grapes (with or without stems), thus making them easier to press. If the stems are left with the berries, the juice will run more freely during pressing. White grape must is often chilled, especially in warmer regions, in order to delay fermentation until after pressing, and in some varieties (*e.g.* Gewürtztraminer) it is held in contact with the skins for extraction of flavours, especially those imparted by monoterpenes that are concentrated just under the skin. Skin contact should take place in specially designed vessels in order to minimise the risks associated with it, which range from over-extraction of astringent phenolics to lowering of acidity. Although crushing ostensibly leads to some oxidation, a few winemakers practise 'oxidative handling', the opposite of protective handling, whereby no attempt is made to protect the must from air. This allows polyphenol oxidase to oxidise phenolics, thus eliminating browning precursors, and gives the wine a greater ageing capacity. Since little, or no, SO_2 is added, malo-lactic fermentation is encouraged. Where white wine is being made from red grapes, crushing is omitted; grapes go whole into the press.

4.2.4   Pressing

White wine grapes/musts are always pressed, with better quality juices being produced when pressing is gentle. Some of the best white wines result from an elongated pressing regime, which, by a combination of very low pressure and a long period of draining, yield juices of extreme delicacy. Such time-consuming practices, however, are not always possible at the height of the harvest, when throughput must be maintained, and it should be emphasised that high-quality wines can be obtained from fast press cycles. A technique originally invoked for the production of premium sparkling wines, but now used for some hot-climate Rieslings and Chardonnays, is that of 'whole-bunch pressing'. It involves loading uncrushed, whole bunches of grapes directly into the press. The resultant juice is very low in phenolics and solids, and possesses fine, delicate flavours.

4.2.5 Settling/Clarification

Juice from the press is drained into holding tanks, where gross material, such as skin, stalk and pip fragments will settle out. The solids content of a juice prior to fermentation can be an important factor in determining the wine style. Traditionally, most French makers have favoured juices with a fairly high level of grape solids (5–10%), while many New World wineries prefer to clarify their juices down to around 0.1% solids, or below. The time-honoured way to clarify a juice is by leaving it to settle, 24 hours usually being sufficient. Settling is invariably carried out at reduced temperatures, despite the fact that warm juices settle more rapidly than cold ones! More modern clarification methods include filtration (usually with diatomaceous earths) and centrifugation.

4.2.6 Fermentation

Although traditionally carried out in oak casks, white wine is now more often fermented in stainless steel vats. Some high-quality whites are still vinified in wood. Temperatures required for white wine fermentations are generally lower (rarely above 20°C) than those used for red wines, so that there is some survival of fruity esters. Hence, temperature control during white wine fermentation is much more critical. Chaptalisation is practised by some white winemakers, but not as frequently as is necessary for red wine production. Many white wines are not fermented out to complete dryness (*i.e.* they contain residual sugar), and this is best achieved by halting the fermentation, by either rapid chilling or yeast removal. After fermentation is deemed to be complete, the winemaker has to decide whether extended lees contact and malo-lactic fermentation are required.

4.3 SPARKLING WINE

Sparkling wines are those that are surcharged with CO_2, such that they contain a residual level of the gas. The addition of CO_2, whether it is by extraneous means or by an internal secondary fermentation, imparts new dimensions to a wine, which *via* all five senses leads to an alteration of human understanding. The dissolved gas changes the taste perceptions of sweetness, acidity and astringency. The escaping gas may contain entrapped aroma molecules, which can become concentrated in the headspace above the wine, and thus affect our sense of smell. In addition, escaping bubbles can be visually attractive (sparkle), and the 'crackling' sound of bubbles escaping can bring aural pleasure. The 'prickly' sensation of dissolved CO_2 in the mouth promotes the feeling of thirst quenching, and is an integral part of the fascination of sparkling wine consumption. Sparkling wines are generally more complicated to make than their still counterparts, and usually have higher acidity, lower alcohol and more delicate flavour than most table wines. Then, of course, there is the 'tingle'!

Most authorities would agree that there are four main methods of making sparkling wine: carbonation, bulk fermentation (the Charmat method), bottle fermentation by filtration (the transfer process) and the classic champagne method (*méthode champenoise*). All but the latter are solely the result of 20th-century technology. Some would suggest that these methods provide a basis for the classification of such wines, but Howe[3] maintains that nearly all of the wines made with residual CO_2 can be classified by eight production variables:

   (i) the type of base wine used;
   (ii) the method of attaining the carbonation;
   (iii) the sugar source for the carbonating fermentation;
   (iv) the vessel used for the carbonating fermentation;
   (v) the amount of time of ageing on fermentation yeast lees;
   (vi) the method of clarifying the wine;
   (vii) the final product container; and
   (viii) the method for obtaining any remaining sugar in the finished product.

A sensible classification of sparkling wines has been adopted by the Australians, where the state-operated Food Acts specify that (1) champagne must be produced by the traditional method of fermentation in a bottle of less than 5-L capacity, and aged on the lees for not less than 6 months; (2) sparkling wine must be made by a complete, or partial, fermentation of the sugar content, and be charged with CO_2. Sugar and/or wine spirit may be added; (3) carbonated wine is that to which CO_2 has been added from any source other than by its own fermentation, in excess of 100 kPa pressure.

The level of carbonation in a sparkling wine can be measured in a number of ways, perhaps the most commonly used being the measurement of the amount of internal pressure in the finished bottle at a given temperature (usually expressed in atmospheres, or bars). The SI unit of pressure is the pascal (the kilopascal (kPa)=1000 pascals). One bar pressure is equivalent to 100 kPa and 0.9869 atmospheres (*i.e.* one atmosphere=101.327 kPa). Methods have now been developed that allow the expression of carbonation to be recorded as grams of CO_2 per litre of wine. Most wine-drinking countries have a legal definition of what is 'sparkling', and what is not. In the USA, for example, a wine must contain more than 3.92 g of CO_2 per litre to be classified as sparkling; anything below this level is legally considered to be still, even though it might taste 'fizzy'. The OIV view things from another perspective, and recommend that, internationally, the maximum CO_2 concentration for still wines should be 2 g of CO_2 per litre. Such a liberal definition is almost certainly designed to allow for worldwide inconsistencies in terms such as 'fizzy' and 'still'.

4.3.1 The Champagne Method

The most complex and most traditional means of gaining carbonation in wine is *via* a secondary fermentation by *Saccharomyces cerevisiae*, and the most notable example is, of course, champagne. Of the great wines of France, champagne is the newest arrival. As a district, Champagne has been making wine since early Roman times, but it was not until the 17th century that the first highly sparkling wines appeared, and not until the early 19th century that 'champagne' became synonymous with sparkling wine. The numerous specialised techniques used today for making champagne have evolved over a period of 200 years, and have involved a large number of makers. Contrary to popular myth, not everything was discovered by Dom Pérignon! Over the years, champagne has benefited from imaginative and energetic marketing, with the result that making objective criticisms about the drink are often difficult, especially in convivial circumstances. As Halliday and Johnson remarked,[4] 'It is true that the greatest champagnes are of extraordinary finesse, balance, and above all else, length of flavour; the intrinsic quality of these wines is on a par with the greatest of the still white or red table wines'. Certainly, the Champagne Appellation has some of the strictest, most exacting standards for growing, producing and labelling in all of the wine world.

Irrespective of how carbonation is achieved, all sparkling wines originate from a 'base wine', which is invariably dry and fermented in a conventional way. The classic grape varieties of Champagne are Pinot Noir, Chardonnay and, to a lesser extent, Pinot Meunier, although in practice many more are employed, especially by New World wineries. The finest champagnes are made from the purest musts/juices, which usually emanate from hand-picked grapes that have been gently bunch-pressed, thus ensuring minimal contact between juice and broken skins. Grapes destined for making champagne are picked earlier in the season than those for still wines, which means that the relatively high level of organic acids will exert a preservative effect during the lengthy processing. The concomitant lower sugar levels will also yield lower levels of alcohol during primary fermentation, but this will be compensated for during secondary fermentation in bottle. Many makers still use traditional basket presses in order to avoid such events. Juice is released from the press and collected in the order of quality, the free-run juice and the first pressing containing the lowest tannin content. Second pressings are used for 'inferior' wines. Skin fragments and other gross impurities can be settled out of the juice at normal cellar temperatures, a process that can be hastened by chilling to $-5°C$ if necessary. The more the material that is settled out at this stage, the more stable the finished wine will be. Bentonite is sometimes used as an aid to sedimentation, but there is a danger of over-fining, which can decrease the flavour complexity of the final product.

Clarified juice is then fermented into base wine. The most renowned (best?) houses would still ferment in oak barrels, but stainless steel vats are now commonly used. Cultured yeasts are being increasingly employed, which eliminates most of the elements of unpredictability during vinification but can decrease the complexity. Many champagne-base wine fermentations require chaptalisation in order to attain the desired ABV, and malo-lactic fermentation is almost always encouraged. After primary fermentation, the wine is settled or fined to achieve some degree of stabilisation. This is preferably carried out at reduced temperature so that further, unwanted yeast enzyme activity is discouraged and tartrate crystal formation prevented.

The wine is then racked off from the sediment and further clarified before being blended (*assemblage*), a process carried out by highly skilled, experienced tasters. *Assemblage* is probably the most critical stage in the convoluted process of making champagne because the blender has to envisage how a series of often hard, acidic and thin base wines will coalesce after blending, and respond to a secondary fermentation before producing a memorable wine. The blenders at Moët & Chandon, for example, often have up to 300 base wines from any one vintage to deal with. The possible permutations are incalculable. Once a suitable blend has been established, called the *cuvee*, the wine needs to be clarified (*collage*) once again, since the combination of so many different base wines invariably causes a deposit to be thrown. Gelatine is the usual fining agent in the Champagne region (sometimes after a light addition of tannin), although isinglass can also be employed. The clarified wine is then cold stabilised before being made available for a secondary fermentation in bottle.

The primary fermentation usually results in a dry base wine[*], so a source of fermentable sugar must first be added to each bottle to accommodate further yeast growth. This is usually effected by addition of a small sample of reserve wine containing sucrose (called *liqueur de tirage*), followed by addition of a suitable strain of yeast (*levurage*). The amount of sugar added is dependent upon the alcohol content and pressure required, and the amount and type of yeast added will vary according to the style of champagne. Yeast for this fermentation should be able to ferment under pressure in an environment of low pH, about 10% ethanol and the presence of SO_2. It should also be able to flocculate after fermentation. Once all components are in place, the bottles are sealed (*tirage*) with crown caps and stored. Bottles are commonly stacked in a horizontal position (*entreillage*) at temperatures of 12–15°C for at least 6 months. In many top houses, secondary fermentation and

---

[*]Occasionally, the primary fermentation does not proceed to completion, and adequate sugar remains in the base wine to accommodate secondary fermentation. In such cases, the wine is transferred to another bottle and a fermentation is allowed to resume to completion. This is known as *méthode rurale* or *méthode ancestrale*.

maturation takes place in deep, cool chalk cellars, and it is not uncommon for ageing to take 20–50 years. The CO_2 liberated during secondary fermentation cannot escape, and the resultant pressure demands that champagne bottles be constructed of thickened glass. During fermentation/ageing, bottles are traditionally stacked and separated by wooden slats (*lattes*) in such a way that if a rogue bottle explodes, the rest will be disturbed as little as possible. Bottled are shaken and re-stacked at intervals to prevent sediment (yeast debris) from adhering to the glass. Once ageing is deemed to be complete, the wine has to be clarified. In olden times this was felt unnecessary, and sparkling wines were decanted and drunk from hollow-stemmed glasses, which allowed the sediment to settle in the stem and the clear beverage to be drunk from the glass above. This is no longer a commercially viable option, and clarification is either *via* filtration under counter-pressure or by traditional riddling (*rémuage*). Riddling is a complex process that used to be carried out manually, but, these days, can be effected automatically by the 'gyropalette' invented by the Spanish house Codorníu, which became popular in the late 1970s. The apparatus considerably shortens clarification time from the minimum 8-week *rémuage* to around 10 days. The gyropalette consists of a cage containing a pallet of inverted bottles that undergoes a pre-set series of movement to bring the yeast deposit down onto the bottom of the crown cap. Manual riddling involves the tilting and turning of bottles in a wooden rack (*pupitre*) by a skilled *rémueur* (a skilled practitioner can turn *ca.* 6000 bottles an hour). Bottles are rotated and fractionally tilted until they assume a neck-down, perpendicular position, an operation that may last several months. During rotation, the sediment and gas bubbles exert a 'scouring' effect on the inner surface of the bottle and promote sediment deposition in the bottle neck. Once the sediment is in this position, the wine may theoretically be stored indefinitely, with the finer champagnes being matured for up to 5 years in upside down position (*sur pointes*). During automated *rémuage*, inverted bottles are placed into cages that are mechanically rotated every 8 h, causing sediment to spiral down to the neck. Automation provides considerable savings in time and labour, without apparently having a negative effect on the finished product.[5] In some wineries, riddling agents such as isinglass, bentonite and alginate may be used.

With the sediment now in the bottle neck, the next stage is to remove it by *dégorgement*, a process that requires extreme dexterity and speed, in order to remove the crown cap and immediately disgorge the sediment. To facilitate this, the neck of the bottle is first frozen (calcium chloride bath at *ca.* −24°C), which causes a small ice plug (containing the sediment) to form. The bottle is then returned to an upright position, ensuring that the ice plug is still in the neck and the crown closure is removed causing the sediment-laden plug to be forcefully ejected. Before freezing was introduced, the operator had to adroitly remove the cap just as the bottle was being reverted to being upright before the sediment had a chance to slide back down the bottle – a method known as 'on the fly'. Although *dégorgement* is essential, it can also be disruptive because it briefly exposes the wine to air. The loss of a small volume of wine during *dégorgement* is compensated for by topping up (*remplissage*) with a considerable quantity (*ca.* 600–750 g L^{-1}) of sugar dissolved in base wine (*liqueur d'expédition*). For a *brut* style of champagne, sugar was originally omitted from the base wine, but nowadays a low dosage (10–20 g L^{-1}) is incorporated. A *sec* style will normally contain a little more sugar, some 20–40 g L^{-1}.

Riddling and disgorging yeast are tedious and time-consuming processes, even when carried out automatically, and over the years experiments have been carried out with the aim of using encapsulated yeast for secondary fermentation. In this method, yeast cells are entrapped in a matrix of calcium alginate, and the resultant 'yeast balls' (called *billes*) allow yeast–liquid interactions across the alginate barrier, but do not permit individual cells to escape into the wine. The technique, which eliminates the necessity for riddling, was developed in France by Moët & Chandon, and first expounded upon by Diviès.[6,7] The enclosure of yeast in an 'immobilised' form in membranous beads, and their use in secondary fermentation has been developed over the last couple of decades.[8] Alginate consists of two basic building blocks, D-mannuronic acid and L-guluronic acid, which form a 'double envelope' around the yeast helping it to reduce 'leakage' of cells from the bead.[9] Such technology can be used for bottle fermentations, tank fermentations or even continuous fermentations, as long as the base wine has been stabilised and filtered. Alginate beads are formed by dripping an alginate solution into dissolved calcium chloride, which transforms the drops into gel-like beads. Because the two building blocks can be found in different sequences, alginates can have a variety of properties useful to the food industry.

Some champagnes are naturally sweet at the end of secondary fermentation and require no sugar addition before reaching the consumer. Most, however, are dry and have low pH values, and thus benefit from sweetening. The addition of sugar at this stage is by means of a syrup, and the process is called *dosage*. Most syrups would consist of a mixture of 65% sucrose dissolved in base wine, usually with SO_2 incorporated. In the USA, grape brandy, grape concentrate, citric acid, ascorbic acid, tartaric acid, fumaric acid and malic acid are permitted additives, as is a syrup of pure sucrose in water, as long as it is at least 60° Brix. The purpose of the final sugar addition is to smooth and protect the final wine, and the interaction of the addition and the wine itself is very complex, since sugar can influence the solubility of some aroma compounds. Sweetness level can also affect the drinkers' perception of acidity, bitterness and viscosity. Where permitted, the addition of brandy and/or acid will affect wine aroma chemistry.

After *dosage*, bottles are now ready for corking (*bourbage*), nowadays often with a conglomerate cork with one or more discs of high-quality cork in contact with the wine. The bottle is then wired (*muselet*) to prevent the cork from being blown out, and then shaken to thoroughly mix in the viscous *liqueur d'expédition*. Finally, the bottle is 'dressed' with foil (*habillage*) and labelled. For a thorough account of the methodologies appertaining to *méthode champenoise*, the reader is recommended to the book by Amerine *et al.*[10]

The secondary (carbonation) fermentation and subsequent events, herald a series of complex changes in the chemical composition of the wine, which radically affect aroma, flavour and foaming ability. These reactions, many of which are enzymatic, follow a definite sequence and involve proteins, amino acids, lipids, polysaccharides and other macromolecules. Perhaps the most obvious change relates to ethanol content, which increases by around 1.3–1.4% as a result of bottle fermentation. This level will decrease fractionally after *dosage* as a result of dilution with *liqueur d'expédition*. There is also a change in total acidity, which decreases slightly due to malic acid breakdown by yeast, and by precipitation of potassium hydrogen tartrate. Again, acidity might change (rise) after *dosage* if citric acid has been incorporated into the *liqueur d'expédition*.

Champagne musts and wines are rich in nitrogenous compounds, especially proteins, since the region naturally yields grapes with a high nitrogen content, and Chardonnay and Pinot Noir are nitrogen-rich varieties anyway. The total nitrogen content of champagnes can vary in the range of 150–600 mg L^{-1}. During secondary fermentation, amino acid levels in the wine decrease since they are required for yeast growth. Some acids are assimilated more rapidly than others. When carbon sources become depleted and the yeast becomes stressed, nitrogen compounds are excreted into the wine. Following this, there is a period of 'nitrogen stability' (normally about 4–6 months after secondary fermentation ceases), after which yeast cells start to autolyse as a result of the activity of intracellular proteases and carboxypeptidases.[11] This releases the likes of proteins, peptides and amino acids back into the wine, and there is a marked increase in the amino nitrogen content of the wine. The pH of *méthode champenoise* wines is usually around 2.9–3.2 – levels that are favourable for proteolytic activities – with amino acid levels reaching a peak after about 12 months on the lees, but low levels of proteolysis can continue for years if the wine is held on lees. Depending on ageing time, up to 25% of yeast protein may be degraded, much of the released material being made available for a wide range of reactions involving compounds such as ethanol, malic acid and tartaric acid. As one would expect, amino acids are the most reactive, and are being continually released, and continually participating in new reactions. They act as precursors of aroma and flavour compounds such as higher alcohols, lactones, polyamines and amino acid esters After a period of time, however, amino acid levels decrease, especially those of alanine and arginine. This is due to them becoming de-aminated, and also because of their participation in other chemical reactions. According to Feuillat,[12] these dynamic fluctuations in amino acid content contribute greatly to the final flavour and aroma of a wine. Larger nitrogen-containing compounds, such as peptides, may adversely affect flavour and aroma by binding with normally volatile wine constituents.

4.3.2   The Tank Method

The tank or bulk process represents a relatively inexpensive means of producing large quantities of sparkling wine. It is also known as the closed tank or Charmat method, in honour of Eugène Charmat, who developed the technique in France during the first decade of the 20th century (supposedly, 1907). Over the intervening years, the practicalities have been modified but the method still differs from *méthode champenoise*, inasmuch as secondary fermentation takes place in a vat, not in a bottle. Modern practice now means that the wine is transferred three times each into air-free tanks, which should be capable of accommodating 6–8 atmospheres pressure. The preferred, albeit rather laborious, way of voiding the tanks of air is to fill them with water, and then replace the water with CO_2. Base wine is pumped into the first pressure tank, where sucrose and yeast suspension are added to give final concentrations of 20 g L^{-1} and 2×10^6 viable cells per ml, respectively. This mixture is then fermented to 'dryness' (*ca.* 1.5 g L^{-1} sugar) at 12–15°C, such that the internal pressure reaches *ca.* 500 kPa. The wine is then chilled to −2 to 0°C, thus causing the pressure to fall to *ca.* 400 kPa. The yeast is then allowed to settle over a few days before being further clarified (centrifugation and filtration) and introduced into a second air-free tank (CO_2 at 200 kPa), which will contain any sweetening material, usually consisting of sucrose, together with ascorbic acid and SO_2. If necessary, citric acid may be added at this point. The amount of sucrose added will depend on the wine style, but will normally be in a region of 10–30 g L^{-1}. This concoction is then mixed and micro-filtered (0.45 μm) into a third pressurised tank (CO_2 at 200 kPa) using CO_2 pressure as the driving force, and from this tank the wine is then bottled under counter-pressure. All transfers are carried out at 0 °C or below so as to retain the desired level of dissolved gas. The Charmat method is often referred to as *cuve close*, and many of the French wines traded under the name of 'mousseux' are manufactured in this way.

The fundamentals of the Charmat method have been used to produce sparkling wines on a continuous basis. The first commercial production of sparkling wine using a continuous method was in 1945 (as reported by Charmat in 1969),[13] but this method was modified considerably by Russian oenologists, who have since been credited for formulating continuous methodology. As Ough and Kunkee[14] remarked, certain difficulties were experienced in reproducing some of the earlier Soviet work, and in the 1970s some important modifications were made, including fermentation at low temperature (4–8°C) and the use of a physical support (*e.g.* oak shavings) for yeast growth. A step in the original Soviet process involved heating the deoxygenated wine to 40°C for 24 hours in order to maximise the flavour and aroma of the wine. This is of interest inasmuch as the original process used by Charmat involved subjecting the base wine to a heat treatment. A successful continuous operation has been functioning in Azeitão, Portugal, since the last decades of the 20th century, and the resultant products are very highly regarded. According to Broussilovsky,[15,16] the original Soviet system could produce 12,000 bottles daily. RISP Ltd., a modern Russian producer of sparkling wines (Sovetskoe Champanskoye), has an annual output of 20 million bottles as a consequence of their continuous process.

4.3.3 The Transfer Method

Like the Charmat process, this seems to have been developed during the early years of the 20th century, but this time in Germany. In some wine-producing countries, such as Australia, most bottle-fermented champagne is made by this method. Cold-stabilised base wine is fermented in bottle *à la méthode champenoise* and stored on its lees for a legal minimum of 6 months. After this time, the internal bottle pressure will be around 11 g L^{-1} CO_2 at 20°C (= 525 kPa). Bottles are then removed from their storage racks and placed on a conveyor, which takes them to a transfer machine. The contents of each bottle is then removed by inserting a hollow spear through the crown seal of the bottle, which is then inverted and 200-kPa CO_2 pressure is applied to force the yeast-laden wine out of the bottle *via* a port in the base of the spear. The wine is chilled in-line and passed to a CO_2-filled tank, where further fermentation and lees storage can occur. The cold wine is then clarified before being passed to a second tank filled with CO_2 to a pressure of 200 kPa and held at 0°C to maintain the gas pressure. Expedition liquor is placed in this tank and stirred in, after which the wine is allowed to stand to permit the bulk of the yeast to settle. The wine is then sterile-filtered and bottled in absence of air. Thus, the transference of the wine into a second bottle avoids the necessity for *rémuage* and *dégorgement*, which vastly reduces production times. The transfer method can reduce the time taken from harvest to bottling to periods ranging from 90 days to 1 year (as opposed to 2–5 years for *méthode champenoise*). Despite the capital cost of equipment, other advantages claimed are that all of the wine bottled has a uniform sugar dosage and a uniform bottle pressure. There are also the more obvious advantages of reductions in manual labour and space, but traditionalists maintain that the wine thus produced cannot be compared with that from *méthode champenoise*. In some countries, it is important to indicate exactly how the champagne was made. In the USA, for example, 'fermented in *this* bottle' refers to *méthode champenoise*, while 'fermented in *the* bottle' means that the wine has been made by the transfer process.

4.3.4 Carbonation

This method is used for producing inexpensive sparkling wines, of what may generally be regarded as the 'spumante' type. A stabilised base wine is prepared and filtered as if it were to be bottled. It is then chilled in a tank to $0°C$, under a blanket of CO_2, before being passed through a carbonator (saturator) under CO_2 pressure. The carbonator is usually a long vertical cylinder in which the chilled wine and CO_2 are mixed counter-currently with internal baffles to induce turbulence and promote contact between gas and wine. The desired level of 'fizziness' is achieved by a combination of flow rate through the carbonator, wine temperature and gas pressure. Wine is then either sterile-filtered and bottled *via* a counter-pressure filler, or passed to a pressurised tank where it is stirred for a few days and then filtered and bottled. The latter protocol aids the retention of CO_2 in the wine. Carbonated wines are often regarded as the 'poor relations' of the sparkling wine family, but this is usually due to the quality of the base wine, rather than any inherent fault in the carbonation process. Certainly, carbonation does not produce the same quality of bubble, as other methods do, for inducing a sparkle.

Next to champagne, the 'cava' wines made in the Penedès region of Catalonia, Spain, are the most widely known of the wines made by *méthode champenoise*. The wine was first produced during the 19th century, and was originally called *champaña*, but this was changed in the 1970s to appease the French! The largest houses, Codorníu and Freixenet, produce huge quantities of this wine style, which, with the EC refusing to allow the wording *méthode champenoise* to appear on the label (only on wines from Champagne), has been elevated to what is essentially the Appellation Contrôlée status. There are now set geographical boundaries, prescribed grape varieties and vinification methodology, the latter being almost identical to that for champagne. All this has resulted in the wine style becoming far more popular worldwide. Cava is not as delicate as champagne, although it abounds in flavour and complexity. The major authorised grape varieties are Xarel-lo, Macabeo and Parellada, which impart 'fruity' and 'earthy' aromas to the wine, and these, together with a warm climate, make it difficult to emulate champagne. Recent years have seen increased plantings of Chardonnay and Pinot Noir for making cava.

Other French wine regions, notably Burgundy and Alsace, also make very respectable sparkling wines known as *crémants* to distinguish them from champagnes, while the sparkling wines of the Loire, based mainly on the Chenin Blanc variety, were almost as famous as those from Champagne during the second half of the 19th century, and some of the great houses such as Bollinger and Alfred Gratien have operations there. Blanquette de Limoux has claim to being the first sparkling wine of modern times, its history dating back to 1531, when monks at the Benedictine abbey of Saint-Hilaire, near Limoux in the Languedoc-Roussillon region of southern France, were noted to be making an 'unusual white wine with a natural sparkle'. The wine, which is bottle-fermented, typically originates from the Mauzac grape (at least 80%) grown in a specialised microclimate in the foothills of the Corbières mountains above Carcassone. Other permitted varieties are Clairette, Chenin Blanc and, more latterly, Chardonnay. Originally, this wine was regarded as *pétillant* and produced by the *méthode rurale*. In Italy, the most venerable sparkling product is probably 'prosecco', made in the Veneto province by the Charmat method. Of a slightly more frivolous nature is 'asti spumante', which is also invariably tank-fermented, and often described as 'grapes in a bottle'.

In France, champagnes are aged for a minimum of 1 year and vintage ones for 3 years. *Crémants* have to be aged on yeast for a minimum 9 months, as do many other European sparkling wines including cava. There appear to be no specific ageing requirements for sparkling wines in the USA. The importance of ageing has been stressed by Feuillat and Charpentier,[17] who commented, 'The important sensory characteristics of bottle-fermented and aged sparkling wines are attributable to the autolysate flavour only obtained by lees contact time'.

Wines with bubbles are associated with festivities and celebrations by many people. In addition to the normal criteria of taste and aroma, sparkling wine quality is assessed on the size of the bubbles – the smaller they are, the better is the wine. Significance is also attributed to their longevity (the more persistent, the better) and how well they are integrated into the wine. Wines made by *méthode champenoise* score on all these counts, being smaller, longer lasting and better integrated than bubbles formed by other production methods, which tend to be less 'creamy' and complex. When decent champagne is poured into a glass, the consumer sees the foam before the liquid phase. The fascinating subject of effervescence in sparkling wines has been extensively reviewed by Casey.[18,19]

4.4 COLD MACERATION

This is a pre-fermentation process, whereby crushed grapes are held in a sealable container at low temperature (0–10°C) for anything from 3 to 10 days, with relatively high levels of SO_2. This facilitates slow aqueous extraction of anthocyanins from the skins, and it is claimed that a different profile of phenols is extracted. The must is then warmed and inoculated for fermentation. Thus, this maceration occurs in the absence of ethanol, and it is a useful way of getting more depth of colour out of low-colour intensity grape varieties. Originally it was a French practice, now has found favour in some New World wineries.

4.5 CARBONIC MACERATION

It is a technique used to produce light, fruity red wines, which have a distinctive aromatic presence in addition to their fruit character. They are low in tannins and varietal character, and are designed for early drinking, and indeed lack the structure for long-term ageing. The technique is mainly used on red grapes, but it is also applicable to white ones. Clusters of uncrushed grapes are placed in a fermentation vessel, which is then sealed. The air in the container is then replaced (purged) with CO_2, thus preventing the grapes from coming into prolonged contact with oxygen. The grapes are then allowed to respire and undergo partial fermentation *via* the grapes' own glycolytic enzymes, the normal transformations being carried out for a few days at 35°C. Some winemakers may add a small volume of fermenting juice to the fermenter to provide initial CO_2. A proportion of the grapes at the bottom of the vessel will get crushed from the weight of the grapes above and release juice that will ferment, but most fermentation occurs while the juice is still inside the intact grape. During the process, the skin cell walls become permeable, allowing pigments and other extractable materials to leak out. After 8–10 days of berry fermentation, the enzymes lose their activity and the grape clusters are pressed, whence the berries liberate their coloured, partially fermented juice (usually *ca.* 1.5% ABV). This juice, minus the skins, is then inoculated with yeast and completely fermented at *ca.* 15–20°C. For most wines made in this fashion, the period between picking the grapes and bottling the wine is rather truncated.

Apart from the conversion of sugar to ethanol, the main chemical reactions undertaken during carbonic maceration are the degradation of almost half of the malic acid present and the formation of the amide amino acids and succinic, fumaric and shikimic acids. Liberation of amino acids serves to increase the nutritional status of the must, and this is said to stimulate malo-lactic fermentation in wines produced by carbonic maceration. Wines resulting from this procedure have a unique aroma, and the enhanced production of four volatiles – benzaldehyde, ethyl salicylate, vinylbenzene and ethyl-9-decenoate – has been shown to be attributable to carbonic maceration, the formation of the first three being a result of their involvement in the shikimic acid pathway.[20]

The development of carbonic maceration has an interesting history, which emanates from research carried out by a French team in the 1930s. They were attempting to overcome the problem of grape deterioration during transport between vineyard and winery. As part of their work, whole bunches of grapes were stored under CO_2 at around 0°C. After 2 months under such conditions, one batch was found to be alcoholic, gassy and generally unfit for sale, but the flavour was found to be quite palatable. As a desperate measure, the grapes were crushed and vinified rather than being discarded, and the resulting product was heralded as 'pleasantly unusual'. The first research into what actually went on during storage under CO_2 was carried out in 1936 by Prof. Michel Flanzy in southern France, and much of the developmental work was carried out by his son, Claude, and co-workers at the Station de

Technologie des Produits Végétaux, Institut National de la Recherche Agronomique (INRA), Montfavet, near Avignon, together with collaborators from Beaujolais, Burgundy and other areas who formed Group de Travail Macération Carbonique. The success of their work can be gauged by the increasing acceptance of the technique throughout the winemaking world. As part of their research, the Avignon group investigated the possibility of using SO_2 as well as CO_2, but came out in favour of the latter gas as being the best way to achieve anaerobiosis.

The work carried out by Flanzy's group was, in fact, the follow-up of an observation made around 60 years earlier by Pasteur, who in 1872 observed that storage of whole grapes under anaerobic conditions enabled them to retain more of their flavour than when they had been stored in air. In his inimitable way, Pasteur predicted that anaerobic storage of grapes and their subsequent vinification may well result in wines with special properties perhaps with commercial possibilities, but he did not investigate any further.

According to Rankine,[2] who notes that carbonic maceration is widespread in Australia, the process can be divided into four phases:

(i) Storage of whole grape bunches in a closed vessel with CO_2. Grapes should not be de-stemmed, and it is usual practice to use grapes with little varietal character such as Gamay and Grenache. The gas can be added extraneously or can result from spontaneously fermenting juice, which is expressed from grapes at the base of the vessel. It is essential to keep oxygen levels to a minimum ($<0.5\%$ by volume) to avoid growth of aerobic spoilage organisms such as *Acetobacter*.

(ii) Maceration phase, in which berries undergo intracellular anaerobic metabolism. This is not the same as alcoholic fermentation of grape juice by yeast, and involves some complex biochemistry. When freshly harvested, grapes are 'alive' and consume their own, 'internal' supply of oxygen within a day or so. This enzymatic process within the grape is called 'intracellular fermentation' or 'anaerobic metabolism', and continues until its demise. Death of the berry is a result of either the accumulation of ethanol or the rupture of skins, and signifies the end of the aroma-generating phase of carbonic maceration. Depending on temperature, it takes

place at least 5 days after the berry has depleted its endogenous oxygen supply. At 35°C, the normal maceration
temperature, death grape will occur in *ca.* 8 days, whereas
at 15°C it may take 2 weeks. During anaerobic metabolism,
the grape initially absorbs CO_2 from its environment, but this
is followed by release of the gas when malic acid is metabolised. Again, at 35°C, around half of the malic acid is
consumed with less at lower temperatures. Tartaric acid
undergoes little, or no, modification at this stage. Characteristically, about 1–2% ethanol by volume is liberated
within the berry, and the other major products are glycerol
(*ca.* 2 g L^{-1}), methanol (*ca.* 50 mg L^{-1}) and acetaldehyde
(*ca.* 20 mg L^{-1}).

(iii) Pressing whole bunches of grapes without crushing. The
juice from pressed grapes is of high quality and is a highly
amenable environment for fermenting yeasts. Rapid fermentation ensues, and a cooling system is necessary at this stage
in order to prevent overheating, which is undesirable in
fragrant wines of this genre.

(iv) Alcoholic fermentation. Pressed juice is seeded with yeast
and fermentation proceeds to dryness. Malo-lactic fermentation occurs either during alcoholic fermentation or slightly
afterwards.

Red wines produced by carbonic maceration have the following
characters: (1) less colour and tannin than those made by conventional vinification; (2) a readily discernible character imparted not
only by anaerobic, intracellular fermentation in the berry, but also
by alcoholic fermentation, and only minimal alcoholic extraction
of grape constituents; (3) a fruity aroma reminiscent of 'cherries'
and 'strawberries' can often be detected. Wineries designed solely
for carrying out carbonic maceration vinifications are more expensive and less adaptable that the traditional winery.

By substituting an atmosphere of nitrogen for CO_2, sugar conversion can be prolonged, and some wines are fermented in this
way until the ethanol level reaches *ca.* 7%, whence the grapes are
crushed, and the must is then subjected to a subsequent 'normal'
fermentation. This strongly suggests that the loss of endogenous
grape enzyme activity is brought about by CO_2. Perhaps the most
renowned wine made by this method is Beaujolais nouveau, where

the period from grape to bottle can be less than 6 weeks. Produced in the Beaujolais (AOC) region of France, 'nouveau' is the most popular *vin de primeur*, a wine in which the grapes are harvested in autumn and the bottles sold before Christmas. Such a timescale would not be possible with a 'normal' fermentation and, of course, has great financial benefits for the winemaker and vintner. By convention, the new 'nouveau' is officially released on the third Thursday in November (often associated with much celebration). The product is very variable and should be consumed within a few months of release. It is reckoned that almost 5,00,000 hL of 'nouveau' is produced annually, around one-third of the output in the region. An informative review of carbonic maceration has been written by Flanzy *et al.*[21]

4.6 THERMOVINIFICATION

This is a pre-fermentation technique applicable to the bulk production of certain red wines in which a higher level of some phenolic and volatile compounds is required. Thermovinification produces soft 'grapey' wines suitable for early consumption, and is also used in the manufacture of purple grape juice. The process is also known as 'hot pressing', and it consists of subjecting de-stemmed grapes to temperatures of 60–87°C, thus rupturing cells and facilitating maximum extraction of coloured compounds, such as anthocyanins, from the skins. The highly coloured juice that results is then fermented in the absence of skins and seeds, as if it were a white wine. The holding time of the heated grape tissue varies from 2 min at 87°C to several hours at lower temperatures. The latter regime can only be used for sound grapes since certain enzymes emanating from diseased grapes (*e.g.* laccase from grapes infected with *Botrytis*) can withstand temperatures of around 60°C. After the chosen heat treatment, the resultant pulp is then pressed and allowed to cool to around 25°C, whence yeast is added and fermentation ensues.

Fischer *et al.*[22] investigated the impact of this fermentation technology on the phenolic and volatile composition of some German red wines made from Pinot Noir, Dornfelder and Portugieser grapes. They found that thermal inactivation of lipoxygenase enzyme systems reduced the level of C_6-alcohols and their esters, but that overall thermovinification yielded wines with high ester content, thus rendering them more 'fruity'. Certain ester levels were markedly increased, the most notable being 3-methyl-1-butyl (isoamyl) acetate and hexyl acetate.

4.7 CHAPTALISATION

This is the addition of sugar (sucrose) to must or wine, and is an important process in colder winemaking areas where grapes often contain insufficient endogenous sugar. It is a common practice in France, but less so in the New World, being prohibited in Australia[†] (except for addition to sparkling wines prior to secondary fermentation) but permitted in New Zealand. The name comes from Jean Antoine Chaptal, Napoleon's minister of agriculture, who formally sanctioned the process in 1801. Chaptalisation is carried out with the sole purpose of improving wine quality, and in most cases it does just that, being considered almost essential for the making of Burgundy regardless of the level of sugar naturally present in grapes. Many Burgundian winemakers consider that it gives their products a required 'mouthfeel' and tend to make small serial sugar additions at the end of fermentation (considered by some to be illegal), which encourages the formation of glycerol. This undoubtedly adds a 'roundness' and 'fatness' to Burgundian wines. In the manufacture of white wines, cane sugar is preferred to beet sugar, but for red wines the source of sugar is not so important. For white wines sugar should be added to the expressed juice, whereas in reds it is a must addition. Either way, it is essential that the sugar be dissolved in water before being added. In practice, in order to raise the level of alcohol by 1%, it is necessary to add sugar at the rate of 17 g L^{-1} to white wines and 19 g L^{-1} to red wines, the difference aimed at compensating for the greater loss of ethanol during red winemaking (*i.e.* through warmer fermentation and pumping-over operations).

In Europe, the EC has discerned five zones and has issued regionally based regulations governing chaptalisation. These prescribe the minimum alcohol potential of the wine before any sugar addition, the maximum sugar addition and the maximum alcohol content. In the coldest, most northerly regions, chaptalisation is

---

[†] But the addition of concentrated grape juice is permitted.

permitted every year; in the central, more moderate zones, it is authorised in poor years, whereas in the warmer, southern zones, it is not permitted at all. In no case may sugar and acid be added to the same wine. Chaptalisation is often referred to as 'sucrage' in France, 'zuckerung' in Germany and 'zuccheraggio' in Italy.

German wines designated as 'Qualitätswein mit Prädikat' (QmP) may not be chaptalised but can have their alcoholic potential increased by *süssreserve*, a technique by which sweet, unfermented juice or must is added to a wine after fermentation. This slightly contentious practice is permissible as long as the addition is made with grapes of the same type and grade as the wine itself.

4.8 USE OF COMMERCIAL ENZYMES IN WINEMAKING

The most widely used enzymes available for commercial use, oenologically, are pectinases, hemicellulases, glycosidases and glucanases, the singularly most important being pectinases, which occur naturally in all fruit, including grapes, and are partly responsible for the ripening process. Grape pectinases are, however, inactive under the pH and SO_2 conditions prevalent during winemaking, and so in modern, automated wineries, it is sometimes necessary to 'help nature' by calling for biochemical assistance and extraneously adding lytic enzymes in order to expedite juice extraction and to make it a much more thorough process. Fungal pectinases, for example, are resistant to the conditions encountered in winemaking, and with certain provisos it is permissible to use commercial preparations from these microbes. In the European Union, the methods used to govern enzyme production for use in winemaking are governed by the OIV, who have decreed that only *Aspergillus niger* and certain *Trichoderma* species may be used as source organisms (*i.e.* have GRAS, 'generally regarded as safe', status). Of the four enzyme groups mentioned above, the first three are produced from *A. niger*, while glucanases emanate from *T. harzianium*. Pectolytic enzymes have been used in the processing of fruit juices since the 1950s in order to improve juice yield and to aid clarification, but pectinase preparations have only commonly been used in the wine industry since the 1970s. Such enzymes are most commonly used to break down the pectin in white grapes, but are occasionally used with red wine grapes in order to enhance the extraction of colours and tannins from the berry. Results with red grapes have been variable, with some studies showing improved colour extraction, some showing no difference and some showing a decrease in colour. Such results have been attributed to impurities (such as phenol esterases) in the enzyme preparations, which can degrade anthocyanins and phenols rather than aiding their extraction. Modern commercial macerating 'pectinase' preparations are rarely homogeneous and often contain traces of other enzymes, such as cellulases and hemicellulases, in order to expedite tissue breakdown.

The substance of the grape berry is a cellular (compartmentalised) structure, and the juice, which is of prime interest to the winemaker, actually originates in an intracellular position, in the vacuole. Before juice is available for vinification, it must be expressed from the fruit in some way, and this has traditionally been effected by crushing and pressing, which serve to disrupt grape tissue and permit the activity of an array of natural lytic enzymes that will rupture cell walls and allow the escape of juice. The grape cell wall consists of a pectin foundation, the middle lamella, which is impregnated with celluloses and hemicelluloses, the whole structure being capable of conferring shape and rigidity. The middle lamella is not actually a part of the cell wall because it is technically intercellular in nature and forms a base upon which cell walls are formed. It is composed of 'pectic substances', which are derivatives of polygalacturonic acid. The primary cell wall is deposited directly onto the middle lamella, and is a relatively elastic layer comprising a framework of dispersed linear cellulose microfibrils with a low degree of crystallinity (paracrystalline) in a matrix of pectin, hemicellulose and a small amount of protein. Grape cell walls have around 30% pectin content. The secondary cell wall, where it is formed, is produced after any necessary cell extension has been completed since it is a much more rigid structure containing cellulose in a higher state of crystallinity, and usually containing lignin (*i.e.* becomes 'woody'). Grape seeds and stems would invariably contain secondary cell wall material, whereas the grape flesh cells show very little secondary thickening. To obtain juice from the berry, middle lamellae must first be broken down and the adjacent cells released, this being the prime object of maceration. Cell walls then have to be degraded in order that the vacuolar sap can be expressed. This can be achieved through mechanical maceration, heat, chemicals or enzymes. One or more of these agencies can be used, or it may be sufficient to rely on the endogenous grape enzymes released upon crushing, along with the heat, ethanol and enzymes produced during fermentation. It should be stressed that the macromolecules that comprise the middle lamella, and primary and secondary cell walls, are quite diverse, and a number of enzymes are required for their complete breakdown.

Pectic substances are essentially based upon linear, $\alpha, 1 \rightarrow 4$ linked polymers of pyranosyl D-galacturonic acid molecules, often referred to as the polygalacturonan or galacturonan backbone. The polygalacturonic acid itself is known as pectic acid, and the carboxyl groups of the uronic acid moiety in the galacturonic acid residues may become partially methylated to form pectinic acid or pectin. Pectin, whose molecular weight varies between 30 and 40 kDa, will normally have between 50 and 75% of its carboxyl groups esterified with methanol, while pectinic acid represents a situation intermediate between pectin and pectic acid. If the degree of methylation is greater than 50%, the material is often referred to as 'high-methoxy' pectin. Grape pectins are normally 44–65% esterified. In addition to methylation, the carboxyl groups of pectic acid may form salts with Ca^{2+}, Mg^{2+} or Fe^{2+}, and these divalent cations then allow bonds to be formed between adjacent molecules of uronic acid. The pectin backbone usually contains some α-1,2 rhamnopyranosyl molecules, and short side chains of up to three sugar residues can occur. D-galactose and L-arabinose are also present in more complex chains. Luckily for the winemaker, the middle lamella pectin is quite soluble and easily degraded enzymatically, but once in solution and left undegraded, it can lead to viscosity problems. Pectin located in the primary and secondary cell walls is not nearly as soluble. The interconversions of pectic materials in the cell wall are of particular importance in the process of fruit maturation. Little soluble pectin is present in immature fruit, and the process of maturation by which the fruit softens and attains 'ripeness' is accompanied by the conversion of 'protopectin'[‡] to soluble pectin, and in particular the loss of pectic acid (pectate) from the middle lamella. The latter is, in large measure, responsible for fruit softening.

[‡] A general term for all insoluble pectic substances, including cross-linked chains of pectic acid as well as their salts.

Pectolytic enzymes can be placed into four categories according to how they act on their target. Pectin methyl esterase (PME) removes methyl groups from high-methoxy pectin, thus yielding methanol and 'low-methoxy' pectin. The origin of this enzyme determines exactly how it goes about its business; higher plant-derived (endogenous) material works systematically along the pectin chain, while that from fungal sources acts in a random fashion. Technically, when methyl units have been cleaved from the polygalacturonic backbone chain, pectin becomes pectic acid (pectate). Polygalacturonase (PG) is a depolymerase and acts by breaking the bonds between non-methylated galacturonic acid units in the chain. Endo-PG acts randomly on the chain, while exo-PG only breaks bonds at the non-reducing ends of chains. PME and endo-PG are often found together in pectinase preparations used for grape processing. Pectin lyase (PL) acts randomly on the chain to depolymerise polygalacturonans, in a way similar to that for endo-PG, but differs in that it cleaves bonds between methylated molecules. Thus, PG recognises galacturonic acid units without methyl groups, whereas PL does not. Pectate lyase (PAL) acts randomly to break glycosidic linkages between non-methylated galacturonic acid molecules in low-methoxy pectin, thus again making it rather similar to endo-PG. To the winemaker, PAL is the least useful of the pectinases because it has an optimum pH of 8–9 and has an absolute requirement for Ca^{2+}. The activity of PME is inhibited, somewhat, by ethanol and polyphenols. It is interesting to note that minute amounts (ppm) of methanol are released into wine as a result of the pectolysis of high-methoxy pectin.

Cellulose is abundant in primary and secondary cell walls, and comprises some 20–30% of the dry weight of most primary cell walls. It is the most abundant natural product on the planet and takes the form of linear chains of $\beta,1 \rightarrow 4$ linked D-glucose molecules. These unbranched chains can form hydrogen bonds with each other along their lengths to give aggregates called microfibrils, which have varying degrees of crystallinity. Microfibrils can also become attached to hemicelluloses *via* hydroxyl groups. The properties of cellulose vary according to the degree of polymerisation and the degree of crystallinity, the less crystalline material being more susceptible to enzyme attack. Cellulose exists in such a highly organised structure that no single enzyme is capable of degrading native cellulose into its constituent glucose or cellobiose units.

'Cellulase' is thus a complex of enzymes comprising endo-glucanase, exo-glucanase and β-glucosidase (cellobiase). Endo-glucanase randomly hydrolyses the $\beta,1 \rightarrow 4$ glycosidic bonds in the chain, while exo-glucanase only breaks the bonds at the non-reducing end, to give either glucose or its dimer, cellobiose. Cellobiase splits cellobiose into two glucose units. Cellobiase strongly inhibits the activity of both endo- and exo-glucanase, and so it is an essential component of the cellulase complex.

Hemicelluloses are made up of four major polysaccharides: arabinans, galactans, xylans and xyloglucans, the first two being the most significant. Arabinans have a backbone of $\alpha,1 \rightarrow 5$ linked arabinofuranosyl residues with branches of further arabinofuranosyl molecules at about every third unit along the chain. Galactans have a backbone of D-galactopyranosyl residues that are either $\beta,1 \rightarrow 4$ or $\beta,1 \rightarrow 3$ linked. There are three forms of arabinase enzyme: arabinosidase A degrades arabinan oligomers to monomers; arabinosidase B degrades branched arabinans to give a linear chain by removing terminal $\alpha,1 \rightarrow 3$ linked arabinofuranosyl side chains. At the same time, it sequentially splits off the terminal arabinofuranosyl group from the non-reducing end of the linear chain. The endo-arabinase enzyme randomly hydrolyses the linear chain, giving rise to oligomeric units that arabinosidase A can act on. The two galactanases are endo-galactanase and galactanase. The former acts randomly on the galactan chain, breaking $\beta,1 \rightarrow 4$ linkages, while galactanase works at random to cleave $\beta,1 \rightarrow 3$ and $\beta,1 \rightarrow 6$ linkages.

Grape juice can be obtained by crushing (mechanical or otherwise) and pressing of grape pulp, which yields a turbid liquid containing cell contents and cell wall fragments. Expressing maximum juice from such a pulp can be difficult if even moderate amounts of pectin are still present. If soluble pectin is present in juice, viscosity problems are likely to ensue downstream, while if pectin persists in the pulp itself, it will be bound to celluloses and hemicelluloses causing liquid retention and a resulting gelatinous mass from which it is difficult to remove the liquid phase. To improve juice yield (*i.e.* 'pressability'), viscosity and clarification, PME and PG – with or without PL – are typically added to degrade pectins and gelled particles.

After crushing, negatively charged pectin molecules form a 'protective' layer around grape solid particles, which have a positive charge. This keeps the solid particles in suspension. When present in sufficient amount, pectinases can fragment this pectin coat, thus exposing some of the grape solid particles underneath (*i.e.* this exposes a positive charge). These positive charges then bind to intact pectin-covered grape solid particles, thus forming larger structures that will then settle out. Settling enzymes, which are for use with grape pulp, are probably the most basic commercial products to be used in winemaking in terms of their composition and mode of action. They work mainly on the soluble pectins in the pulp and comprise a mixture of PL, PME and PG. If very ripe grapes have been used in processing, then it is necessary to incorporate higher levels of PG into the enzyme preparation in order to prevent settling problems.

Grape skins contain much more insoluble pectin (more side chains), and so a 'skin contact' enzyme preparation will be more concentrated, and besides the basic 'settling' components will contain 'side chain activity'. Skin contact enzymes are used in white grape processing to extract not only more juice, but also more flavour. Many grape flavour compounds, such as pentanones, norisoprenoids and terpenols, are concentrated in the skin and will be released into must after enzyme action. Nitrogen compounds are also released, and this is advantageous except when heat-labile proteins are liberated. If skins are exposed to enzyme action for too long, over-maceration can occur, and this results in a settling problem whereby there is clear juice at the top of the tank, compact lees at the bottom of the tank and a metre or so of 'fluff' above the lees. This fluff consists of very fine fragments of grape skin that remain in suspension. Skin contact enzymes for use with red grape varieties must extract and protect compounds responsible for colouring, as well as liberating flavours and more juice. For red grapes, higher enzyme concentrations have to be attained than are necessary for white grapes, and this is partly due to the fact that the tannins present bind to enzymes and thereby partially inactivate them. It is also partly due to more extended period of skin contact needed in the extraction of red wine grapes. In pectinase preparations intended for extended skin contact with red grapes, it is critically important that there should be no 'anthocyanase' activity. Grape anthocyanins, which are more concentrated in the skin, are stabilised by forming covalent linkages with a molecule of glucose (glycosidic form; anthocyanin proper). They become unstable and colourless when these linkages are broken and the aglycone form (anthocyanidin) is liberated. Anthocyanases, which are essentially glycosidases, have the capacity to remove the glucose units and cause colour stability problems. The use of enzymes for skin maceration in red winemaking has shown that they actually increase colour stability, especially when used with the more 'difficult' varieties, such as Pinot Noir, which naturally have a low phenol and anthocyanin content. This increase in stability is attributed to the enzymic extraction of tannins, which have a protective effect on the red pigments. As would be expected, anthocyanase activity does not generally present problems in enzymes intended for white wine production.

As we have just seen, glycosidase impurities in pectinases can be deleterious to the colouration of red wines, but the presence of these sugar-removing enzymes is not always undesirable. Glycosidases can be of commercial use for grape varieties that contain flavour groups attached to sugar residues, such as Muscat, Gewürztraminer and Riesling. Such molecules are flavour/aroma precursors in essence, while they are non-volatile in this bound form. When the sugar moieties are removed, the flavour becomes volatile and can contribute to the aroma. Examples of these flavour groups are monoterpenes and C_{13}-norisoprenoid derivatives. In *V. vinifera* grapes, monoterpene precursors are normally di-glycosidic, which means that the hydrocarbon is bound to glucose and one other carbohydrate residue, such as arabinose, rhamnose or apiose. Grapes contain endogenous glycosidases (such as glucosidase, rhamnosidase and apiosidase), which are capable of slowly releasing the volatile flavour components from their non-aromatic precursors, but under winemaking conditions, they are not very efficient because they work most effectively around pH 5. Fungal glycosidases, on the other hand, are fully operative at wine pH. β-Glucosidase on its own is ineffective in releasing the aromatic flavour compounds from di-glycosidic precursors because sugar removal is sequential, and the 'other sugar' must be removed first before glucose can be removed. In addition, fungal β-glucosidases are repressed by glucose, which means that, in practice, preparations should be added to a finished wine or to a wine with less than 50 g L^{-1} residual sugar. Normally, monoterpenes are quite stable in wine and are hydrolysed over a period of time, but if exogenous enzymes are added, much of the flavour emanating from them will be released all at once, which is not necessarily what the winemaker wants, since it gives rise to an unbalanced flavour profile. To counteract this, it is usual to add enzyme to only a part of a blend so that the other fractions can slowly supply their volatiles to the final product.

Commercial preparations of glucanases for wine production were developed in the 1980s, with the industrial enzyme coming from *T. harzianium*. As we have seen from the biology of *Saccharomyces*, β-glucans are major component of the cell walls of many fungi, and, accordingly, are secreted by *B. cinerea* into infected grapes, whence they survive in their juice causing clarification and filtration problems. These processing problems are a consequence of the high molecular mass of the polysaccharide secreted by the fungus, which has been identified as a β-(1,3–1,6) glucane. The polysaccharide is impossible to remove conventionally (filtration, centrifugation, *etc.*) and is unaffected by endogenous enzymes. It can be removed by *Trichoderma* β-glucanase. A more recent, alternative use for these enzymes is to enhance yeast autolysis, which can save some of the time for which wine is left on the lees. Other enzymes occasionally used in winemaking include proteases and ureases, the latter breaking down the urea formed by yeast and preventing it from being converted into ethyl carbamate.

The pulp of a number of grape varieties is rich in pectic compounds, and their incomplete hydrolysis by endogenous enzymes may cause problems during processing. After pressing, grape juice is turbid and has a viscosity due to pectins, the content of which is dependent on grape variety, degree of ripeness and the methods of harvesting and pressing. The viscosity interferes with clarification processes, and often extends the time necessary for filtration. During the production of certain white wines, juice viscosity can be rapidly reduced and its stability enhanced by enzyme addition. During the manufacture of red wines, extraction of phenolics usually occurs during the pulping of the must during alcoholic fermentation. The efficiency of this extraction is dependent on how they were pressed, *etc.* Improved anthocyanin extraction can be effected *via* enzyme addition to the grapes at the onset of pulping. The enhanced colour remains stable during maturation phases, and it has been shown that enzyme treatments have a positive effect on organoleptic parameters, such as 'fruitiness'.

The potential problems likely to be encountered by using 'impure' enzymes can be illustrated by referring to the flavour and aroma problems that have been experienced by users of certain pectinase preparations, which have been shown to contain traces of cinnamyl esterase (CE). This enzyme catalyses the hydrolysis of the ester linkage in the tartrate esters of hydroxycinnamic acids (hydrocinnamoyl tartaric derivatives) in must liberating, for example, ferulic, caffeic and *p*-coumaric acids. During fermentation, these hydroxycinnamic acids can then be subjected to cinnamyl decarboxylase produced by certain yeast strains (referred to as Pof+, phenyl off-flavour positive), with the result that volatile vinyl-phenols (most notably vinyl-4-phenol) are produced. These compounds have a rather pleasant (*e.g.* 'clove-like') aroma at low concentration, but when more concentrated cause off-flavours described as 'phenolic' or 'medicinal'. The effect of vinyl-phenols in red wines is ameliorated by the fact that they react covalently with tannins to produce compounds that are non-volatile. The activity of CE is inhibited by ethanol, and so hydroxycinnamic acids are only formed during the pre-fermentation phase and in the early stages of fermentation.

REFERENCES

1. M. Krstic, G. Moulds, W. Panagiotopoulos and S. West, *Growing Quality Grapes to Winery Specifications*, Winetitles, Adelaide, 2003.
2. B.C. Rankine, *Making Good Wine*, Macmillan, Sydney, 2004.
3. P. Howe, in *Sparkling wines*, 2nd edn, A.G.H. Lea and J.R. Piggott (eds), *Fermented Beverage Production*, Kluwer Academic, New York, 2003.
4. J. Halliday and H. Johnson, *The Art and Science of Wine*, Mitchell Beazley, London, 1992.
5. G. Hardy, *Revue des Oenologues*, 1993, **69**, 23.
6. C. Diviès, *Bull. O.I.V.*, 1981, **54**, 843.
7. C. Diviès, On the utilisation of entrapped microorganisms in the industry of fermented beverages, in C. Cantarelli and G. Lanzarini (eds), *Biotechnology Applications in Beverage Production*, Elsevier, London, 1989, pp. 153–165.

8. K. Yokotsuka, M. Yajima and T. Matsudo, *Am. J. Enol. Vit.*, 1997, **48**, 471.
9. A. Lallement, *Revue des Oenologues*, No. 58, 1991, 29.
10. M.A. Amerine, H.W. Berg, R.E. Kunkee, C.S. Ough, V.L. Singleton and A.D. Webb, *The Technology of Wine Making*, 4th edn, AVI, Westport, CT, 1980.
11. M. Sato, Y. Suzuki, K. Hanamure, I. Katoh, Y. Yagi and K. Otsuka, *Am. J. Enol. Vit.*, 1997, **48**, 1.
12. M. Feuillat, *Vigneron Champenoise*, 1981, **102**, 340.
13. P. Charmat, *Wines Vines*, 1969, **50**, 40.
14. C.S. Ough and R.E. Kunkee, *Appl. Microbiol.*, 1966, **14**, 643.
15. S.A. Broussilovsky, *Bull O.I.V.*, 1963, **36**, 209.
16. S.A. Broussilovsky, *Bull O.I.V.*, 1963, **36**, 319.
17. M. Feuillat and C. Charpentier, *Am. J. Enol. Vit.*, 1982, **33**, 6.
18. J.A. Casey, *Aust. Grapegrower Winemaker*, July 1988, **295**, 19.
19. J.A. Casey, *Aust. Grapegrower Winemaker*, June 1995, **378a**, 37.
20. V. Ducruet, *Lebensm. Wiss. Technol.*, 1984, **17**, 217.
21. C. Flanzy, M. Flanzy and P. Benard, *La Vinification par Macération Carbonique*, INRA, Paris, 1987.
22. U. Fischer, M. Strasser and K. Gutzler, *Int'l. J. Food Sci. Technol.*, 2000, **35**, 81.

CHAPTER 5

Lactic Acid Bacteria and Malo-Lactic Fermentation

5.1 LACTIC ACID BACTERIA

The very name encompasses a very diverse group of prokaryotes, which have sufficient characters in common for some generalisations to be made, and occur by design, or by chance, in many drinks and foodstuffs. It is impossible to do justice to these bacteria in such a concise allotment of space, and this section will primarily attempt to document some of the aspects of their metabolism likely to be of interest to the winemaker. A general description of them would be that they are Gram positive, non-sporing, non-respiring cocci or rods, which produce lactic acid as the major endproduct during the fermentation of carbohydrates. Lactic acid bacteria (LAB) are more than merely lactic acid producers, and lactic acid production is a reflection of an underlying mode of metabolism that is far more complex and adaptive than was first imaginable. They have become highly adapted to their particular natural environments, which are often rich in nutrients and energy sources, and have consequently evolved intricate nutritional requirements. Consequently, growth of LAB in 'normal', chemically defined laboratory media, is, at best extremely slow, a sure indication that they have adapted to environments that are nutritionally rich. Metabolically, they have dispensed with their biosynthetic capabilities, and are devoid, for example, of the Krebs cycle, and a typical terminal electron transport system, a reflection of their environment, which is often low, or completely lacking in oxygen.

203

It should be noted, however, that the genetic material for, now defunct, biosyntheses is still present in some strains (*e.g. Lactobacillus plantarum*). In order to adapt to nutrient-rich niches, LAB have evolved very effective transport systems, which enable them to take up the required solutes. In particular, they are often highly saccharolytic, the genome of *L. plantarum* for example, containing 25 complete sugar phosphotransferase systems (PTS).

The overriding feature of LAB metabolism is that of highly efficient carbohydrate fermentation coupled to substrate-level phosphorylation. The ATP so generated is subsequently used for biosynthetic purposes. As a group, LAB have an enormous capacity to degrade a variety of carbohydrates, and related compounds, with the predominant end products being lactic acid ($>50\%$ of sugar carbon is generally converted into the acid), and it is manifest that these bacteria can adapt to a wide range of environmental conditions, and modify their metabolism accordingly. Nutrient-rich environments are, of course, ideal for supporting the growth of a plethora of microbes, and so LAB have therefore developed strategies to enable them to compete successfully with them. One such strategy is to produce acid, and a concomitant tolerance to the acid environment. Many LAB are also highly salt tolerant. Some of the more unusual features of LAB metabolism are undoubtedly important to them in their natural habitats. Their metabolic versatility may be gauged by the fact that in the genome of *L. plantarum* strain WCFS1 (the second LAB genome to be mapped[*]), there is a large region of the chromosome containing genes encoding several nutrient utilisation systems and extracellular functions. Kleerebezum *et al.*[1] have called this the 'lifestyle adaptation region'.

In terms of the food industry, they can metabolise many constituents in their particular environment, sometimes by unusual means, thereby modifying flavours, often by introducing unique ones. In foodstuffs that contain large quantities of organic acids, their metabolism by LAB can be important in reducing acidity, and this is particularly the case in winemaking where they can affect the final product in a favourable (or, in a few cases, unfavourable) way. The major metabolic products of LAB are lactic acid, CO_2 and ethanol, while minor ones are acetic acid, diacetyl and acetaldehyde.

---

[*]The first LAB genome to be mapped was from *Lactococcus lactis* (Bolotin *et al.*[2]).

The term 'lactic acid bacteria' was first coined at the end of the 19th century to describe 'milk-souring organisms', one of their early names being 'Milchsäurebazillus', and it is claimed that the first pure culture of a bacterium was obtained by Joseph Lister in 1873, and recorded as '*Bacterium lactis*' (probably *Lactococcus lactis*). The first serious attempt to classify these bacteria is to be found in Orla-Jensen's classic monograph,[3] which, although revised substantially, still forms the basis for their taxonomy in the 21st century. Orla-Jensen used the following characteristics as a basis for his classification: Gram stain (all positive); morphology (cocci, rods or tetrads); mode of glucose fermentation (homo-, or hetero-fermentative); growth at specified 'cardinal' temperatures (*e.g.* 10 and 45°C), and range of sugar utilisation. As a result of using these criteria, the view emerged that the LAB comprised four genera, *Lactobacillus*, *Leuconostoc*, *Pediococcus* and *Streptococcus*. During the last 20 years, with the advent of modern taxonomic methods, many new genera have been described, most of them comprising strains previously included in one of the four above (*e.g.* *Oenococcus*). Orla-Jensen regarded the LAB as a 'great natural group' and believed that they were phylogenetically related and separate from other groups of bacteria. At that time, only phenotypic characters could be evaluated, whereas nowadays we have the means to evaluate cell macromolecules, particularly nucleic acids, which are believed to be far more accurate in defining relationships and phylogenetic positions. Close relationships at species and subspecies level can, and have, been determined using DNA–DNA homology studies, and this method is still used for defining what constitutes a 'species' in the prokaryotic world. For determining phylogenetic positions of species and genera, ribosomal RNA (16S rRNA) gene sequencing is more suitable, since the sequence contains both well-conserved and less-conserved regions. Using polymerase chain reaction (PCR) technology, it is now a relatively easy task to determine the sequence of rRNA from bacteria, and the technique is now considered an important tool in the classification of LAB. Chemotaxonomic markers, such as cell fatty acid composition, soluble protein patterns and cell wall constituents, are also useful in classification. With taxonomic revisions, the LAB, in their broadest sense, would now encompass some 20 genera, the principal of which are the following: *Aerococcus*, *Carnobacterium*, *Enterococcus*, *Lactobacillus*, *Lactococcus*,

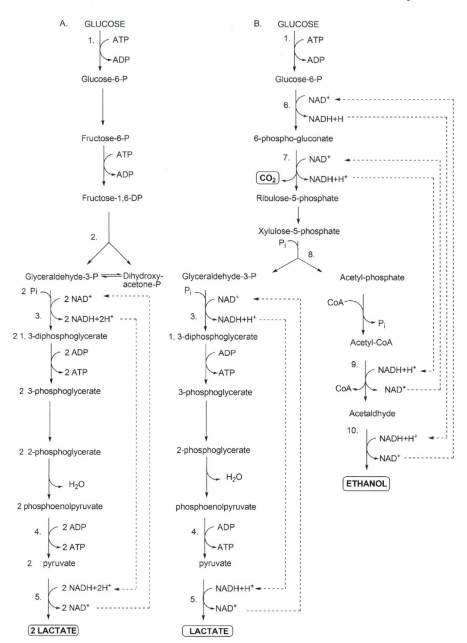

Leuconostoc, Oenococcus, Pediococcus, Streptococcus, Tetrageno-coccus, Vagococcus and *Weissella*. Because of their variety and versatility, some cynics might say that is only their Gram-positive character that cannot be challenged, and that they do not constitute a 'natural' group of bacteria, *sensu* Orla-Jensen. For a review of LAB taxonomy prior to the advent of modern taxonomic methods, the reader is directed towards Ingram,[4] while later classifications are summarised in Wood and Holzapfel,[5] and Axelsson.[6] The review by Stiles and Holzapfel[7] is also very useful, and, at the time of writing, the online publication '*The Prokaryotes*' contains useful chapters on these important industrial bacteria.[8]

Two main sugar (glucose) fermentation pathways may be distinguished among LAB, and these have been usefully employed in taxonomy. Glycolysis (the Embden–Meyerhof–Parnas (EMP) pathway) results almost exclusively in lactic acid as an end product under standard conditions, and is referred to as a homolactic fermentation. Conversely, the 6-phosphogluconate/phosphoketolase (6-PG/PK) pathway results in significant amounts of other end products, such as ethanol, acetate and CO_2, in addition to lactic acid, and the metabolism is referred to as heterolactic fermentation. The two pathways are outlined in Figure 1.

The EMP pathway is used by all LAB except leuconostocs, group III lactobacilli, oenococci and weisellas, and is characterised by the formation of fructose-1,6-diphosphate (FDP), which is split by an FDP aldolase into dihydroxyacetonephosphate (DHAP) and glyceraldehyde-3-phosphate (GAP). The latter (and DHAP *via* GAP) is then converted to pyruvate in a metabolic sequence that involves substrate-level phosphorylation at two sites. Under 'normal' conditions (*i.e.* plentiful sugar and limited O_2), pyruvate is reduced to lactic acid by a NAD^+-dependent lactate dehydrogenase (nLDH), thereby re-oxidising the NADH formed by

---

Figure 1 *The two main glucose fermentation pathways in lactic acid bacteria: (A) homolactic fermentation (glycolysis); (B) heterolactic fermentation (6-phosphogluconate/phosphoketolase pathway). Selected enzymes are numbered: 1. Glucokinase; 2. fructose-1,6-diphosphate aldolase; 3. glyceraldehydes-3-phosphate dehydrogenase; 4. pyruvate kinase; 5. lactate dehydrogenase; 6. glucose-6-phosphate dehydrogenase; 7. 6-phosphogluconate dehydrogenase; 8. phosphoketolase; 9. acetaldehyde dehydrogenase; 10. alcohol dehydrogenase (after Axelsson[6])*
(Reproduced by kind permission of the Taylor and Francis Group LLC)

the earlier steps in glycolysis. A redox balance results, and lactic acid is the only end product of the metabolism.

The other main fermentation pathway has been variously referred to as the pentose phosphate pathway, the pentose phosphoketolase pathway, the hexose monophosphate shunt and the 6-phosphogluconate pathway. It is designated here as the 6-phosphogluconate/phosphoketolase (6-PG/PK) pathway, thus recognising the phosphoketolase split, a key step in the metabolic sequence, that invokes 6-phosphogluconate as an intermediate. This pathway is characterised by initial dehydrogenation steps, and the formation of 6-phosphogluconate, with subsequent decarboxylation to pentose-5-phosphate. The remaining pentose-5-phosphate is split by phosphoketolase into GAP and acetyl phosphate. GAP is metabolised as per the glycolytic pathway, but if no additional electron acceptor is available, acetyl phosphate is reduced to ethanol *via* acetyl CoA and acetaldehyde. This pathway results in substantial amounts of ethanol and CO_2 being formed.

In most LAB, hexose sugar (glucose) is transported into the cell as the free sugar, and is then phosphorylated by an ATP-dependent glucokinase before entering either of the two abovementioned pathways. A few species use the phosphoenolpyruvate:sugar phosphotransferase system (PTS), in which phosphoenolpyruvate is the phosphoryl donor. In either case, therefore, a high-energy phosphate bond is required for activation of the sugar. It must be noted that, under certain conditions, glycolysis can lead to a heterolactic fermentation, yielding significant amounts of end products other than lactic acid. In a similar sort of vein, some normally homolactic strains will use the 6-PG/PK pathway when metabolising certain substrates. Theoretically, homolactic fermentation of glucose produces 2 moles of lactic acid, and a net gain of 2 moles of ATP per mole of glucose consumed, while the fermentation of glucose *via* the 6-PG/PK route gives 1 mole each of lactic acid, ethanol and CO_2, and 1 mole of ATP per mole of glucose fermented. Other hexoses, such as fructose, galactose and mannose, are fermented by many LAB, the sugars usually entering the major pathways at the level of glucose-6-phosphate, or fructose-6-phosphate after isomerisation and/or phosphorylation. In some strains, galactose proves to be an exception because a PTS system is used for its uptake.

Pentose sugars are fermented by many LAB, with specific permeases being the means by which they enter cells. Once inside, pentoses are phosphorylated and converted to ribulose-5-phosphate, or xylulose-5-phosphate by epimerases or isomerases. These phosphate derivatives are then metabolised as per the bottom half of the 6-PG/PK pathway. This would seem to imply that only heterolactic fermenters can utilise pentoses, but this is not the case, because all genera of LAB, with exception of Group I lactobacilli, are pentose positive. Homofermenters that utilise pentoses usually do so in the same fashion as heterofermenters, the phosphoketolase of these strains being induced by substrates fermented by the 6-PG/PK pathway, and repressed by glucose. Fermentation of pentoses by heterolactic strains results in a different spectrum of end products as compared to glucose fermentation. No CO_2 is formed, and since no dehydrogenation steps are necessary to reach the intermediate xylulose-5-phosphate, the reduction of acetyl phosphate to ethanol becomes a superfluous reaction. Instead, acetyl phosphate, under the influence of the enzyme acetate kinase, participates in a substrate-level phosphorylation yielding acetate and ATP. As a consequence, fermentation of pentoses leads to the formation of equimolar quantities of lactic acid and acetic acid.

A range of disaccharides can be fermented by LAB, and, depending upon the mode of transport, they will enter the cell either as free sugars, or as sugar phosphates. In the former instance, the disaccharide will be split by hydrolase enzymes to monosaccharides, which then enter one of the two major fermentation pathways. In the latter case, *i.e.* when sugar PTSs are involved, specific phosphohydrolases split the disaccharide phosphates into one part free monosaccharide and one part monosaccharide phosphate.

It is possible to divide LAB into three groups according to their metabolic characteristics, the genus *Lactobacillus* having species in all three camps, hence its division into groups I–III. The first 'metabolic group' contains those organisms that are obligately homofermentative (*i.e.* sugars can only be fermented *via* the EMP pathway), and includes the group I lactobacilli, and a few other species from other genera. Organisms in the second category are obligately heterofermentative (only the 6-PG/PK pathway is available for sugar fermentation), and includes leuconostocs, oenococci, group III lactobacilli and weissellas. The third category includes organisms in an intermediate position with regard to their

fermentation, and would include the group II lactobacilli, lactococci, pediococci, streptococci, tetragenococci, vagococi and most
species of enterococci. They are best considered as 'facultatively
heterofermentative'. At an enzyme level, the major difference between the first two categories is the presence, or absence, of the key
enzymes of the glycolytic and 6-PG/PK pathways, FDP aldolase
and phosphoketolase, respectively. Obligate homolactic fermenters
possess a constitutive FDP aldolase and lack phosphoketolase,
whereas the opposite is true for heterolactic fermenters. This
difference accounts for the obvious variation in the end products
of glucose fermentation, and explains why group I lactobacilli
cannot possibly use pentose sugars. The facultative heterofermenters resemble the obligate homofermenters inasmuch as they contain a constitutive FDP aldolase, resulting in the use of glycolysis
for hexose sugar fermentation, and these LAB are thus homofermentative in respect of hexose sugars, and heterofermentative
with regard to pentoses and some other substrates.

During the fermentation of sugars, different LAB species may
produce exclusively L-lactic acid, exclusively D-lactic acid, approximately equal amounts of each optical isomer, or a predominance of
one form but measurable amounts of the other. Exactly what is
formed is governed by the presence, and degree of activity, of
specific NAD^+-dependent lactate dehydrogenases (nLDH), and,
so, if only D-lactic acid is produced, for example, D-nLDH would
be responsible. If a mixture of D- and L-lactic acid are formed, one
D-nLDH and one L-nLDH would generally be responsible, but a
few species produce a racemase that converts L-lactic acid into
D-lactic acid. What usually happens here is that the organism
initially produces L-lactic acid, which then stimulates the production of the racemase, which subsequently converts some of the
L- into the D-form. Why some LAB should want/need to produce a
mixture of D- and L-lactic acid is unclear at present.

The well-documented ability of LAB to alter their metabolism in
response to environmental conditions, and thus synthesise a variety
of end products not produced under 'normal' circumstances, can
ultimately be attributed to their ability to modify pyruvate metabolism. The essential feature of most bacterial fermentations is the
oxidation of a substrate to generate energy-rich intermediates,
which can subsequently be used for ATP production by substrate-level phosphorylation. The oxidation results in the formation

of NADH from NAD^+, which has to be regenerated in order for the cells to continue fermentation. Pyruvate holds a key position in many fermentations by serving as an electron (or hydrogen) acceptor for this regeneration step, and, as can be seen from Figure 1, this certainly applies to the two major fermentation pathways of LAB. Under certain environmental conditions, LAB use alternative ways of utilising pyruvate, and do not merely reduce it to lactic acid. Alternative fates for pyruvate are shown in Figure 2, which represents a summary of what may be found in LAB as a group, and is based on the work of Kandler.[9]

Different species may use differing pathways, depending on their environment and enzymic content. Some of these reactions may be operational even under 'normal' conditions (*i.e.* glucose fermentation), but they will then serve in an anabolic role (*e.g.* acetyl CoA may be required for lipid biosynthesis).

In reactions with NADH oxidases, oxygen can act as an external electron acceptor, and be of considerable benefit to some LAB. NADH oxidases appear to be widespread among the group, and the systems are often O_2-induced, the products of the reactions being either NAD^+ and H_2O_2, or NAD^+ and H_2O, depending on whether the enzyme mediates a two-, or four-electron transfer. Most LAB also possess a NADH peroxidase, which uses H_2O_2 as an electron acceptor (with concomitant formation of H_2O). In homofermentative LAB, NADH oxidases may compete efficiently with nLDH and create a pyruvate surplus, which becomes available for metabolism through the diacetyl/acetoin pathway. In heterofermentative LAB, NADH oxidases do not compete with nLDH, but with acetaldehyde dehydrogenase and alcohol dehydrogenase, the enzymes of the 'ethanol branch' of the 6-PG/PK pathway. When this happens, a pyruvate surplus is not created, and lactic acid remains the main product of pyruvate metabolism. External electron acceptors, such as oxygen, can play an important role in energy conservation in some LAB. As an example, those LAB that ferment glucose by the 6-PG/PK pathway use acetyl CoA as an electron acceptor in addition to pyruvate. In practice, this is a waste of acetyl phosphate, which could better be used in substrate-level phosphorylation, thus producing ATP. Such a waste can be avoided by using external electron acceptors. The active role of oxygen as an electron acceptor in the metabolism of some LAB can be amply demonstrated by the fact that certain

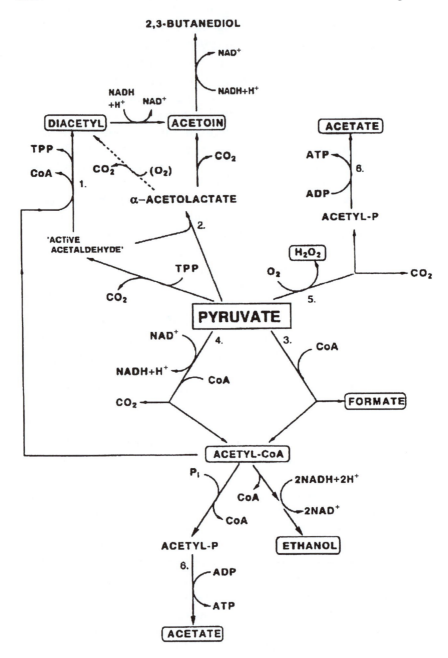

substrates, such as glycerol and mannitol, are fermented only under aerobic conditions, and that some strains of heterofermentative lactobacilli, such as *Lactobacillus brevis*, find it almost impossible to ferment glucose anaerobically. The latter circumstance is due to a lack of acetaldehyde dehydrogenase in the 'ethanol branch' of the 6-PG/PK pathway, creating an absolute requirement for an external electron acceptor, if glucose is to be fermented. Some LAB can even ferment lactate, the end product of 'normal' metabolism, to acetate and CO_2. Where this occurs, the pathway involves NAD^+-dependent, and/or NAD^+-independent LDH, pyruvate oxidase, and acetate kinase, and the energy yield is at the rate of one ATP per mole of lactate consumed.

A key feature of LAB, as a group, is their inability to synthesise porphyrin groups (*e.g.* haem), and this means that they are devoid of a 'true' catalase, and cytochrome system, when they are grown in 'normal' laboratory media. Since they lack an electron transport chain when grown on such media, LAB are forced to rely on fermentative substrate-level phosphorylation for their energy. One of the preliminary diagnostic tests for bacteria is the 'catalsase test', which uses H_2O_2 as the test reagent. Problems can be encountered with some LAB because catalase-like activity can be mediated by a non-haem 'pseudocatalase'. If haemoglobin is incorporated into the growth medium, then, in some species, a true catalase, and even cytochromes, may be synthesised, which, in some instances, leads to respiration through a functional electron transport chain. Thus metabolism changes from a fermentative to a respiratory mode, and this is accompanied by an increase in ATP production.

Because of the abovementioned deficiencies, LAB do not have the same ability as 'genuine' aerobes to protect themselves against the toxic effects of oxygen. Being generally devoid of catalase, some strains may suffer from the toxic effects of the build up of H_2O_2, produced by NADH oxidases. Superoxide dismutase is present in a few LAB, and some strains of *Lactobacillus* (which are devoid of

---

Figure 2 *Pathways for the alternative fates of pyruvate. Dashed arrow denotes a non-enzymatic reaction. Important metabolites and end products are framed. Selected enzymatic reactions are numbered: 1. diacetyl synthase; 2. acetolactate synthase; 3. pyruvate-formate lyase; 4. pyruvate dehydrogenase; 5. pyruvate oxidase; 6. acetate kinase (after Axelsson[6])*
(Reproduced by kind permission of the Taylor and Francis Group LLC)

superoxide dismutase) protect themselves against superoxide by accumulating high levels (30–35 mM) of Mn^{2+}, which have a scavenging effect on the radical.

The absence of an electron transport chain (unless haem is provided in a growth medium) means that LAB are constrained to produce ATP by a modified method. Before detailing how this is done, it is worth summarising the mechanism by which ATP is produced in organisms with an electron transport chain. ATP (or high-energy compounds inter-convertible with ATP), being the universal energy carrier in all living cells, is necessary for the thermodynamically unfavourable business of building cells. The synthesis of macromolecules, and transport of essential solutes against a concentration gradient are possibly the most significant energy-sapping events in the living cell. According to the chemi-osmotic theory, cellular metabolism leads to an electrochemical proton gradient across the cytoplasmic membrane, and the best way of creating a proton gradient is *via* the membrane-linked electron transport chain present in respiring organisms (the 'respiratory chain'). The flow of electrons through the system, *via* a series of carriers, pumps protons out of the cell. The proton gradient across the membrane is composed of two entities, an electrical potential ($\Delta\Psi$), inside negative, and a pH gradient (Δ_{pH}), inside alkaline. $\Delta\Psi$ and Δ_{pH} exert an inwardly directed force, called the proton-motive force (PMF). In organisms with an electron transport chain, this force is sufficiently great for it to be converted into chemical energy, that is, ATP. This is effected by a membrane-bound enzyme, ATP synthase. The energy of the reversed flow of protons 'through' the enzyme, into the cell, is used to form ATP from ADP and phosphate. Without the electron transport chain, LAB cannot make ATP in this way, and have to resort to substrate-level phosphorylation for this energy source, as do other fermenting organisms. LAB do, however, possess H^+-ATPase, an enzyme very similar to ATP synthase, but its major role is to hydrolyse ATP, not to synthesise it. When ATP is broken down, protons are pumped out of the cell and a PMF is established, which is sufficient to drive energy-consuming metabolic reactions (*e.g.* transport of metabolites). The difference between the H^+-ATPase of LAB and the ATP synthase of respiring organisms is a reflection of two very different modes of metabolism. To illustrate the similarity of the two enzymes, it is worth noting that, under certain conditions, the H^+-ATPase of *Lactococcus lactis* is capable of acting as an ATP synthase, and, in a different context, it has been shown that the H^+-ATPases of certain LAB (such as *Lactobacillus casei*) have the same basic structure as the ATP synthase from mitochondria, chloroplasts and respiring eubacteria.

There is some evidence that, in some LAB at least, H^+-ATPase has a major role to play in the maintenance of a satisfactory cytoplasmic pH, and this may in part explain how wine LAB tolerate such a harsh environment. LAB differ from other bacteria inasmuch as they can tolerate a wider and lower range of cytoplasmic pH (pH_i) than other bacteria; lactobacilli being particularly 'tolerant' in this respect by dint of having developed mechanisms by which they can function normally, even with a pH_i of 4.2–4.4. Most other LAB can barely tolerate a pH_i of 5.0. Because of the very nature of their habitats, and their metabolism, the acid released by some LAB often means that the external pH (pH_o), often falls below threshold pH_i levels. To counteract such pH homeostasis problems, the extrusion of protons by the H^+-ATPase, and the electrogenic uptake of K^+, maintain pH_i at a higher level than pH_o, and, where this system is known to operate (*Enterococcus faecalis* being the most extensively studied), both the activity and the synthesis of the H^+-ATPase are regulated by pH_i. The maintenance of a pH_i above threshold level, and the generation of a PMF exhaust a substantial amount of the ATP generated by substrate-level phosphorylation, making less ATP available for other biosyntheses. Any means of generating and/or maintaining the PMF, other than by the H^+-ATPase system, would save the cell energy, and some LAB have evolved alternative energy-saving systems.

A general scheme for PMF-driven solute transport is shown in Figure 3A, where the inwardly directed gradient of protons is the driving force for the influx of solute X, which enters the cell in tandem with a proton (called proton symport). By reversing this mechanism, that is by directing a metabolic end product out of the cell, in tandem with a proton (*i.e.* a net charge), a PMF can be created. Such a system is illustrated in Figure 3B, in which solute Y (which may represent an entity such as lactate) is effluxed. In the case of lactate, it is usually found that more than one proton per lactate molecule has to be exported in order to obtain an electrogenic efflux. Such a mechanism has been referred to as the 'energy

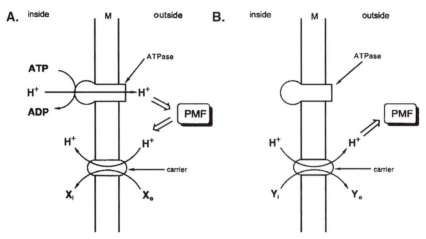

Figure 3 *Schematic representation of proton-motive force (PMF) formation by a H⁺-ATPase and PMF-driven transport (A), and electrogenic end product (B). M denotes the cytoplasmic membrane (after Axelsson[6])* (Reproduced by kind permission of the Taylor and Francis Group LLC)

recycling model'.[10] Conditions for maximum rates of lactate efflux are low external lactate concentration, and high pH_o. An energy recycling system has been shown to operate by acetate efflux in some LAB.

Another mechanism for creating a PMF has been demonstrated in some MLB, as a direct result of trying to ascertain exactly what benefits MLF has for the cell. In 1991 it was found that MLF in *L. plantarum* and *L. lactis* functions as an indirect proton pump.[11,12] The precursor, malate, is exchanged for a product, lactate, with a higher charge, thus making the overall reaction electrogenic. Of key significance here, is the compartmentalisation of the pathway, that is the decarboxylation occurs inside the cell, and consumes one proton. The PMF formed solely by MLF is sufficiently high to drive ATP synthesis by H⁺-ATPase, but under conditions, such as the co-fermentation of sugar and malate, be an energy-conserving process, rather than one of energy generating. Such a mechanism is referred to as the 'electrogenic precursor/product exchange reaction', and a generalised scheme is illustrated in Figure 4, and as the figure indicates, the exchange reaction may be direct, and mediated by one exchange protein (Figure 4A), or, indirect, *via* one precursor uptake protein and one product exit protein (Figure 4B). It has

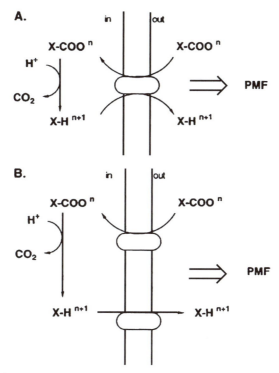

Figure 4 *Schematic representation of electrogenic precursor/product exchange with intracellular decarboxylation. The precursor (X-COO) with a charge (n) is transported into the cell and decarboxylated to yield the product (X-H) with higher charge (n+1). The product can be exported via a precursor/product antiport (A) or by a separate carrier (B). The reaction contributes to the formation of the PMF (after Axelsson[6])* (Reproduced by kind permission of the Taylor and Francis Group LLC)

been shown that the energy benefits of citrate metabolism in *L. lactis* can be attributed to the same mechanism.

By entering into co-fermentations, LAB have developed metabolic systems that will permit them to derive more energy from a nutritionally rich medium than merely from carbohydrates. By using an otherwise non-fermentable substrate as an electron acceptor during carbohydrate fermentation, they can indirectly derive energy from that substrate that would, in other circumstances, be unavailable. Under anaerobic conditions, several organic compounds can act as external electron acceptors, especially in heterofermentative species, which, energetically, have more to gain by employing such entities. Even in circumstances of 'normal' glucose fermentation, the theoretical amount of ATP produced can be doubled if the cell resorts to an external electron acceptor. The intermediate, acetyl phosphate, occupies a key position in the 6-PG/PK pathway, and the presence, or absence, of an external electron acceptor will determine whether ethanol (0 ATP), or acetate (1 ATP) is formed. Lucey and Condon[13] have suggested that the 'ethanol branch' of the 6-PG/PK pathway is, in reality, a 'salvage route', facilitating growth when an external electron acceptor is unavailable. In heterofermenters, the use of these acceptors invariably signals a major alteration in end product profile, something that does not apply to homofermenters, where only minor modifications are observed. Examples of organic external electron acceptors that can be used by LAB are: acetaldehyde, citrate, fructose, α-keto acids, fumarate and glycerol.

The citrate ion is not used directly as an external electron acceptor, but acts as a precursor to one, when it is cleaved by the enzyme citrate lyase, to form acetate and oxaloacetate, which are further metabolised by some LAB. For the enologist, the decarboxylation of oxaloacetate, yielding pyruvate, and the latter's entry into the diacetyl/acetoin pathway, is probably the most significant. Actively growing cells of heterofermentative LAB metabolising citrate in a co-fermentation with a carbohydrate source, do not produce problematic amounts of diacetyl or acetoin, but reduce excess pyruvate to lactic acid. Such an event spares the energetically wasteful reduction of acetyl phosphate to ethanol, and results in the formation of more prolific amounts of ATP through the acetate kinase reaction. This provides the cell with a more efficient means of utilising glucose, and, according to Cogan,[14] results in an increased rate of growth. Using four strains of *Leuconostoc*, Cogan studied the growth, substrate utilisation and end product formation from the metabolism of glucose, citrate and a mixture of both. He found that citrate was not used as an energy source, but that it was rapidly metabolised when glucose was present as well. Predictable amounts of lactate and ethanol were produced when glucose was the carbon and energy source, but in some strains the co-metabolism of both glucose and citrate resulted in the stimulation of growth, decreased uptake of glucose, increased acetate and lactate production and suppressed ethanol production.

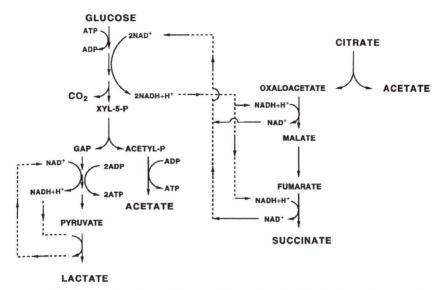

Figure 5 *Proposed pathway for succinic acid production in heterofermentative lactobacilli growing on glucose and citrate (after Axelsson[6])*
(Reproduced by kind permission of the Taylor and Francis Group LLC)

Diacetyl, and the other compounds in the acetoin pathway, were not detectable. Differing LAB may use the citrate lyase breakdown products as external electron acceptors in a variety of reactions, as, for example, in the anaerobic fermentation of mannitol by *L. plantarum*, and the production of succinic acid from citrate, which is a fairly common phenomenon among heterofermentative lactobacilli. Citrate is also used as an electron acceptor in an anaerobic degradation of lactate in some strains of *L. plantarum*, the products of this co-metabolism being succinic acid, acetate, formate and CO_2. This would indicate the use of both the succinic acid pathway, and a pyruvate-formate lyase. The differences between LAB regarding their use of citrate may ultimately depend upon the presence, or absence, of the enzyme oxaloacetate decarboxylase. If the enzyme is present, citrate utilisation results in an increase in the pyruvate pool, which may lead to an altered end product profile (*e.g.* acetoin). If oxaloacetate decarboxylase is absent, an alternative route would be *via* the succinic acid pathway, a proposed scheme for which is shown in Figure 5.

The reaction may be summarised thus:

1 glucose + 1 citrate + 2 ADP + 2 P_i → 1 lactate + 2 acetate + 1 CO_2
+ 1 succinate + 2 ATP

Some heterofermentative lactobacilli (*e.g. Lactobacillus buchneri*) can use glycerol as an electron acceptor in an anaerobic co-fermentation with gluscose. The end products are lactate, acetate, CO_2 and 1,3-propanediol. The NADH formed during glucose fermentation is not re-oxidised by the ethanol pathway, but rather by using glycerol as electron acceptor. Some strains of *L. brevis* will ferment glucose very weakly under anaerobic conditions, but will do so far more vigorously if glycerol is added to the growth medium.

It has been known for many years that the fermentation of fructose by heterofermentative LAB results in the production of mannitol, and it is now known that such a reaction provides an example of a pathway where a single compound acts as the growth substrate, and an electron acceptor. Fructose is fermented by the 6PG/PK pathway, but some sugar is reduced to mannitol by a NAD:mannitol dehydrogenase. As in the use of other external electron acceptors, this enables cells to produce ATP through the acetate kinase reaction. Assuming that no ethanol is formed, the overall equation for fructose fermentation would be:

3 fructose + 2 ADP + P_i → 1 lactate + 1 acetate + 1 CO_2 + 2 mannitol + 2 ATP

In terms of ATP production per unit of sugar consumed, this is less efficient than glucose fermentation, but, idiosyncratic as they are, some heterofermentative lactobacilli show a higher rate of growth on fructose than on glucose.

As a general rule, LAB have a very limited capacity to synthesise amino acids from inorganic nitrogen sources. They are, therefore, mostly dependent upon pre-formed organic nitrogen sources in their growth media, although some strains of *L. lactis* are photo-trophic for most amino acids. The growth requirement for a particular amino acid in a growth medium may be the result of mutation(s) in the gene(s) for its biosynthesis, and/or the down-regulation of those genes, or the enzymes involved. Studies on the entry of nitrogen compounds into the cells of LAB have thus far been more or less confined to the milk-acidifier, *L. lactis*, in which the protein casein looms large in the environment. What evidence we have, suggests that proteins play little, or no, part in the nutrition of wine LAB, and that, in such organisms, single amino acids, or, at the most di-, or tripeptides, are taken up by the relevant transport system and passed across the cell membrane. In *L. lactis*, there is an oligopeptide transport system (Opp), which accepts 4- to 8-residue peptides and takes them across the membrane.

The fastidious nature of most LAB, as exemplified by their requirement for amino acids, nucleotides, *etc.*, demands that they have evolved efficient uptake and transport systems for these essential nutrients, and we find that the transport of solutes is closely linked to their cellular bioenergetics. LAB use any of three broad categories of transport system, according to the form of energy used in the translocation process: (1) primary transport (using chemical energy), that is ATP-driven; (2) secondary transport (using chemiosmotic energy), that is PMF-driven, and (3) group translocation (chemical modification, concomitant with transport), that is phosphoenolpyruvate: sugar phosphotranferase system (sugar PTS). The most common class of primary transporters in LAB are members of the ATP-binding cassette (ABC)-transporter family, the most studied examples being the glutamate/glutamine transporter, and the Opp. The PMF-driven transport systems are perhaps the most common and general among LAB (as in most bacteria). A specific, membrane-bound protein translocates the solute across the menbrane in tandem with one proton. It is thought that many sugars are transported in this way, and, in lactococci, at least, most amino acids are transported by PMF-driven symport systems. In the group translocation category, the abovementioned PTS is a complex enzymic machinery, whose main function is to translocate a sugar across a membrane, and phosphorylate it at the same time. For a thorough review of this topic, see the paper by Postma *et al.*[15]

A variation of the secondary transport system is the 'precursor: product antiport mechanism', which does not appear to involve a PMF. Many LAB can derive energy from substrate-level phosphorylation through the metabolism of arginine (the ADI pathway, see below). For reasons unknown, most arginine-metabolising LAB cannot use the amino acid as their sole energy source, but catabolise it simultaneously with a fermentable carbohydrate.

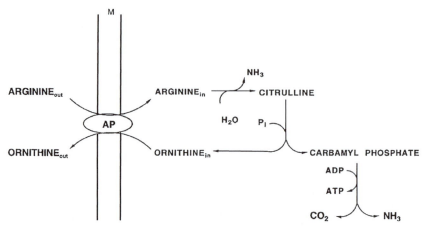

Figure 6   *The arginine deiminase (ADI) pathway with the arginine/ornithine anti-porter (AP). M denotes the cytoplasmic membrane (after Axelsson[6])* (Reproduced by kind permission of the Taylor and Francis Group LLC)

When this occurs, one of the end products, ornithine, is excreted. In lactococci, it has been shown that the driving force for arginine uptake, and for ornithine excretion, is the concentration gradient of these two compounds. The stoichiometry for the arginine:orni-thine interchange is 1:1, and is mediated by a single membrane-bound protein, the arginine/ornithine antiporter. A summary of the mechanism is illustrated in Figure 6. Production of the anti-porter is repressed by glucose, and induced by arginine.

Of prime importance to the wine maker is *Oenococcus oeni*, an organism at one time included in the genus *Leuconostoc*, which had been previously described as being 'heterofermentative, coccoid LAB, producing only D-lactic acid from glucose, and not produc-ing ammonia from arginine'. It was not difficult to confuse leuconostocs with some coccoid rods of heterofermentative lacto-bacilli, and phylogenetic analysis of leuconostocs revealed a con-siderable heterogeneity within the genus. Phylogenetic studies in the late 1980s revealed that the so-called 'wine leuconostocs', allotted to the species *L. oenos*, were only distantly related to other leuconostocs, and that this warranted their placement in a separate genus, and *Oenococcus* was duly proposed.[16] Oenococci are easily distinguishable by their extreme acid and ethanol tolerance. In evolutionary terms, the distance from the 'leuconostoc-like LAB' (*i.e. Leuconostoc sensu stricto* and *Weisella*) to the genus *Oenococcus* is a large one, and it is assumed that, during its adaptation to the highly unusual wine environment, extensive changes have necessarily occurred in the genome, which have resulted in a rapid divergence from related species.

The metabolic versatility of some LAB may cause problems for the winemaker. As an example, fermentation of excessive amounts of fructose by heterolactic fermenters may result in the formation of excess acetic acid (volatile acidity, VA) and mannitol. High VA is considered to be undesirable in wine, and an abundance of mannitol may encourage the growth of spoilage lactobacilli, such as *L. plantarum*, after MLF has been completed. According to Amerine and Kunkee,[17] this is known as 'mannitol spoilage'. Furthermore, citrulline excreted into wine during arginine catabolism by *O. oeni* (see page 235), may support the growth of wine spoilage heterofermenters, such as *L. buchneri* and *L. brevis*.

Although reduction of acidity is the most obvious result of the growth of LAB in wine, their metabolic activity can also significantly modify wine aroma, flavour and mouthfeel. Such changes can be good, or bad, depending on which strains predominate. Some strains of *Lactobacillus*, for example, impart 'sweaty' and 'mousy' taints, the latter being, arguably, the most unpleasant of all wine aroma faults. The role of LAB in the quality improvement and depreciation of wine has been reviewed by Lonvaud-Funel.[18]

5.2 MALO-LACTIC FERMENTATION

This is a bacterial fermentation, rather than one carried out by micro-fungi. It was well over 100 years ago that Pasteur found that bacteria were involved in some, so-called, 'wine diseases', but it was not until the early years of the 20th century that the importance of LAB in wine-making was clearly demonstrated. Pasteur proved that lactic acid was produced by some bacteria (or 'new yeast', as he called them). He also noticed that 'sound' wines would lose acidity upon storage, and maintained that this phenomenon could be attributed to bacterial activity. Some other workers at that time attributed this rise in pH to yeast activity, and it was not until 1900 that Alfred Koch[19] first isolated, what were to be known as, malolactic bacteria (MLB), from wine sediment, and actually 'induced' a malo-lactic fermentation (MLF) with them. It was around this time, also, that Koch noticed that the fall in wine acidity could be correlated with the disappearance of malic acid from the wine, and proved, by adding salicylic acid (which inhibits bacterial growth, but not yeast growth) to wine, that yeasts could not possibly be responsible for the observed loss of acidity. This fact had been strongly suggested in 1891 by Müller-Thurgau,[20] who was the first to attempt to control acid reduction in wine. Seifurt[21] also isolated bacteria that would induce loss of acidity in wine, and provided both a description, and a microscopic illustration of their form (paired cocci). He named his isolate *Micrococcus malolacticus*, and showed that the loss of acidity was due to the conversion of malic acid to lactic acid. During the next decade, or so, four more strains of bacteria capable of MLF were isolated, and in 1913, Müller-Thurgau and Osterwalder[22] provided a detailed account of these organisms, and constructed a taxonomic key to all non-spoilage bacteria that had been isolated from wine (and other alcoholic beverages).

The predominant organic acids in grape must are malic acid and tartaric acid, both of which are metabolised by certain strains of LAB. Malic acid ('malic' is derived from *malum*, the Latin for apple) is the most widespread organic acid in fruit (whereas tartaric acid is peculiar to grapes) and is generally present in grapes in concentrations in the range of 2–4 g L^{-1}, and often constitutes around half of the grapes' acidity. In small berries growing in cool climes, the level can reach 6 g L^{-1}, whereas the acid will be nearly absent from overripe grapes growing in hot locations. Levels of malic acid in berries will vary with grape cultivar, cultural conditions and berry density within a cluster. The isomer found in grapes is the L(+) form, and it is synthesised from glucose *via* pyruvic acid. L-Malic acid can be almost totally converted by some bacteria to L-lactic acid and CO_2, an event generally referred to as MLF. The D-isomer of malic acid is not involved in the reaction. In addition, malic acid may be partially removed by the yeast during alcoholic fermentation, and by treatment with calcium carbonate, since its calcium salt is sparingly soluble. MLF occurs not only in grape wine, but also in cider, and other fruit wines, and in the manufacture of soy sauce and sauerkraut. According to Halliday and Johnson,[23] it is 'a process now applied to all red, but not all white wines'.

The pathway for MLF (*i.e.* malic acid to lactic acid and CO_2) was first elucidated by Möslinger in 1901,[24] who realised that lactic acid arose both from the decomposition of malic acid, and by fermentation of a carbohydrate source (lactic acid fermentation). With the information available at that time, MLF could be summarised thus:

$$COOH - CH_2 - CHOH - COOH \rightarrow CH_3 - CHOH - COOH + CO_2$$

As the equation shows, the reaction essentially involves the decarboxylation of a dicarboxylic acid, and it was found to occur under the influence of, what was called, the 'malic enzyme', the exact nature of which was not fully understood until 1950,[25] when a sample was isolated from *L. arabinosus*. It was later shown that the 'malic enzyme' catalysed a different reaction (the decarboxylation of malate to pyruvate), and that the enzyme catalysing the 'true malo-lactic reaction' required the presence of Mn^{++} and NAD+, the latter not actually being involved in the reaction as a substrate. Accordingly, the enzyme concerned was renamed L-malate: NAD carboxylyase (the 'malo-lactic' enzyme), the new name distinguishing it from the 'malic enzyme', and from malate dehydrogenase. The presence of the malo-lactic enzyme, in nature, was shown to be restricted to a few genera of LAB, which consequently became known as MLB. It is rather curious that, although NAD^+ is a prerequisite for enzymatic activity, no NADH can be detected, indicating that hydrogen exchange reactions occur within the malo-lactic enzyme, and that oxaloacetate, and/or pyruvate may be enzyme-bound intermediates.

A few LAB (*L. casei* is an example) can use malate as the sole energy source, the NAD^+-dependent malic enzyme catalysing the decarboxylation of malate to pyruvate, and CO_2 as follows:

$$malate + NAD^+ \rightarrow pyruvate + CO_2 + NADH$$

Pyruvate is then converted to acetate, ethanol, CO_2 and ATP, presumably *via* the acetate kinase reaction.

The pathway for malate dissimilation was first believed to be *via* a sequence of reactions involving pyruvate as a free intermediate. In this scheme, pyruvate would subsequently be reduced to lactic acid by an LDH. That this was not the whole story was indicated by the fact that malo-lactic fermenting leuconostocs only possess a D-LDH (which means that they form D-lactic acid from glucose), while producing exclusively L-lactic acid from malate. It is now

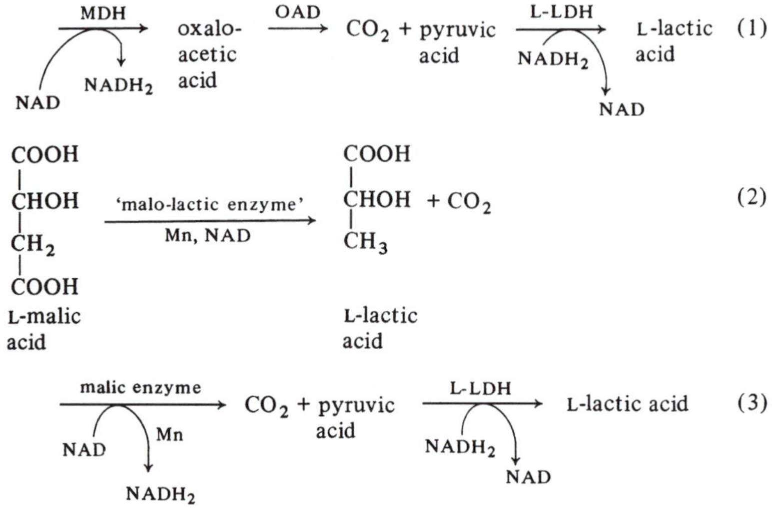

$$\begin{array}{ccccccc}
& \text{MDH} & & \text{OAD} & & \text{L-LDH} & \\
& & \text{oxalo-} & \longrightarrow & CO_2 + \text{pyruvic} & \longrightarrow & \text{L-lactic} \quad (1) \\
& \text{NADH}_2 & \text{acetic} & & \text{acid} & \text{NADH}_2 & \text{acid} \\
\text{NAD} & & \text{acid} & & & & \\
& & & & & & \text{NAD}
\end{array}$$

$$\begin{array}{ccc}
\text{COOH} & & \text{COOH} \\
| & & | \\
\text{CHOH} & \xrightarrow[\text{Mn, NAD}]{\text{'malo-lactic enzyme'}} & \text{CHOH} + CO_2 \qquad\qquad (2) \\
| & & | \\
\text{CH}_2 & & \text{CH}_3 \\
| & & \\
\text{COOH} & & \\
\text{L-malic} & & \text{L-lactic} \\
\text{acid} & & \text{acid}
\end{array}$$

$$\begin{array}{ccccc}
\xrightarrow[\text{NAD} \quad \text{Mn}]{\text{malic enzyme}} & CO_2 + \text{pyruvic} & \xrightarrow[\text{NADH}_2]{\text{L-LDH}} & \text{L-lactic acid} \quad (3) \\
\text{NADH}_2 & \text{acid} & & \text{NAD}
\end{array}$$

Figure 7 *Theoretical pathways of the conversion of malic acid to lactic acid (after Radler[26]) (OAD, Oxaloacetate decarboxylase; LDH, lactate dehydrogenase; MDH, L-malate dehydrogenase)*
(Reproduced with kind permission of Reed Elsevier)

known that among the LAB, as a whole, there are three enzymatic pathways for the conversion of L-malic acid to L-lactic acid and CO_2 (summarised by Radler[26,27]): (1) by the malo-lactic enzyme directly to lactic acid; (2) by the malic enzyme to pyruvic acid and subsequent reduction by L-lactate dehydrogenase to lactic acid and (3) reduction by malate dehydrogenase to oxaloacetate, then decarboxylation to pyruvate (with the aid of oxaloacetate decarboxylase) and finally reduction to lactic acid. These pathways are outlined in Figure 7.

Bacteria undergoing MLF in co-fermentation with a source of carbohydrate generally benefit from an increased growth rate, and a higher yield from glucose (Y_{gk}), when compared to growth on glucose alone. This is difficult to explain, because no potential electron acceptors, such as oxaloacetate or pyruvate, are produced by the reaction, although it has been shown by some workers that the reaction is not stoichiometrically complete, and that small amounts of pyruvate and NADH are released, so that the reaction can supply additional electron acceptors. As far as we know, all wine LAB use the true malo-lactic enzyme to decarboxylate malic to lactic acid, without production of free intermediate compounds,

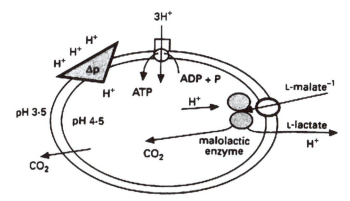

Figure 8 *Energy conservation during malo-lactic fermentation (after Henick-Kling[28])*
(Reproduced by kind permission of the Society for Applied Microbiology)

and produce energy by linking the transport and decarboxylation of malic acid to the production of a proton gradient across the cell membrane. Investigations of the PMF and ATP charge of the malo-lactic bacterial cell have shown that MLF is an energy-liberating pathway (Figure 8), which is based on the electrogenic transport of malate and the malo-lactic enzyme, and the efflux of lactic acid.[28]

This energy can be used for transport processes or converted into ATP *via* membrane-bound ATPase. In wine, with its low pH (<4.5) and limited amounts of fermentable sugar, the additional ATP allows increased growth yields of *ca.* 25–50%. As Figure 8 shows, malate enters the cell, *via* an energy-requiring active transport system, where it is decarboxylated by the intracellular malo-lactic enzyme; lactic acid and CO_2 leaving the cell. No active transport system has been identified for lactate excretion, but there is evidence for a mechanism facilitating malate/lactate interchange. The optimum temperature for malate-degrading activity of whole cells of *O. oeni* is 30–35°C, while the optimum growth temperature of the organism is *ca.* 25°C. The MLF does not contribute any carbon for the synthesis of new cell material, and so it is essential that, if a MLB starter culture is inoculated post-fermentation, there is sufficient sugar present for cell syntheses. The major sugars in wine are glucose and fructose (anything from 10 to <0.5 g L^{-1}). Although both sugars are utilised by many wine LAB, some strains of *O. oeni* only ferment glucose very weakly or not at all. Other sugars, such as ribose, arabinose, xylose, mannose and galactose may be present in concentrations up to 1.3 g L^{-1}, but evidence suggests that the growth of most MLB in wine is normally at the expense of glucose and fructose.

MLB in wine are not only able to metabolise malate to lactate, they can also convert the small amounts of citrate present in wine to pyruvate, lactate, acetate, ethanol and diacetyl.[29] The production of diacetyl is particularly important, since it can directly influence flavour, by introducing 'buttery' notes, an excess of the compound imparting 'caramel' and 'rancid butter' characters. To ensure that the correct level of diacetyl is imparted into a wine such as Chardonnay, producers will blend their base wines in such a way that only a certain percentage of them have undergone MLF. The amount of diacetyl produced is largely determined by the amount of citrate present, and the redox potential in the wine, and levels of SO_2. Small quantities of diacetyl are also formed by certain strains of yeast during primary fermentation. The inability to predict, or control, elevated levels of diacetyl, acetic acid and other unwanted components, in certain wines, gives ammunition to those who aver that MLF is not universally beneficial. For a summary of aroma compounds formed during MLF, the reader is recommended to Bertrand *et al.*[30]

Considering the importance of MLF, the initiation and control of this secondary fermentation was, for many years, very much a 'chance' affair. The practice of inoculating wine, with a specific bacterial culture, in order to initiate the bio-transformation of malate, was not reported until Webb and Ingraham's paper in 1960,[31] and since that time there have been numerous studies on the organisms concerned and their growth requirements. Since the 1970s, some red winemakers have inoculated their musts with bacterial cultures, and, in modern wineries, it is not uncommon for musts to be inoculated with yeast and MLB simultaneously, so that MLF can ensue at an early stage. Strains of *O. oeni* are now widely used as starter cultures, and the factors affecting the induction of MLF by this organism in red wines has been studied at length.[32]

A variety of MLB have been isolated from wines (Henick-Kling, 1993[33]), mostly of the genera *Lactobacillus, Pediococcus* and *Leuconostoc*, although organisms in the latter genus which can be isolated from the wine environment are now generally referred to as *Oenococcus oeni* (ex-*Leuconostoc oenos*), as recommended by Dicks *et al.*[16] The lactobacilli can be both homo- and hetero-fermentative, while the pediococci are homofermentative, and *O. oeni* is heterofermentative. Which species predominates in the fermentation is largely dependent upon pH and ethanol content of the must/wine. Fortunately, the oenologically desirable *O. oeni* generally dominates a post-fermentation MLF, as it has greater tolerance of low pH and high levels of ethanol. For a review of the biology of *O. oeni*, see the 1993 paper by van Vuuren and Dicks.[34]

Depending upon the wine style, the location of the vineyard, and the method of production, MLF may be regarded as beneficial or deleterious, and the causative organisms encouraged, or discouraged accordingly. The presence of non-grape odours and flavours in white wines is generally more unacceptable than in red wines, hence MLF is only permissible in a few white styles. In general, wines that benefit from MLF are champagnes, full-bodied dry whites (classic Burgundian whites, such as Chardonnay) and the medium- to full-bodied reds. The reaction is unwanted in wines such as Riesling, and in most wines made from German white grape varieties, which, with their delicate flavour characters, can easily be overwhelmed by the endproducts of MLF activity. As a rule, grapes grown in cool regions, such as Canada and New Zealand, tend to be high in acidity, much of which is attributable to malic acid. Even the Champagne region of France is sufficiently cold to render base wines very acidic (often with a pH of <3). For wines produced from such grapes, bio-deacidification by MLB, results in a more balanced wine. White wines from the Minho region of northern Portugal are traditionally very acidic, not from the influence of climate, but from time-honoured viticultural practices. Vines in this region are very lofty and difficult to train (they are often grown up trees), and grapes are usually over-cropped, making it difficult to control quality. Grapes are hand-harvested, and the wines produced, *vinhos verdes* (green wines) are named after the unique fruit rather than their colour. Such wines *have* to undergo MLF.

In wines with a low intrinsic acidity, an additional rise in pH can reduce quality, but if MLF is necessary for its other attributes, the acidity can be corrected by addition of acid (usually tartaric; sometimes citric). MLF is also said to enhance the body and flavour-retention capability of a wine, giving greater palate 'softness' and 'roundness'. According to the strain of MLB employed, MLF will remove strong 'vegetal/herbaceous' aromas, enhance 'fruity' and 'floral' aromas, and add 'yeasty', 'nutty' and 'buttery' aromas. Any wine sample, with low SO_2 content, and kept at 20°C, or above, is likely to undergo spontaneous MLF, especially those with low intrinsic acidity, in which it is not always wanted. Control of MLF arises as a direct result of the control of MLB, and a number of methods can be employed where MLF is considered undesirable; examples being: early removal of grape skins (red wines); low cellar temperature; high SO_2 levels; early clarification; pasteurisation, and sterile filtration. With the exception of SO_2, addition of permitted anti-microbial agents to must/wine have little effect upon MLB growth. Spontaneous MLF can be managed successfully if a minimum of SO_2 ($<$10 ppm free, and $<$40 ppm bound) is used during processing, and if a moderate temperature (18°C) is maintained. The effect of the gas on MLB (especially *O. oeni*) growth and activity has been thoroughly studied by Wibowo *et al.*,[35] who found that free SO_2 at 10 ppm kills *O. oeni*; bound SO_2 at $>$30 ppm delays growth, and lowers the final cell density attainable in a wine, while bound SO_2 at $>$50 ppm can totally inhibit growth. In terms of the malo-lactic activity of the cell: 20 ppm bound SO_2 reduces activity by 13%; 50 ppm by 50% and 100 ppm inhibits it totally.

In a synthetic wine medium at pH 3.5, lactic acid at a concentration of 0.5 g L^{-1} reduces the growth of MLB, and at concentrations over 3.0 g L^{-1} growth is inhibited. In California, fumaric acid, a widely used acidulating agent in foods and beverages, has been used in that capacity in wines for over 40 years, and, in 1969 it was shown that it could prevent, or delay, MLF.[36]

The MLF normally occurs after the completion of alcoholic fermentation, usually within a year of primary yeast activity, but it can take place during ethanolic fermentation. Growth conditions for MLB are completely different during ethanolic fermentation, as compared to those when it has subsided. In respect of the latter situation, table wines present a rather harsh environment, and the bacteria must be able to routinely tolerate a rather low pH (3–4), and a high ethanol content (*ca.* 10–14%). Having said that, such conditions generally seem to delay the onset of MLF, rather than to prohibit it completely. It is generally agreed that when MLF occurs naturally alongside fermentation by yeast, it proceeds more rapidly and there is less likelihood of unwanted by-products being liberated, whereas unsolicited MLF in primary fermented wine is less rapid and less consistent. It should be remembered that MLB are fastidious organisms, with complex nutritional requirements, and that during red winemaking these are far more likely to be met by the components of a must than those of a primary fermented wine. It is essential to ensure that MLF does not occur too late in the life of a wine, especially when it is in bottle. Not only does this give the consumer the impression that the wine is still fermenting, but it also causes the wine to lose 'fruitiness' and take on an intrusive 'lactic' flavour. It can also impart turbidity to the product. If MLF occurs while a wine is being aged in wood, winemakers maintain that a superior integration of 'oak' and 'fruit' characters can be achieved. The practical implications of MLF in wine have been reviewed by Davis *et al.*,[37] and the energetics of the reaction have been summarised by Henick-Kling.[28]

Malo-lactic fermentation is deemed to be complete when the malic acid level is reduced to $<0.1g\ L^{-1}$. The reaction should then be stopped by clarification and SO_2 addition, except if diacetyl levels are too high, in which case one has to wait until bacterial activity has reduced the compound to an acceptable level.

There is one school of thought that recommends that spontaneous (natural) MLF should be discouraged at all costs, since the characteristics of most indigenous wine LAB are, at best, imperfectly known. There is some foundation for this approach, because, as we shall see, several of these organisms have the ability to degrade arginine and excrete undesirable intermediates into wine.

5.2.1 Urethane (Ethyl Carbamate), and Arginine Metabolism

Ethyl carbamate (EC), the ethyl ester of carbamic acid, has achieved a notoriety (somewhat exaggerated by the media) as a proven carcinogen for some animals.[38,39] Its effects on humans, however, are imprecisely known, although Zimmerli and Schlatter[40] contended that the average levels found in alcoholic beverages could well increase the lifetime risk of cancer in imbibers. Carbamic acids are very unstable, whereas their esters are stable, and non-volatile. According to Ough,[41,42] the ester is a naturally-occurring micro-component of all fermented foods and beverages, and has been recorded in detectable quantities in wine, beer, orange juice, soft drinks, bread, soy sauce and yoghurt. Although it is synthesised in minute amounts by some fermenting microorganisms, its levels increase if the foodstuff is subjected to heat, either intentionally (*e.g.* during the production of sherry), or unintentionally (*e.g.* in transit, or during storage). In context of the latter, it has been shown that, as a result of the shipment of wine during hot weather, the EC level doubles with each 14 °F rise in temperature, although Stevens and Ough[43] showed that the compound can be formed at low to normal wine storage temperatures. EC has attested soporific qualities, and has been used as a food preservative in some countries (not now!), and its concentration in wine is now controlled by legislation. In 1985 the Canadian government introduced a limit of 30 ppb for table wines, and 100 ppb for fortified wines, while in 1988, the American wine industry and the FDA agreed a limit of 15 μg L^{-1} (15 ng g^{-1}) in table wines.[44] At the other end of the scale, some bourbons can contain several hundred ppb, and some brandies have tested as high as 12,000 ppb.

Since the role of EC as a suspected human carcinogen was mooted, the origin and levels of the compound in wine (and other beverages) have received much attention.[45] The formation of EC is a spontaneous chemical reaction involving ethanol and EC precursors, which include urea, citrulline, carbamyl phosphate, *N*-carbamyl α- and β-amino acids and allantoin, the main source in wine being by ethanolysis of urea.[46] The involvement of urea in EC formation in wine was conclusively demonstrated in 1989 by Monteiro *et al.*,[47] who employed radio-labelled precursors. Urea is an intermediate in the metabolism of L-arginine by yeast and the extent of the production and release of EC is dependent upon the initial level of arginine (and other nitrogenous compounds), and the yeast strain used for fermentation.[48]

The concentration of arginine in grape juice ranges from a few hundred mg L^{-1} to 2.4 g L^{-1},[49,50] and has been recorded as high as 10 g L^{-1}. Although arginine in grape juice is mostly metabolised by yeast,[51] the amino acid is still present in wine in amounts varying from 100 mg L^{-1} to 2.3 g L^{-1},[52] and so there is still plenty available for metabolism by LAB during subsequent MLF.

The development of EC during the ageing of wines seems to be directly linked to the presence of urea, although, as we shall see, some heterofermentative LAB can produce EC from arginine by a different pathway. Urea in the base wine has been shown not to be

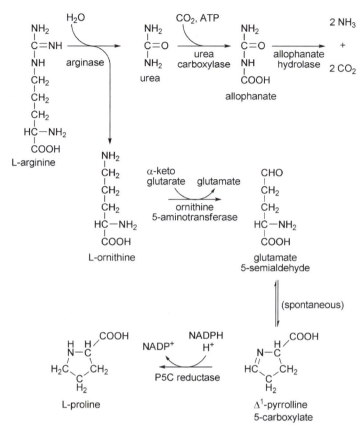

Figure 9 *Pathway of arginine degradation by yeast*
(Reproduced by kind permission of Springer Science and Business Media)

the precursor of the EC found in brandies, and, similarly, if EC has been formed prior to the distillation process, then, being non-volatile, it will not be present in the spirit distillate. In most brandies, EC is rather more attributable to hydrocyanic acid.

The catabolism of L-arginine by yeast is summarised in Figure 9, the first step being catalysed by arginase and yielding ornithine and urea. Many microbes produce urease, an enzyme which breaks down urea into ammonia and CO_2,[53] but *S. cerevisiae* possesses no urease activity, and, instead, urea is broken down by a bifunctional enzyme, urea amidolyase. In its first manifestation, urea carboxylase, the amidolyase yields allophanate, which is then broken down to equimolecular quantities of ammonia and CO_2 when it acts as allophanate hydrolase.[54] As can be seen from Figure 9, the pathway involving urea carboxylase involves energy (ATP) consumption. It also requires the presence of the vitamin biotin, and, as a consequence, urea will not be metabolised by *S. cerevisiae* under conditions of low energy and/or vitamin status. This being the case, the relatively non-toxic urea is regarded by the cell as being a surplus nitrogen source, and is excreted. Excretion is *via* a urea-facilitated diffusion permease, and, from the cell's point of view, this mechanism is more preferable than excreting surplus nitrogen as the potentially more toxic NH_4^+. As we have said, during fermentation, yeast may rapidly, and irreversibly, deplete must of nitrogenous compounds, even though there might be a 'surplus of nitrogen' inside the cell. The ability to secrete urea gives the cell a mechanism by which it can regulate its internal nitrogen levels, since the cell can re-absorb urea when metabolic conditions dictate.

Ornithine, the other result of arginase activity, can be further degraded to proline (by glutamate semialdehyde), or can be converted to polyamines. It has a pronounced inhibitory effect on the growth of wild yeasts, such as *Hansenula minuta*, and it is thought that high concentrations of ornithine produced from arginine catabolism may well confer microbiological stability to wine. Cooper[55] has reported that genetic analysis of proline degradation has revealed that ornithine breakdown proceeds *via* proline as an intermediate. Proline is formed from ornithine in the yeast cytoplasm, transported to the mitochondria, and then converted to glutamate. Since proline cannot be metabolised during anaerobiosis, ornithine cannot form glutamate in the absence of oxygen, and this is reason why it is possible for wine to contain higher levels of proline than its original must.

The ability of some wine LAB to degrade L-arginine has been demonstrated by a number of workers, many of who have proposed that the event occurs *via* the arginase–urease pathway, which produces urea as an intermediate,[56,57] *viz*:

$$\text{L-arginine} \xrightarrow{\text{arginase}} \text{L-ornithine} + \text{urea}$$

$$\text{urea} \xrightarrow{\text{urease}} CO_2 + 2NH_3$$

Certainly, as far as many of the workers were concerned, this was a supposition and was based on the stoicheiometry of arginine conversion to ornithine and ammonia (*i.e.* 1 mole of ornithine and 2 moles of NH_3 from each mole of arginine metabolised), since the relevant enzymes had not been unequivocally identified. For many years, the degradation of arginine by wine LAB was attributed to the combined activity of arginase and urease, and the mechanism was accepted until the work of Liu *et al.*,[58] who showed that, during MLF in grape must or wine, some heterofermentative LAB may completely degrade arginine *via* the arginine deiminase (ADI) pathway (see Fig. 6), resulting in the production of ornithine, ammonia, CO_2 and energy. The same workers had already reported that the utilisation of arginine by this pathway resulted in the excretion of citrulline, a precursor of EC.[59] The ADI pathway may be summarised thus:

$$\text{L-arginine} + H_2O \xrightarrow[\text{arginine deiminase (ADI)}]{} \text{L-citrulline} + NH_4^+ \quad (1)$$

$$\text{L-citrulline} + P_i \xrightarrow[\text{ornithine transcarbamylase (OTC)}]{} \text{L-ornithine} \\ + \text{carbamyl-P} \quad (2)$$

$$\text{carbamyl-P} + \text{ADP} \xrightarrow[\text{carbamate kinase (CK)}]{} \text{ATP} + CO_2 + NH_4^+ \quad (3)$$

Here again, 1 mole of arginine is converted into 1 mole of ornithine and 2 moles of ammonia. All three enzymes, ADI, OTC and CK could be detected, and it was shown that their levels increased significantly over time in the presence of arginine. On the other hand, arginase and urease activities were undetectable in all cultures grown in the presence of arginine, indicating that, in some strains of wine LAB at least, arginine breakdown is not *via* the arginase–urease pathway. The operation of the ADI pathway indicates a different potential route for EC formation, namely L-citrulline, which can be chemically detected in wine, and carbamyl phosphate. It is not precisely known why citrulline should be excreted, but, according to Russell and Cook,[60] it may be due to energy uncoupling or overflow metabolism between catabolic and anabolic reactions. The intermediate, citrulline, is re-assimilated and catabolised by some strains of wine LAB (notably *L. buchneri* CUC-3), but not others, and so, in certain circumstances, it is unavailable for EC formation. Re-assimilation of excreted citrulline occurs only after the exhaustion of arginine from the growth medium. In a survey of commercial cultures of MLB, Mira de Ordũna et al.[61] showed that all strains tested excreted citrulline during arginine catabolism, the amount of citrulline produced being strain specific. The same research group studied the kinetics of arginine metabolism in selected wine LAB,[62] and have summarised the practical implications of the products of the pathway proposed by Liu and Pilone.[63] Arginine has a bitter, musty taste and its degradation is deemed to be desirable for overall wine quality.

As can be seen from Equation (3), the complete breakdown of arginine leads to ATP formation, which implies that the presence of energy-rich substrates, such as fermentable sugars, in must would suppress arginine catabolism, and, thus, citrulline excretion. It has, therefore, been suggested that it would be oenologically advantageous to conduct MLF in grape must with high sugar concentrations (180–240 g L^{-1}), instead of inoculating bacteria after the completion of ethanolic fermentation. Unfortunately, it has been observed by Mira de Ordũna et al.[62] that higher rates of conversion of arginine to citrulline occur in media with high glucose concentrations. Strains of wine LAB that are able to derive energy from arginine catabolism may be more competitive, in the stressful environment provided by wine, than those unable to do so.

Formation of ammonia can be undesirable because it is liable to increase wine pH, and render it more susceptible to spoilage organisms, although the extent of the pH rise will vary according to the amount of arginine available and the buffering capacity of the wine. A moderate pH rise will help desirable, acid-sensitive wine LAB, but, if it goes too far, spoilage organisms may proliferate. In a recent study of commercial MLB, all strains tested degraded arginine and excreted citrulline, and it was shown that there was a linear relationship between the two events. Some MLB, such as *L. buchneri*, have a greater capacity for utilising arginine than others, and show an enhanced rate of growth as arginine levels in the culture medium are increased. This, of course, leads to increased citrulline and EC formation. Fortunately, *O. oeni* does not respond to arginine in this way, and is regarded as a non-arginine-degrading organism. Based on this sort of evidence, oenococci will be the microbes of choice for initiating MLF, while most lactobacilli and pediococci should be considered undesirable because of their potentially adverse effects on flavour.

EC concentrations in wine may be linked to grape variety, and to how they are grown, particularly with respect to the use of nitrogenous fertilizers. It is also evident that certain vinification processes can elevate EC levels, notably high temperatures during maceration, and the non-removal of grape stems. In addition, wines that have undergone a MLF always have slightly higher EC levels than those that have not; a fact attributable to the MLB involved. In response to an FDA request, the wine industry in the USA has prepared an *Ethyl Carbamate Preventative Action Manual*,[64] which encourages lower EC levels in wines by recommending the use of modified agricultural and winemaking practices.

5.2.2 Bacteriophages

Bacteriophages (phage) are bacterial viruses which infect bacterial (host) cells to replicate and in the process destroy their host. Work carried out many years ago in the dairy industry indicated that bacteriophages may attack and destroy essential LAB, with the result that commercial milk fermentations have been ruined. This has resulted in deleterious economic consequences, and the results of an extensive appraisal were reported by Huggins in 1984.[65] A scan of the literature reveals that there has been little mention of a similar phenomenon in respect of wine LAB, particularly those responsible for MLF. As far as I can ascertain, the first report of the presence of bacteriophages in wine was penned by Sozzi *et al.* in 1976,[66] who were investigating Swiss wines, and this group were responsible for much of the early work in this field. Characteristic phage particles were observed in electron micrographs of concentrated wine, and it was shown later that these particles were lytic towards *L. oeonos*, and that this caused abnormalities in the MLF. Some difficulties in MLF of wines have been ascribed to high levels of phage, and malic acid degradation has been shown to terminate in both spontaneous and induced fermentations. What was then known about the problems of bacteriophages in wine was reported by Sozzi and Gnaegi in 1984.[67] The second group to conclusively identify phages in wine were led by Graham Fleet in Australia, who demonstrated the fact by a direct, cultural procedure.[68] The group showed that the phages encountered did not lyse all strains of *O*.

oeni, and that strain susceptibility was correlated, to some extent, with colony morphology on solid growth media. Phages survived in wines at pH values greater than 3.5, but were inactivated in wines of higher acidity, and by the addition of SO_2 and bentonite. Interestingly, a known sensitive strain of *O. oeni* to its specific phage was found to be resistant to that phage when both were cultured in filter-sterilised wine. Since the phage remained viable in the wine, it was speculated that its adsorption onto the surface of the bacterium was inhibited by some property in the wine, or that the bacterial cell wall did not produce the required phage receptor sites when grown in wine.

Malo-lactic bacteria are most susceptible to phage attack when they are in their growth phase, and one way to prevent infection is to inoculate large numbers into a wine after primary fermentation. This is because the phage particles attach themselves to the bacterial cell wall, and this is at its most 'sensitive' during active growth. Sanitation, and the use of phage-resistant starter strains also lower the likelihood of phage infection. The rotation of starter cultures is also a deterrent.

REFERENCES

1. M. Kleerebezum *et al.*, *Proc. Natl. Acad. Sci. U.S.A.*, 2003, **100**, 1990.
2. A. Bolotin, P. Wincker, S. Mauger, O. Jaillon, K. Malarme, J. Weissenbach, S.D. Ehrlich and A. Sorokin, *Genome Res.*, 2001, **11**, 731.
3. S. Orla-Jensen, *The Lactic Acid Bacteria*, Host & Son, Copenhagen, 1919.
4. M. Ingram, The lactic acid bacteria – A broad view, in J.G. Carr, C.V. Cutting and G.C. Whiting (eds), *Lactic Acid Bacteria in Beverages and Food*, Academic Press, London, 1975.
5. B.J.B. Wood and W.H. Holzapfel, *The Genera of Lactic Acid Bacteria*, Chapman & Hall, London, 1995.
6. L. Axelsson, Lactic acid bacteria: Classification and physiology, in S. Salminen, A. von Wright, A. Ouwehand (eds), *Lactic Acid Bacteria: Microbiological and Functional Aspects*, Marcel Dekker, New York, 2004.
7. M.E. Stiles and W.H. Holzapfel, *Int. J. Food Microbiol.*, 1997, **36**, 1.

8. M. Dworkin (ed), *The Prokaryotes An Evolving Electronic Resource for the Microbiological Community*, Springer-Verlag, Berlin, 2000 (http://link.springer.de/link/service/books/10125/).
9. O. Kandler, *Antonie van Leewenhoek*, 1983, **49**, 209.
10. P.A.M. Michels, J.P.J. Michels, J. Boonstra and W.N. Konings, *FEMS Microbiol. Lett.*, 1979, **5**, 357.
11. E.B. Olsen, J.B. Russell and T. Henick-Kling, *J. Bacteriol.*, 1991, **173**, 6199.
12. B. Poolman, D. Molenaar, E.J. Smid, T. Ubbink, T. Abee, P.P. Renault and W.N. Konings, *J. Bacteriol.*, 1991, **173**, 6030.
13. C.A. Lucey and S. Condon, *J. Gen. Microbiol.*, 1986, **132**, 1789.
14. T.M. Cogan, *J. Appl. Bacteriol.*, 1987, **63**, 551.
15. P.W. Postma, J.W. Lengeler and G.R. Jacobsen, *Microbiol. Rev.*, 1993, **57**, 543.
16. L.M.T. Dicks, F. Dellaglio and M.D. Collins, *Int. J. Systematic Bacteriol.*, 1995, **45**, 395.
17. M.A. Amerine and R.E. Kunkee, *Ann. Rev. Microbiol.*, 1968, **22**, 323.
18. A. Lonvaud-Funel, *Antonie Van Leeuwenhoek.*, 1999, **76**, 317.
19. A. Koch, *Weinbau Weinhandel*, 1900, **18**, 395.
20. H. Müller-Thurgau, *Ber. XII Deut. Weinbaucongr., Worms.*, 1891, 3–27.
21. W. Seifert, *Zeits. Landwirts, Versuch. Öst.*, 1901, **4**, 980.
22. H. Müller-Thurgau and A. Osterwalder, *Zentr. Bakteriol. Parasitenk. Abt. II.*, 1913, **36**, 129.
23. J. Halliday and H. Johnson, *The Art and Science of Wine*, Mitchell Beazley, London, 1992.
24. W. Möslinger, *Z. Untersuch. Nahrungs Genussmittel.*, 1901, **4**, 1120.
25. S. Korkes, A. Del Campillo and S. Ochoa, *J. Biol. Chem.*, 1950, **187**, 891.
26. F. Radler, The metabolism of organic acids by lactic acid bacteria, in J.G. Carr, C.V. Cutting and G.C. Whiting (eds), Lactic Acid Bacteria in Beverages and Food, Academic Press, London, 1975.
27. F. Radler, *Microbial Biochem.*, 1995, **42**, 884.
28. T. Henick-Kling, *J. Appl. Bact. Sym. Suppl.*, 1995, **79**, 29S.
29. J.C. Nielsen and M. Richelieu, *Appl. Environ. Microbiol.*, 1999, **65**, 740.

30. A. Bertrand, C. Smirou-Bonnamour and A. Lonvaud-Funel, Aroma compounds formed in malo-lactic fermentation, in L. Nykanen and P. Lehtonen (eds), *Flavour Research of Alcoholic Beverages*, Foundation for Biotechnical & Industrial Fermentation Research, Helsinki, 1984, 39–49.
31. R.B. Webb and J.L. Ingraham, *Am. J. Enol. Vitic.*, 1960, **11**, 59.
32. D. Wibowo, G.H. Fleet, T.H. Lee and R.E. Eschenbruch, *J. Appl. Bacteriol.*, 1988, **64**(5), 421.
33. T. Henick-Kling, Malo-lactic fermentation, in G.H. Fleet (ed), *Wine Microbiology & Biotechnology*, 1st edn, Harwood Academic Publishers, Chur, Switzerland, 1993.
34. H.J.J. Van Vuuren and L.M.T. Dicks, *Am. J. Enol. Vitic.*, 1993, **44**, 99.
35. D. Wibowo, R. Eschenbruch, C.R. Davis, G.H. Fleet and T.H. Lee, *Am. J. Enol. Vitic.*, 1985, **36**, 302.
36. A. Tchelistcheff, A.G. Peterson and M. van Gelderen, *Am. J. Enol. Vitic.*, 1971, **22**, 1.
37. C.R. Davis, D. Wibowo, R. Eschenbruch, T.H. Lee and G.H. Fleet, *Am. J. Enol. Vitic.*, 1985, **36**, 290.
38. A.W. Pound, *Aust. J. Exp. Bio. Med. Sci.*, 1967, **45**, 507.
39. S.S. Mirvish, *Adv. Cancer Res.*, 1968, **11**, 1.
40. B. Zimmerli and J. Schlatter, *Mut. Res.*, 1991, **259**, 325.
41. C.S. Ough, *J. Agric. Food Chem.*, 1976, **24**, 323.
42. C.S. Ough, *Bull. Soc. Med. Friends Wine*, 1993, **25**, 7.
43. D.F. Stevens and C.S. Ough, *Am. J. Enol. Vitic.*, 1993, **44**, 309.
44. B.J. Canas, F.L. Joe Jr., G.W. Diachenko and G. Burns, *J. AOAC Int.*, 1994, **77**, 1530.
45. P.A. Henschke and V. Jiranek, Yeast-metabolism of nitrogenous compounds, in G.H. Fleet (ed), *Wine Microbiology & Biotechnology*, 1st edn, Harwood Academic Publishers, Chur, Switzerland, 1993.
46. C.S. Ough, E.A. Crowell and B.R. Gutlove, *Am. J. Enol. Vitic.*, 1988, **39**, 234.
47. F.F. Monteiro, E.K. Trousdal and L.F. Bisson, *Am. J. Enol. Vitic.*, 1989, **40**, 1.
48. C.S. Ough, E.A. Crowell and L.A. Mooney, *Am. J. Enol. Vitic.*, 1988, **39**, 243.
49. Z. Huang and C.S. Ough, *Am. J. Enol. Vitic.*, 1991, **42**, 261.
50. W.R. Sponholz, Nitrogen compounds in grapes, must, and wine, in *Proceedings of the International Symposium on*

Nitrogen in Grapes and Wine, American Society for Enology and Viticulture, Seattle, WA, 1991, 67–77.

51. F.F. Monteiro and L.F. Bisson, *Am. J. Enol. Vitic.*, 1991, **42**, 199.
52. P. Lehtonen, *Am. J. Enol. Vitic.*, 1996, **47**, 127.
53. P.J. Large, *Yeast*, 1986, **2**, 1.
54. R. Sumrada and T.G. Cooper, *J. Biol. Chem.*, 1982, **257**, 9119.
55. T.G. Cooper, Transport in *Saccharomyces cerevisiae*, in J.N. Strathern, E.W. Jones and J.R. Broach (eds), *The Molecular Biology of the Yeast* Saccharomyces: *Metabolism and Gene Expression*, Cold Spring Harbor, New York, 1982, 399–462.
56. U. Kuensch, A. Temperli and K. Mayer, *Am. J. Enol. Vitic.*, 1974, **25**, 191.
57. W.R. Sponholz, *Biologia Oggi.*, 1992, **6**, 15.
58. S.-Q. Liu, G.G. Pritchard, M.J. Hardman and G.J. Pilone, *J. Appl. Bacteriol.*, 1996, **81**, 486.
59. S.-Q. Liu, G.G. Pritchard, M.J. Hardman and G.J. Pilone, *Am. J. Enol. Vitic.*, 1994, **45**, 235.
60. J.B. Russell and G.M. Cook, *Microbiol. Rev.*, 1995, **59**, 48.
61. R. Mira de Ordũna, S.-Q. Liu, M.L. Patchett and G.J. Pilone, *FEMS Microbiol. Lett.*, 2000, **183**, 31.
62. R. Mira de Ordũna, S.-Q. Liu, M.L. Patchett and G.J. Pilone, *J. Appl. Microbiol.*, 2000, **89**, 547.
63. S.-Q. Liu and G.J. Pilone, *J. Appl. Microbiol.*, 1998, **84**, 315.
64. C.E. Butzke and L.F. Bisson, *Ethyl Carbamate Preventative Action Manual*, U.C. Davis, CA, 1997.
65. A.R. Huggins, *Food Technol.*, 1984, **38**, 41.
66. T. Sozzi, R. Maret and J.M. Poulin, *Experientia*, 1976, **32**, 568.
67. T. Sozzi and N. Gnaegi, The problems of bacteriophages in wine, in T.H. Lee and T.C. Somers (eds), *Advances in Viticulture and Oenology for Economic Gain, Proceedings of the 5th Australian Wine Industry Technical Conference*, AWRI, Adelaide, 1984.
68. C. Davis, F. Neliane, A. Silveira and G.H. Fleet, *Appl. Environ. Microbiol.*, 1985, **50**, 872.

CHAPTER 6

Clarification, Stabilisation and Preservation

6.1 CLARIFICATION

Wine is one of the few consumer products that can improve with age, but it can also deteriorate drastically if it is not kept under appropriate conditions. A bottle, once purchased, should be kept unopened and stored with care until its point of use. Wine closures, traditionally the cork, are intended to provide a consistent, easily removable seal that prevents the wine from coming into harmful contact with oxygen. Wine should be stored with due consideration paid to light, temperature and humidity. If wine is exposed to strong sunlight, is kept at too high or too low a temperature or in conditions that encourage the cork to dry out, it will deteriorate. Having said this, it is a part of the job of the winemaker to endow his/her products with as much longevity as possible. To this end, there are a number of standard, often simple, procedures that can be undertaken. If a wine is stored long enough it will eventually 'stabilise', *i.e.* it will reach a stage at which it is unlikely to throw a precipitate or undergo any further chemical change. Unfortunately, the winemaker can rarely wait that long, and so more rapid stabilisation processes have evolved, some of which can lower wine quality if not carried out carefully (*i.e.* one can 'over-stabilise' a wine). Clarification, stabilisation and preservation protocols must be designed to give a wine sample a reasonable shelf-life, something like a minimum of 1 year in bottle. Some natural changes in a bottle of wine can lead to phenomena that are acceptable or even coveted, the classic example being the precipitated pigment or 'bottle crust' in an expensive red wine, which provides evidence of extended bottle ageing.

Before a bottle is opened, wine clarity is one of the most important consumer quality requirements. The visual assessment of a bottled wine often forms the most important part of the consumer's first contact with the product, and will enhance, or otherwise, the assessment of palatability. Turbidity is a major negative factor when assessing a wine, and results from an optical phenomenon called the Tyndall effect. Named after the great Irish scientist John Tyndall (1820–1893), whose main interest was the interaction of light with matter, the effect is caused by the presence of particles in suspension that deflect light from its normal path. A freshly prepared wine will contain two fractions that are likely to give rise to turbidity: (1) gross particles such as grape debris and yeast cells, which cause obvious opacity. These particles can be removed by filtration; (2) macromolecules such as proteins and tannins, which are not normally large enough to cause cloudiness or be removed by filtration. The latter are generally referred to as 'colloids' and can be divided into two groups: stable and unstable. Stable colloids are not normally a problem, remaining invisible in wine, but unstable forms have to be removed because they are likely to cause cloudiness after bottling. The main causes of colloidal instability in wines are proteins, pectins, gums, metallo-colloids and polyphenol degradation products.

Sound wines often settle 'bright' on their own accord if left standing, especially if they are in small containers, but this usually takes a considerable time. Clarification can be expedited by adding fining agents to wine, which helps to precipitate any particles held in suspension. Clarification can also be achieved by filtration, an entirely different process. Fining generally facilitates more than the removal of particulate matter since the agents used invariably help to stabilise the wine as well. Filtration serves, first and foremost, to clarify, not to stabilise. The object of stabilisation is to ensure long-term clarity and prevent deposits forming under whatever conditions the wine is stored. Wines may be clarified in the short term by eliminating suspended particles prior to bottling, but the clarity need not necessarily be permanent because of the many naturally occurring, turbidity-forming reactions that may occur during this time in the bottle. The chemical and biological mechanisms likely to cause turbidity and deposits are now pretty well known, and most may be predicted by laboratory tests. As a result, efficient treatments are now available for stabilising wines before they are bottled. The mechanisms of fining can be quite complex, and the oft-quoted (simplest) explanation is the one involving electrostatic attraction, whereby the fining agent carrying a particular charge reacts with a wine constituent bearing the opposite charge, the resultant combination being sufficiently bulky to precipitate out. This is part of the story, but not all of it because chemical bond formation and/or absorption and adsorption also occur. What is universal is that fining is a surface phenomenon, and so hydration becomes important. Most fining reactions are empirical, with the reaction between fining agent and wine constituent being logarithmic rather than linear. Because of this and other considerations, each fining operation should be preceded by a laboratory assessment to establish the fining rate. It should be remembered that many fining agents can exert an adsorptive effect on some flavouring compounds in wine, especially if used in excess. Also, a proteinaceous substance such as gelatin will induce a protein haze in wine if used over-enthusiastically. In the latter instance, the situation then has to be addressed by counter-fining. Most winemakers would agree that fining should only be practised when really necessary and only at the minimum effective rate of addition.

The ancient practice of clarifying wine has employed a wide variety of agents throughout the ages, some of which may raise a few eyebrows in our modern, over-protective society. Bull's blood, 'fish-guts' and white of egg seem to be three of the more innocuous examples. From the limited evidence available, it seems as though powdered, fired clay was used to treat wine in ancient times. The main aims of such additions have always been to prevent haze and to remove some of the tannins to improve wine balance, but there are several other aims as well.

The first theoretical approach to fining wine was carried out in the 1920s and reported upon by Rüdinger and Mayr,[1-3] who showed, by means of electrophoresis, that gelatin particles were positively charged at the pH of wine, and that the particles responsible for wine turbidity were negatively charged. It was deemed, therefore, that the fining action was the result of an interaction between these opposite charges. Subsequent work showed that many fining mechanisms are more complicated than

Table 1 *Summary of methods available for clarifying and stabilising wines (not all of these agencies are permissible in all countries (e.g. use of potassium ferrocyanide is forbidden in the US))*

Oenological intention	Agencies available
Clarification	Sedimentation by racking, fining (e.g. proteins, earths), filtration, centrifugation
Biological stabilization	SO_2, sorbic acid, fatty acids, heating
Prevention of oxidation	SO_2, ascorbic acid, blanketing with inert gas
Prevention of KHT precipitation	Cold stabilisation, electrodialysis, ion-exchange, metatartaric acid, mannoproteins
Prevention of protein hazes (white wines)	Bentonite, tannins, cold stabilisation, heating
Prevention of turbidity due to colouring matter (red wines)	Cold stabilisation, bentonite, gum arabic
Prevention of metallic casse	
(1) Copper	Bentonite, gum arabic, potassium ferrocyanide, heating
(2) Iron	Citric acid, gum arabic, ascorbic acid, potassium ferrocyanide, calcium phytate
Improving colour and aroma	Charcoal, casein

this and can be divided into two stages: (1) flocculation-produced interactions between tannins and proteins and (2) the precipitation of suspended matter, which effects clarification.

A number of treatments are available for the clarification and stabilisation of wine, and these together with the agents for treatment are illustrated in Table 1.

Fining agents may be classified into eight groups according to their composition:

   (i) Proteins such as gelatin, isinglass, casein, albumen and yeast proteins
  (ii) Earths such as bentonite and kaolin
 (iii) Polysaccharides, most notably alginates and gum arabic
 (iv) Carbons
  (v) Synthetic polymers such as nylon and polyvinyl poly-pyrrolidone (PVPP)
 (vi) Silicon dioxide gel
(vii) Tannins
(viii) Miscellaneous, e.g. metal chelators, enzymes and blue finings

Each fining agent has its own unique properties and will be selected for making a specific improvement to juice and/or wine quality.

6.1.1 Proteins

The use of egg white to clarify wine goes back to thousands of years, and the Romans noticed that wine and beer stored in containers made of animal stomachs were much more translucent than samples stored in other forms of container. Thus, the use of natural products to enhance wine quality has a long and honourable history. Some products such as gelatin and albumen are still widely used, but others such as bovine blood have been prohibited in many countries.

All proteins have a pH at which they carry no net charge when they are in solution, and this is the isoelectric point (p*I*). Although most proteins are least soluble at their p*I*, certain examples such as gelatin and ovalbumin have significant amounts still in solution when they attain their p*I*. Conversely, other proteins such as casein precipitate because they have no residual solubility when brought to their p*I*, even in an acidic medium such as wine. The significance of this is that gelatin and ovalbumin will form stable solutions when their charges are neutralised, whereas casein will flocculate. When at their p*I*, proteins will not migrate in an electric field because their molecules carry an equal number of positive and negative charges.

Proteinaceous fining agents are principally used to adjust the polyphenol content of wines, and the mechanism of interaction is by hydrogen bonding between the phenolic hydroxyl and the carbonyl oxygen of the peptide bond. Since the hydrogen bonds formed are weak, the efficacy of a protein-fining agent is partially a function of the number of potential bonding sites, and the combination of phenolic and fining agent, which provides the strongest total hydrogen bonding, usually occurs preferentially. This means that the larger polyphenols with more available hydroxyl groups will react first. The weakest bonds will be formed with monomeric phenols. Since the larger, polymeric phenols are associated with astringency, and the monomeric forms are mainly associated with bitterness, the selective removal of the former by proteinaceous fining agents can lead to undue bitterness in a wine. An extended treatment of the theory of fining wine by proteins has been provided by Ribéreau-Gayon *et al.*[4]

6.1.1.1 Gelatin. Gelatin has been used since Roman times to clarify wine, but with alternative products now available, it is mostly used to reduce the harshness (astringency) of red wines. It is also used to improve juice clarity prior to fermentation. Gelatin is produced by the hydrolysis of the multi-stranded polypeptide collagen, the main structural protein of skin and bones and the most common protein in the animal kingdom. On acidic, alkaline or enzymic hydrolysis, the polypeptide is unravelled and broken down to yield an interesting combination of amino acids. The amino acid composition of gelatin is variable, particularly for the minor constituents, and depends on the source of the raw material and process used to isolate it; approximate values by weight are as follows: glycine 21%, proline 12%, hydroxyproline 12%, glutamic acid 10%, alanine 9%, arginine 8%, aspartic acid 6%, lysine 4%, serine 4%, leucine 3%, valine 2%, phenylalanine 2%, threonine 2%, isoleucine 1%, hydroxylysine 1%, methionine and histidine <1% with tyrosine <0.5%. The proportion of glycine, proline and hydroxyproline is higher than that found in most other proteins. Although a useful protein, gelatin is not a 'complete' one in terms of mammalian nutrition because it lacks essential sulfur-containing amino acids and tryptophan. The p*I* of gelatin used in winemaking is pH 4.7, well above the normal pH of grape juice or wine, which means that it will carry a positive charge in such media and will bind with negatively charged molecules such as phenolics. Gelatin is classified as a food ingredient rather than an additive and it is generally regarded as safe (GRAS).

Industrial preparation of gelatin dates from the early 18th century, and several different preparations are available according to how the material is prepared. Only certain grades of gelatin are suitable for winemaking, and manufacturers are usually required to make special oenological preparations, which have a neutral odour and little colour. The size of the peptide chain varies, but gelatin used for oenological purposes should have a molecular weight between 15 and 140 kDa. It should be remembered that the peptide bond has considerable aromatic character, and hence gelatin shows an absorption maximum at *ca.* 230 nm. Gelatin is an amphoteric protein with an isoelectric point varying between 5 and 9 depending on raw material and method of manufacture.

There are two main types of gelatin, Type A and Type B, and the product is sold with a wide range of special properties, like gel strength, to suit particular applications. Industrial gelatins are classified according to their purity and by their gelling power, which is expressed in 'Bloom' units. The Bloom number (Bloom strength) indicates the ability of a gelatin sample to absorb water, which can vary from six to eight times its weight. The higher the Bloom number, the greater is the binding capacity of the gelatin. Bloom strength is the force in grams required to press a 12.5-mm diameter plunger 4 mm into 112 g of a standard 6.66% w/v gelatin gel at 10°C. Several penetrometer-type instruments have been adapted to determine Bloom strength. Gelatin recommended for fining wine should have values of 80–150 Bloom. Some workers maintain that pI is a more important criterion than Bloom. Type A gelatins are usually derived from acid pre-treatment of pigskin and have pIs between 6 and 9, with the high gel strength (Bloom strength) gelatins having the higher pI and the low Bloom strength gelatins having a pI closer to 6. Type B gelatins are the result of an alkaline pre-treatment (usually lime) of collagen and have isoelectric points ranging from 4.8 to 5.2. The significance of pI is, of course, that the higher the pI, the greater is the cationic charge on the molecule in a beverage of, say, pH 3.6. In other words, at pH 3.6, all gelatins would be positively charged but the charge density would be far higher for high-pI gelatins. A vegetable-derived 'version' of gelatin is now available for those who want nothing to do with animal products.

The commonly available form of gelatin is as a powder, but it can also be purchased in sheet and liquid forms. The latter are usually prepared by hydrolysis, which lowers the molecular weight and prevents gelling at higher concentrations. Such preparations must be stabilised with SO_2 and/or benzoates. To prepare a solution ready for fining wine (from powder), the material is mixed in cold water, warmed to dissolve and added to the wine with thorough mixing. The fining solution should always be freshly prepared before use (*i.e.* do not store). Addition levels vary according to the purpose. Larger doses are required for fining juices, where heavy-press juice may need up to 48 g hL^{-1} to reduce colour and astringency. Addition to red wines is normally in the range

4.8–10.0 g hL^{-1}, and for white wines 0.75 g L^{-1} normally suffices. The quantity of gelatine that needs to be added to a wine in order to achieve clarity may reduce the astringency of that wine to unacceptably low levels, especially if it is white. To counteract this, it is usual to add small quantities of tannin or silica gel before the addition of gelatin. This is especially the case with some French white wines, where gelatin is added in conjunction with silica gel or siliceous earth. In a kind of role reversal, gelatin is often added as a counter-fining agent to wines that have been stabilised with bentonite and have a residual haze.

6.1.1.2 Isinglass (Ichthyocolla: 'Fish Glue'). Like gelatin, this ancient fining material consists mainly of collagen, but here the traditional source is the swim bladder of fish of the sturgeon family, and its use in winemaking followed in the wake of its successful performance in breweries. The name 'isinglass' comes from the German, *hausenblase*, 'living blister', a reference to the swim bladder. Another ancient name for the material is *Colla piscium*, 'fish glue'. One of its main drawbacks is that its preparation from the raw material is protracted and laborious, involving, as it does, air-drying, cleaning, bleaching, shredding and acid treatment (cutting). The material then has to be made up into a 'solution' before it can be added to wine. Once in liquid form, isinglass begins to denature even at relatively low temperatures (*ca.* 10°C), and so it should be stored in a refrigerator. Isinglass has a molecular weight of *ca.* 140 kDa and a pI of 5.5, so that it will carry a net positive charge in wine. Denaturation results in a reduction in molecular weight and a concomitant weakening of fining ability. On slow hydrolysis in solution, it liberates gelatin (and behaves like gelatin). Isinglass is regarded by many as the best of the protein finings because it (1) can be used in white wines without removing colour; (2) produces small amounts of lees; (3) removes less-condensed tannins, but more leucotannins than, say, gelatin or casein; (4) gives a 'brilliant' and 'softer' wine; (5) is effective at low fining rates (usually *ca.* 0.02–0.1 g L^{-1}), with red wines needing a slightly higher pitching rate (*ca.* 0.5 g L^{-1}); (6) will capture some non-particulate material in wine by entrapping them in the main phenol–protein complex; and (7) does not require extensive counter-fining measures. In short, isinglass is more efficient than other protein finings, and it has a much more 'gentle' action. One disadvantage is that, if care is not taken, denatured fining samples can impart a 'fishy' odour to wine. In addition, the small volume of lees formed can be rather non-compact and 'fluffy', making the deposit difficult to handle. This can be counteracted by using bentonite.

Isinglass is principally used in white wines to accentuate or unmask the fruit character without significantly altering tannin levels. Because it is not aggressively antagonistic towards condensed tannins, it has no untoward effects on the 'body' of a wine. Isinglass is used as a riddling agent in *méthode champenoise*. The BATF and OIV impose no limit on the use of isinglass. The problems associated with obtaining a sufficient supply of sturgeon swim bladders have meant that more prosaic sources are now often used, such as the waste from fish canneries.

6.1.1.3 Egg White (Albumen).   This is one of the oldest of fining agents, and is still an important material in the preparation of some of the better red table wines, where it is used to remove the harsher tannins thus reducing astringency. A few of the great houses still use nothing else. Albumen is recommended for softening wines with a high tannin content and excessive astringency. It has to be used with care because over-enthusiastic use can render wines unnecessarily thin. Albumen is not recommended for use with white table wines, even though it removes less 'fruit character' than something like gelatin. Albumen is colloidal in nature and has a positively charged surface that attracts negatively charged tannins. Neither BATF nor OIV regulate the quantity of egg white used for fining wine.

The main active constituent in egg white, the cytoplasm of the egg, is water-soluble egg albumin, which consists of several proteins and represents some 12.5% of the weight of a fresh egg. The predominant components are the serpins, ovalbumin (54%) – with a molecular weight of 45 kDa – and conalbumin (13%), together with ovomucoid (11%) and lysozyme (3%). There are also two globular proteins (globulins G2 and G3), which comprise some 8% of the matrix, and a few other proteins are present in smaller amounts. Albumins are soluble in water, whereas globulins are soluble in neutral, dilute salt solutions, and it was a common practice to add small amounts of common salt in order to aid solubility of egg white. In the USA, sodium chloride additions are prohibited by BATF, so potassium chloride is used instead. In fact,

BATF suggests the following protocol for the use of albumen finings: dissolve 2 lb (907 g) egg albumen and 1 oz (28.35 g) potassium chloride in 1 gallon (3.8 L) water and use at a rate not exceeding 0.25 L hL^{-1}. As a rule of thumb, 1 g albumin precipitates 2 g tannin. When fresh egg white is used, the recommended rate of addition is 3–8 whites to a barrique (225 L) of wine. Again, as a rule of thumb, it is maintained that one egg white corresponds to 4 g of dry protein. When egg whites are added, it is essential to avoid the frothing that results from over-mixing. Egg albumen is denatured by heat, so avoid warming to aid its dissolution in wine.

Egg albumen is available in solid form prepared by desiccating fresh egg white with a little sodium carbonate added to aid solubility. In this powdered form, some of the higher molecular weight proteins are lost during processing, which slightly reduces fining efficacy.

6.1.1.4 Casein. Casein is a heterogeneous phosphoprotein, which is present in milk in calcium form. It can be obtained by coagulating milk. The term 'casein' represents the principal (*ca.* 80%) group of proteins found in fresh milk and consists of four main fractions, α(s1)-, α(s2)-, β- and κ-caseins, all of which have low solubility at pH 4.6. The conformation of caseins is much like that of denatured globular proteins. The high number of proline residues in caseins causes particular bending of the protein chain and inhibits the formation of close-packed, ordered secondary structures. There are also no disulfide bridges, and as a result, they have little secondary or tertiary structure, and because of this they do not easily denature, so that they are not coagulated by heat. Within the group of caseins, there are several distinguishing features based on the number of proline residues present, their charge distribution and their sensitivity to calcium precipitation:

(i) α(s1)-casein: (molecular weight 23 kDa; 199 residues, 17 proline residues). Two hydrophobic regions containing all the proline residues separated by a polar region, which contains all but one of eight phosphate groups. It can be precipitated at very low levels of calcium.

(ii) α(s2)-casein: (molecular weight 25 kDa; 207 residues, 10 proline residues). Concentrated negative charges near N-terminus and positive charges near C-terminus. It can also be precipitated at very low levels of calcium.

(iii) β-casein: (molecular weight 24 kDa; 209 residues, 35 proline residues). A highly charged N-terminal region and a hydrophobic C-terminal region. Very amphiphilic protein that acts like a detergent molecule. Self-association is temperature dependent and will form a large polymer at 20°C but not at 4°C. Less sensitive to calcium precipitation.

(iv) κ-casein: (molecular weight 19 kDa; 169 residues, 20 proline residues). Very resistant to calcium precipitation, stabilising other caseins.

Casein is found in milk as a suspension of particles called micelles, which are held together by calcium ions and hydrophobic interactions. In fact, as it exists in milk, it is effectively a calcium salt. Without a tertiary structure, there is considerable exposure of hydrophobic residues, rendering the molecule relatively hydrophobic and thus poorly soluble in water. It is available as either a purified 'milk casein' that is soluble in alkaline solution, or a sodium or potassium caseinate that is water-soluble. Caseinates will dissolve in wine with some difficulty, but it will only dissolve at a pH above 8.0, and will require to be made alkaline before being dissolved. Even caseinates are often added in conjunction with potassium carbonate, which makes them more soluble. At wine pH casein flocculates quite naturally, and the resultant precipitate adsorbs and mechanically removes any suspended material as it does so.

Casein is most useful for decolourising white table wines and sherries, and the decolourising power of a product is directly related to its 'formol titration' value, which is a measure of the free amino groups present. It is often used as a replacement for decolourising charcoal (sometimes called 'vegetable carbon') to modify the colour of juice and to remove the 'cooked' flavour from sherries. It is not as effective as carbon, but does not catalyse oxidative degenerative reactions. It has also been shown that casein reduces the levels of copper (by up to 45%) and iron (by *ca.* 60%) in wines The protein can also be used in such wines to remove phenolic bitterness, off-flavours and odours (especially 'reduced' odours).

Several types of 'milk fining' materials are available to the oenologist: skimmed milk, whole milk, lactic casein and caseinates, with lactic casein being the material of choice. Unfortunately, it is the most expensive form of casein, especially since it is needed in a purified form. Casein is the only proteinaceous wine fining that can be used in high doses without the risk of over-fining and concomitant protein haze. When fining with milk or casein, it is essential to ensure that wine and fining are thoroughly mixed (an injection pump is often used). The fining material should be added slowly to prevent pre-mature coagulation, which, if it occurs, results in gross material either floating to the surface or precipitating rapidly; either way, fining is ineffectual. Addition levels to wine range from 1.25 to 24.0 g hL^{-1}, depending on wine type and purpose.

Fining with whole milk is not permitted in the EU, but is allowed in some winemaking countries. The effectiveness of whole milk lies partially in its fat content. Skimming milk reduces its adsorption capacity, but intensifies its clarifying capabilities. It is calculated that one litre of cow's milk contains approximately 30 g casein and 10–15 g of other proteins, and it is the latter that are likely to provide any risk of over-fining, if too high a dose is used (*i.e.* >0.2–0.4 L hL^{-1}). BATF authorises the addition of pasteurised milk to white table wines and sherries. One well-known proprietory brand of casein finings is 'Casesol', which is prepared from food-grade casein and is 'totally and immediately soluble' in wine; thanks to 'a basic additive'. In addition to the standard claims, Casesol will prevent the 'progressive development of 'Madeira' taste', 'replenish colour' and 'revive old wines'.

6.1.2 Polyvinyl Polypyrrolidone

This is unlike any other fining agent for wines, inasmuch as it is a synthetic, high-molecular-weight and insoluble plastic that has been milled into small particles. It was the first synthetic fining material to be used in winemaking. PVPP was developed in 1961, and its manufacturing involves an alkaline polymerisation reaction. The important feature of this polyamide is that it has available carbonyl oxygen atoms on its surface. It has been called a 'protein-like' fining agent, but unlike the soluble protein fining agents, PVPP contacts relatively few reactive groups. It has a high affinity for low-molecular-weight phenolics, which conform to the PVPP molecule, and its main use in winemaking, therefore, is to bind with and remove the smaller, phenolic species such as the monomeric catechins and leucoanthocyanins, which are precursors of 'browning' and 'pinking' oxidation reactions in white wines and browning and bitterness in red wines. There is also much affinity for dimeric phenolics. PVPP is based on 2-pyrrolidone, a five-membered ring system containing a nitrogen atom subtended by a vinyl side chain. The polymer consists of many of these units linked *via* their side chains. The mode of action of PVPP involves the formation of hydrogen bonds between the carbonyl group on a polyamide residue and the hydrogen atom of the phenol. Compare this to the proteinaceous finings that react with polyphenols by interacting with their numerous hydroxyl groups. It is most beneficial as a post-fermentation treatment, but can be added to juice and then removed by settling prior to fermentation. PVPP removes more anthocyanins than gelatin, and, therefore, has to be used with care in red wines, or else a wine can be stripped of its complexity. Its main benefit in reds wines is to soften exces sively tannic samples. PVPP has an advantage over other stabilis-ing materials in that it can be used in conjunction with other processing agents, such as enzymes and gels. Either on its own, or combined with casein, PVPP is used to prevent 'maderisation' by eliminating tannins, oxidisible hydroxycinnamic acids and quin-ones formed when they oxidise. When used to reduce the colour in white wines, PVPP is often used in conjunction with activated carbon.

In the laboratory, PVPP may be used to differentiate the free anthocyanin concentration (as opposed to bound to tannins) in wines. A wine sample is adsorbed onto a PVPP column, rinsed with water and eluted with a dilute solution of ethanol. This releases free anthocyanins, while the combined ones remain on the column. The eluate is evaporated and brought up to the initial volume, analysed and assayed using SO_2 to obtain free anthocyanin concentration. This value is compared with a SO_2-bleached intact sample.

The normal method of use is to slowly add the appropriate amount of PVPP powder (as determined by trial) to a stirred wine and leave for about 1 h. After all turbulence has ceased, the PVPP will settle out and the wine can be racked off. The material can be incorporated into sheet form, which enables the wine to be treated continuously without having to add the fining and then the residue can be collected. Doses used in white wines to prevent browning are of the order of 20–30 g hL^{-1}, and at these levels, there are no negative organoleptic effects. Levels of usage for other purposes vary from 12 to 72 g hL^{-1}.

Adsorbed phenolics can easily be removed from PVPP by treatment with weak, warm caustic solution, thus re-generating the material that is inherently expensive. This practice is widely used in Europe, but not so much in the USA. A widely used commercial preparation is Polyclar®, which is insoluble in water, alcohol, acid and alkali, and is removed from wine after fining to allow the product an additive-free 'clean label'. BATF insists that wines that have been treated with PVPP must be filtered free of the agent before bottling. The material should be regarded as a processing aid rather than an additive.

6.1.3   Bentonite

Bentonite is a natural, clay-like material of volcanic origin, which is geologically a form of the smectite clay montmorillonite, a complex, hydrated sodium, calcium, aluminium silicate with a 2:1 expanding crystal lattice structure and a general composition conforming to $Mg \cdot Ca \cdot Na \cdot Al_2O_3 \cdot 5SiO_2 \cdot nH_2O$. A 'typical' empirical analysis might well be as follows: K_2O 1.85%, Na_2O 2.10%, CaO 3.05%, MgO 5.33%, Al_2O_3 18.00%, SiO_2 59.30%, Fe_2O_3 2.17%, TiO_2 0.23%, CO_2 1.18% and H_2O 6.79% with a molecular weight estimated at 504. Montmorillonite, a term first used in 1847, was named after the French town Montmorillon, near Poitiers, where it was first mined. Its origin can be traced to ancient volcanic eruptions whereby fine volcanic ash particles were carried by winds and deposited in discrete layers, which metamorphosed over time from the 'glassy' state to claystone. The term 'bentonite' was first applied in 1898 to a particularly highly colloidal clay found in the Cretaceous beds near Fort Benton, Montana, USA, and the product is still mined in this state. Some of the largest and highest quality deposits of bentonite occur in neighbouring Wyoming, from which the product takes on the name 'Wyoming clay'.

Bentonites are usually composed of about 90% montmorillonite with the residue consisting of feldspar, gypsum, calcium carbonate, quartz and traces of heavy metals, and it is these metallic impurities that impart any colour to the mineral. In the pure state montmorillonite is almost white. The source of bentonite has a major influence on its properties, the main differences being in the proportions of magnesium, calcium and sodium in the lattice. Not all bentonites are suitable for treating wines; some are too coarse and impart off-flavours, while others have insufficient adsorption capacity. Bentonites from Europe are predominantly 'calcium' forms, while those from the USA are mostly 'sodium' forms. When dispersed in water, bentonite exists as a series of minute, flat, rectangular plates or sheets, *ca.* 500-nm wide by 1-nm thick, and these are arranged in typical 'house of cards', fashion. The plates consist of layers of atoms in a defined order, and their composition, particularly with respect to silicon, aluminium and magnesium, differs with bentonite type. Each plate consists of two rows of tetrahedra chained together. These tetrahedra have oxygen atoms at their nodes and a silicon atom in the centre. Between these two rows, there is a series of octahedra linked together by oxygen atoms or hydroxyl radicals, which are primarily Al_2O_3 or $Al_2(OH)_6$. Some octahedra contain magnesium. The difference in charges between the central layer and the two rows of silica tetrahedra creates a negative charge on the surfaces between the plates. This keeps the plates apart and creates a space that varies according to the origin of the clay. The net negative charge attracts exchangeable cations such as sodium, calcium or magnesium, which form a layer around each plate. These cations are not part of the permanent clay structure, but they confer an ion-exchange capacity to the mineral. When bentonite is soaked in hot water, the plates separate to form a homogeneous colloidal suspension, and in this dispersed state, an enormous surface area for adsorption is formed in the case of sodium bentonite, something in the order of 750 $m^2 g^{-1}$ of powder, which is about 50 times that of kaolin and some 5000 times that of silica flour. Plates of calcium bentonite have more of a tendency to clump, with a subsequent reduction in surface area. In addition, this material has a greater propensity for settling out of suspension, thus further reducing fining potential. As Rodriguez *et al.*[5] found, the calcium form has a basic spacing between plates in the order of 10 Å and adsorbs proteins primarily on external surfaces, while the sodium form expands to give spacings of *ca.* 100 Å and protein adsorption occurs over the entire surface. Thus, sodium bentonites swell more and have a higher protein adsorbing capacity, roughly twice that of calcium forms, but either form will exhibit maximum adsorption capacity after being soaked in water for 48 h. Even though it is less efficient, calcium bentonite is sometimes preferred to the sodium form because it settles more rapidly and produces a less voluminous deposit. The calcium form is more widely used in

Europe, especially as a riddling aid in the production of champagne (see page 179). In general, European law is more concerned about sodium levels in wine and this is another reason for favouring the calcium form. New World winemakers generally favour sodium bentonites.

When a 5% w/v suspension of bentonite is made up with warm water, a pale, green–grey milky liquid results, which, if the concentration of bentonite reaches 10–15% w/v, forms a gelatinous paste. These thicker slurries will form a gel when left to stand but will liquefy upon being agitated; *i.e.* they are thixotropic. This is due to the presence of positive charges on the edges of the plates, which are electrostatically drawn to the negatively charged faces of adjacent plates and thus form a contact at an angle. This means that although bentonite primarily acts by an adsorptive interaction between the flat, negatively charged platelet surfaces and positively charged proteins, a certain amount of binding with negatively charged wine constituents may occur. This cation exchange capacity is very limited, which means that it is difficult for bentonite to fine negatively charged and neutral wine proteins, and, in addition, the clay is essentially inert as far as wine phenolics are concerned (except for cationic anthocyanins). There is little effect of temperature on the fining activity of bentonite, but its adsorption capacity is almost three times higher at pH 3.0 than it is at 4.5.

Bentonite is now widely used for the adsorption of proteinaceous material from wines but was originally employed for clarifying vinegar, as reported by Saywell in 1934,[6] who thought that the mode of action was to remove the iron involved in metal–protein hazes, rather than the removal of protein fractions *per se*. Its use as a means of conferring heat stability to white wines was soon reported by the same author, and its use in Australia was reported soon after. Bentonite was not used in Europe until after World War II, when German producers used it in an attempt to prevent white wine heat hazes brought about by the hot summer of 1947. Because of wide variations in the chemistry of wine proteins, the differences in bentonite structures and the different protocols adopted by winemakers for mixing and adding the fining agent, there are few universal recommendations for achieving protein stability with bentonite. It is essential, therefore, that each individual case be assessed by fining trials and the effective dosage then discerned. This should always be the lowest level possible since it is known that addition of bentonite to wine increases its aluminium content.[7] From this Australian study of juices and wines at various stages of winemaking, it was observed that the addition of bentonite was a major source of aluminum contamination with increases of 100% common after bentonite treatment. Pick-up of iron and some heavy metals has also been suggested as a consequence of bentonite use. The adsorption of protein by bentonite in a model wine solution was studied in depth by Blade and Boulton.[8]

There are disadvantages in using bentonite, the most obvious one being the copious quantity of lees sediment produced, which can result in significant wine losses when it is removed. Calcium bentonites produce less, more compact lees than their sodium counterparts. Bentonites can also be responsible for stripping some of the 'fruit' aromas and flavours from a white wine.

A number of other clays that have the silica–alumina matrix with exchangeable cations have been used as alternatives to bentonite in wine clarification including kaolin and Spanish earth, but they generally have a lower adsorption capacity. Kaolin is also a silicaceous clay, which is refined from the mineral kaolinite. It is more compact than bentonite, and has less than 10% of the hydrative and adsorptive properties of that mineral. The surface area provided by kaolin on hydration and swelling is only 20–40 m^2 g^{-1} of material, which makes it far less effective for fining. At least a ten-fold increase in kaolin addition is required in order to achieve fining comparable with sodium bentonite. While the use of kaolin is permitted by both BATF and OIV, it is now not used very frequently.

6.2   TARTARIC ACID, TARTRATES AND WINE STABILITY

Despite the many recent advances in our knowledge of wine chemistry and the implementation of modern technology, the major non-biological instability in bottle wines is still caused by the precipitation of the salts of tartaric acid (2,3-dihydroxy-butan-dioic acid). Tartrate crystals in a bottle of wine, variously described by the intended consumer as 'broken glass' or 'sugar crystals', are probably the biggest cause of complaint in the wine trade, and it is unfortunate that these crystals can still be produced after the best intentioned efforts of the winemaker. It is interesting to note that, as far as I am aware, no supplier is prepared to give a warranty that

Table 2 *Water solubility (at 20°C) of L-tartaric acid and its main salts present in wine*

Substance	Formula	Solubility $(g\ L^{-1})$
L(+)-tartaric acid	$C_4H_6O_6$	4.90
Potassium bitartrate	$KHC_4H_4O_6$	5.70
Calcium tartrate	$CaC_4H_4O_6 \cdot 4H_2O$	0.53

his/her products will never throw a deposit of tartrate crystals. In reality, these crystals are entirely natural and harmless and do not affect the drinking quality of the wine in any way, but the average consumer does not want to see them. As we shall see, when wine is chilled the solubility limit of some tartrate salts is exceeded and they come out of solution; this is often the root cause of the problem. The earlier a wine is bottled, the greater is the likelihood of a crystal deposit being thrown.

Tartaric acid (H_2T) is, of course, grape-derived and, in the presence of K^+ and Ca^{2+} and at wine pH, can exist in the following possible forms: potassium bitartrate (KHT), potassium tartrate (K_2T), calcium tartrate (CaT), potassium calcium tartrate and calcium tartromalate. The first three represent the major species, while the latter two tartrates are formed and remain stable at a pH in excess of 4.5. Calcium tartromalate is relatively insoluble and crystallises out, while potassium calcium tartrate is highly soluble. The solubilities of tartaric acid and its two main salts in water at 20°C are shown in Table 2.

In unripe grapes, tartaric acid can be present in levels up to 15 g L^{-1}, while ripe grapes from warmer climes would normally be expected to have the acid at levels of 2–3 g L^{-1}. Ripe grapes from cooler (more northerly) vineyards often have tartaric acid levels in excess of 6 g L^{-1}. The tartaric acid composition of grape musts ranges from 2.0 to 10 g L^{-1}, and varies according to growing region (climate, soil), cultivar, stage of grape maturity and viticultural practice. Naturally occurring tartrates in unfermented must/juice are completely soluble because they are present in levels that are below the concentrations at which their solutions become saturated. Some of the KHT present in must and juice precipitates out during fermentation as a deposit known as 'argols', a crude, pigmented form of tartar. Both during and after fermentation, these salts become less soluble because of the presence of ethanol,

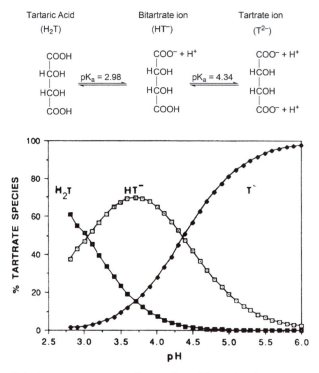

Figure 1 *Relative concentration of tartaric acid species in aqueous solution at different pH values (after Zoecklein et al.[29])*
(Reproduced by kind permission of Springer Science and Business Media)

and their solubilities now exceed their solubility limits and so they become supersaturated.

In grapes and wines, tartaric acid is found in its ionised forms, bitartrate (HT^-) and tartrate (T^{2-}), and depending on pH, the ratios of H_2T:HT^-:T^{2-} can vary greatly, and thus significantly influence the potential for the precipitation of insoluble salts. The relative distribution of each of these three species as a function of pH is given in Figure 1.

Like tartaric acid itself, potassium levels in grape juices display a wide variation, whereas Ca^{2+} levels remain relatively constant. The levels of these three entities in any particular juice depend primarily on the cultivar, the growing season and the level of maturity. The tartaric acid concentration in a given cultivar is essentially independent of temperature during the growing season and seems to be

Table 3 *Solubility of potassium bitartrate (g L^{-1}) in model solutions (after Berg and Keefer[9])*

Temperature (°C)	Ethanol content (% v/v)				
	0	10	12	14	20
0	2.25	1.26	1.11	0.98	0.68
5	2.66	1.58	1.49	1.24	0.86
10	3.42	2.02	1.81	1.63	1.10
15	4.17	2.45	2.25	2.03	1.51
20	4.92	3.08	2.77	2.51	1.82

determined only by its biosynthesis and the degree of berry expansion during maturation (Phase III, see page 87). The potassium content of berries is more dependent on growing conditions (*e.g.* soil), and unless something like ion exchange has occurred during processing (in which case they will be reduced), the potassium content of a wine will reflect that of its grape/must. Calcium levels, on the other hand, are often increased during processing owing to 'pickup' during various treatments (such as 'liming') or if concrete tanks are used for fermentation and/or storage.

As Figure 1 shows, the bitartrate ion (HT^-) is at its highest concentration at pH 3.7, and at this point maximum precipitation of KHT will normally occur. Precipitation depends on (1) the concentration of the salt and any other components that may be involved; (2) the presence of nuclei to initiate crystallisation and (3) the presence or absence of complexing factors (usually grape-derived macromolecules) that might impede crystal growth. As a rule, an adequate level of KHT supersaturation[*] is necessary before nucleation can occur. Once nucleation has occurred, further crystal growth will ensue until a point is reached whereby KHT is precipitated. During alcoholic fermentation, KHT becomes increasingly insoluble (see Table 3), and this promotes supersaturation.

The solubility of potassium bitartrate in model wine solutions of different alcoholic strength is shown in Table 3.

The solubility of KHT shows a 60% decrease as the temperature is decreased from 20°C to 0°C in 12% ethanolic solution, and the overall effect of ethanol is to reduce the solubility by almost 40% for each increase of 10% v/v at 20°C. Compare this with the

---

[*]Supersaturation is the condition in which a solution has a concentration above its solubility limit.

Table 4   *Solubility of calcium tartrate (g L^{-1}) in model solutions (after Berg and Keefer[10])*

Temperature (°C)	Ethanol content (% v/v)				
	0	*10*	*12*	*14*	*20*
0	1.56	0.65	0.54	0.46	0.27
5	1.82	0.76	0.64	0.54	0.32
10	2.13	0.89	0.75	0.63	0.38
15	2.48	1.05	0.88	0.75	0.45
20	2.90	1.24	1.04	0.88	0.53

solubility of calcium tartrate under a similar set of conditions (Table 4), which shows almost a 50% decrease from 20°C to 0°C in 12% v/v ethanol. In addition, the effect of ethanol on calcium tartrate is to reduce its solubility by almost 30% for each rise of 10% v/v.

The time-honoured way of achieving KHT stability in wine is by conventional cold stabilisation (chill-proofing), whereby wine temperature is decreased in order to decrease KHT solubility. In 1977, Perin[11] calculated that the optimum temperature required for bitartrate stabilisation could be deduced from

$$\text{Temperature (°C)} = \frac{(\text{ethanol \% v/v})}{2} - 1$$

KHT precipitation occurs in two stages. During the initial stage, the concentration of crystal nuclei increases as a result of chilling; this is then followed by crystal growth and development. Without the formation of crystal nuclei, crystal growth and subsequent KHT precipitation cannot occur. Naturally occurring tartrates in unfermented musts are completely soluble because they are below the concentration at which their solution becomes saturated. As soon as fermentation commences, they become far less soluble because of the emerging presence of ethanol, and their concentration now exceeds the solubility limit (supersaturation) and they start to crystallise out. Colloids present in the wine can act as protective agents and prevent crystallisation by coating crystal nuclei and preventing subsequent deposition of more KHT. This protective effect is only temporary because over a period of time these compounds denature and lose their protective effect. This results in renewed crystallisation and precipitation of KHT. Unfortunately, this often happens in bottle. For this reason, wines should be fined before any stabilisation process is attempted. This will minimise the level of protective colloids.

Calcium is present in wines at levels between 6 and 165 g L^{-1} and may complex with tartrate and oxalate anions to form crystalline precipitates. Under certain conditions, calcium mucate can produce a crystalline deposit, although this is rather rare. The presence of significant levels of calcium in wine can be attributed to a number of causes: vineyard soil, practices such as 'liming', fermentation and/or storage in concrete tanks and use of calcium-containing fining material and filter pads. It may also be caused by the use of calcium carbonate as a deacidifying agent. Calcium tartrate instability is often regarded as a particular problem because the insolubility of the salt is less temperature sensitive than KHT, and it often occurs many months (sometimes years) after bottling. It is a notoriously difficult phenomenon to predict. At best, CaT is a rather insoluble salt, being ten times less soluble than KHT and this causes problems for the winemaker because wines that have been stabilised in respect of KHT may still throw a precipitate of CaT, especially since crystallisation of KHT does not induce that of CaT. Interesting though, crystallisation of CaT may induce that of KHT. Fortunately, CaT precipitation is not terribly frequent in table wines and is rather more confined to sparkling wines and fortified products.

According to McKinnon,[12] the mechanism of CaT precipitation in model wine solutions is a two-step process: the formation of soluble, unionised CaT followed by the nucleation of soluble material to form a precipitate. The particular problem of CaT in fortified wines can be understood when it is realised that, as with KHT, the solubility of CaT decreases as the ethanol level increases. MacKinnon maintains that fortification may decrease the time needed for CaT precipitation by a factor of 12. CaT precipitation is also highly pH-dependent: the higher the pH, the lower is the degree of solubility. Consequently, any activity that leads to a reduction in acidity, such as malo-lactic fermentation, will increase the likelihood of a precipitate being thrown. McKinnon also reported that all untreated Australian wines are saturated with respect to CaT, and that wines containing 80 mg L^{-1} or more of calcium are at risk of precipitation, a fact reiterated by Boulton *et al.*,[13] who maintained that many wines are naturally close to CaT instability with concentrations of 80–100 mg L^{-1} calcium. In some sherries, CaT precipitation can occur with calcium levels as low as 50 mg L^{-1}, with approximately 30% of the precipitate being in the form of calcium oxalate. Agitation decreases the time required for a CaT precipitate to be thrown, and this is amply demonstrated during champagne manufacture when CaT is precipitated just after disgorgement (see page 180). Agitation promotes the formation of nucleation sites. The crystallisation kinetics of CaT was the subject of a study by Abgueguen and Boulton,[14] and from their work it has been possible to quantify conditions that favour this particular instability and improve treatment accordingly.

6.2.1 Static Cold Stabilisation

In this traditional process, also known as slow cold stabilisation, wine is chilled to just above its freezing point (–4°C for a 12% product; –8°C for a fortified wine) and then stored in insulated tanks for anything from 7 to 14 days. The process then relies on crystal nuclei being formed by any minute particle that happen to be present in the tank. Once small crystals have been formed, they gradually 'grow' until they precipitate to the bottom of the tank. This makes the process inefficient since, once on the bottom, the only way that crystals will be in contact with the wine is by convection current. The method is also expensive in terms of capital outlay for equipment and energy for chilling. In addition, there is an expensive stock holding and the method seldom reduces KHT to 'safe' levels. Slow cold stabilisation often causes a loss of colour in both white and red wines, and this effect can be mini-mised by adding a small quantity of KHT to induce nucleation.

6.2.2 Contact Cold Stabilisation

The inherent weaknesses in the static process led to the notion of encouraging the formation of crystal nucleation sites by seeding wine with tartrate crystals. The duly modified cold stabilisation process that has evolved is quicker, cheaper and more effective. In this method, wine is brought into contact with finely ground (<40 μm) crystals of KHT, which act as nucleation sites and expedite crystallisation. Only KHT can be used since other sharp crystals do not produce the desired effect. It is necessary to add a large volume of very small crystals in order to give a sufficient surface area for crystallisation, and 4 g of finely powdered KHT – the sort of dose used – has a crystal surface area of approximately 3 m^2. The enormous surface area provided by the powdered KHT reduces or eliminates the energy-consuming induction of nuclei and stimulates immediate crystal growth. Crystallisation is rapid and complete in about an hour, because in theory the KHT molecules in solution in the wine only have to travel around 0.3 mm before they encounter a crystal face on which to grow. The wine still has to be chilled, but not as low as for above; 0°C is usually sufficient. Once the crystals have been added to the chilled wine (normally at a rate of 4–6 g L^{-1}), the mixture is stirred vigorously for 1–2 h.[15] This keeps the crystals in suspension and promotes contact between wine and KHT crystals. At the end of this period, the cold wine is filtered and crystals are cleaned, powdered and saved for future wine-seeding purposes. The procedure normally takes place in a vessel with a conical bottom called a 'crystalliser', which is equipped with a drainage port to remove excess crystals. Because polymeric inhibitors of nucleation have little effect on the contact cold stabilisation process, tartrate stability *via* this means is relatively easy to predict. For contact cold stabilisation of red wines, it is necessary to chill the wine beforehand in order to remove pigments and tannins, which may be unstable at low temperature. Recycling of KHT crystals from red wine treatments is not possible because the crystals become coated with phenolics and colouring material, which lessens their effectiveness for further use. Care must be taken to avoid oxygen pick-up during the handling of cold wine because of the high solubility of the gas at low temperature. To minimise this, all transfer lines should be purged with nitrogen before use. Because of the potential for the increased absorption of oxygen in wines held at low temperatures, alternatives to cold stabilisation treatments have been sought. Such methods have involved ion-exchange (removal of K$^+$), electro-dialysis, reverse osmosis and addition of crystal inhibitors, such as metatartaric acid (MTA) and carboxymethyl cellulose.

Rankine[16] has summarised the basic refrigeration requirements for cold stabilisation processes, and notes that if a wine is to be cold stabilised by initially chilling it through a heat exchanger, the chilling requirement (in kilowatts) is the product of the wine flow rate (in litres per hour) and the temperature drop (in °C), divided by 850. To this must be added the minor heat loads of the electrical power of the wine pump and the heat gain from the surroundings.

If the wine is to be chilled in tank, the chilling load (in kilowatts) is the product of the volume of wine (in kilolitres), the required temperature drop (in $^{\circ}$C) and 1.15, divided by the time (in hours) required to cool the wine. Adjustments then have to be made for tank insulation and, if the tank in uninsulated, for the surroundings (people, lights, fans, *etc.*).

There is now a 'dynamic' or continuous contact process, whereby KHT crystals are packed into the conical base of a vertical tank/column through which cooled wine is pumped upwards. Stabilised wine is drawn off from the top of the tank. Periodic shutdown is necessary in order to remove much of the enlarged mass of crystals.

6.2.3 Ion-Exchange Stabilisation

Ion-exchange treatment for prevention of KHT and CaT precipitation in wine was developed in Australia, where it has been used since 1955. Conveniently, the same resins and equipment that suffice for the acidification of juices and wines can be used. According to Rankine,[16] the resins used in Australian wineries are highly acidic, uni-functional cation exchangers, with a polystyrene base and sulfonic acid groups that contain exchangeable hydrogen atoms. Wine can be stabilised by ion exchange in the following three ways:

   (i) By replacing K^+ with Na^+ with a cation-exchange resin in the sodium cycle, thus forming soluble sodium bitartrate.
  (ii) By replacing the T^{2-} anion with OH^- or another anion with an anion-exchange resin.
 (iii) By replacing K^+ and T^{2-} with H^+ and OH^- with a cationic and an anionic resin, respectively, in effect exchanging a portion or all of the KHT for water.

In practice, the sodium cycle is the most suitable because wine can be stabilised without affecting its flavour and with only a slight reduction in its acidity. Unfortunately, the technique increases the level of Na^+ in wine, which is usually considered undesirable. If necessary, the acidity of a wine can be increased by preparing a resin in a mixed Na^+ and H^+ form. This is particularly desirable when treating low-acid dessert wines since the pH of the wine can be lowered to any desired level without stimulating KHT precipitation. It should be borne in mind that when exchanging K^+ for Na^+ the resin will actually have a greater affinity for Ca^{2+} and Mg^{2+} than for K^+, and that if care is not exercised a situation can arise whereby KHT precipitation is promoted. For example, when wine is passed down the ion-exchange column, a top zone of 'exchanged' Ca^{2+} and Mg^{2+} will overly a zone of 'exchanged' K^+, while the remainder of the column (lower down) will be in the 'un-exchanged' Na^+ form. As more wine is introduced, the 'K^+ zone' is moved downwards, until a point is reached whereby no more Na^+ remains on the resin. At this stage, the 'K^+ zone' will be at the base of the column and the resin will be in the combined K^+, Ca^{2+} and Mg^{2+} form. If wine flow continues, K^+ will be displaced from the base of the column and the wine will be prone to KHT precipitation.

6.2.4 Estimation of Cold Stability

There is no universally accepted definition of cold stability, although it is a valuable parameter for the winemaker. Three methods of assessment are commonly used: the freeze tests, the conductivity test and the calculation of concentration product. Freeze tests rely on the ability of KHT to form crystals when a 'suspect' wine is held at low temperature for a specified period (*e.g.* $-2°C$ for 7 days). Quite often, a sample is frozen and then thawed to determine the development of crystals, and whether they re-solubilise or not. The absence of crystal formation or re-solubilisation indicates KHT stability. One inherent problem is that when the sample is frozen, the removal of water effectively concentrates all the wine components, including ethanol, and this enhances nucleation and crystallisation. Despite their deficiencies, freeze tests are probably the most widely used predictors of cold instability. The conductivity test is a far more accurate means of predicting cold stability, and involves seeding the wine sample with finely ground KHT crystals (*ca.* 1 g L^{-1}), which causes the electrical conductivity to change. If the change in conductance from the beginning to the end of the treatment is less than 5%, the wine may be considered stable (although some winemakers adopt a change of 3% or lower as their criterion). Samples passing the test are only stable at or above the test temperature, which is usually $0°C$ for white and $5°C$ for red wines. Thanks largely to the work of Berg and Keefer,[9,10] one can calculate the solubility of KHT by measuring the concentration of K^+, total tartrates, pH and ethanol, and from these calculate the concentration product (CP) of the salt.

The relationship between solubility products and long-term tartrate stability was the subject of a report by De Soto and Yamada,[17] and as a result of the findings of these workers and Berg and Keefer, the CP for KHT can be calculated from

$$CP = [K^+ \text{ mole } L^{-1}] \text{ [total tartrate mole } L^{-1}] \text{ (% bitartrate)}$$

The values for percentages of bitartrate and/or tartrate at measured pH and ethanol levels are taken from tables prepared by Berg and Keefer. If we consider an example of a 'typical white wine' with the following determinants, $K^+ = 1100$ mg L^{-1}; $Ca^{2+} = 76$ mg L^{-1}; tartrate$=1600$ mg L^{-1}; pH $= 3.74$; ethanol $= 12\%$ (v/v), then the CP calculation is as follows:

$$CP = \frac{1100 \times 10^{-3}}{39.1} \times \frac{1600 \times 10^{-3}}{150} \times 0.661 = 19.8 \times 10^{-5}$$

According to DeSoto and Yamada,[17] the suggested minimum CP for a dry white wine is 16.5×10^{-5} at $0°C$, and so our theoretical sample is likely to precipitate KHT crystals.

As stated previously, it is unlikely that CaT stabilisation can be brought about simply by reducing the temperature of the wine, and so chill tests are not effective stability predictors for this salt. Traditionally, CaT stability has been determined by CP values, which, in view of the lack of the sort of data that is available for KHT, can be calculated from the following equation:

$$CP = [Ca^{2+}] [T^{2-}]$$

where Ca^{2+} and T^{2-} represent the free or ionised calcium and tartrate concentrations. Total calcium concentration is usually measured by atomic absorption, and free calcium should be measured by specific ion electrode. Since a large proportion of the total calcium may be bound, analysis of 'total' rather than 'free' leads to abnormally high CP values. Muller et al.[18] reported the use of a minimum contact conductivity method for predicting CaT stability. In this procedure, a wine is stirred in the presence of CaT crystals. A decrease in conductance indicates CaT crystal growth.

6.2.5   Prevention of Crystallisation

The first attempts by winemakers to prevent the formation of tartrate crystals involved the use of the sequestering agent sodium hexametaphosphate (SHMP), which worked reasonably well, except that high doses were required to treat some wines and this inevitably resulted in phosphate take up by the wine, which in turn led to the formation of a ferric complex that caused instability on contact with air. SHMP is strictly speaking a hexamer, with the composition $(NaPO_3)_6$, and as well as its use in the food industry, is widely used in detergents and water softeners (*e.g.* Calgon). The next substance to be used for the same purpose was metatartaric acid (MTA), which was first used in Europe in 1955. MTA is a polyester resulting from the intermolecular esterification of tartaric acid, and the compound (labelled E353) is used to prevent KHT precipitation when wine is chilled. In effect, it acts as an inhibitor of nucleation. When finely ground tartaric acid is heated at 170°C, it melts and water vapour is released. In addition to this dehydration process, esterification and polymerisation also occur and the resulting substance is an amorphous, deliquescent, off-white, high-molecular-weight mass known as a hemipolylactide, with the molecules being linked in a polymer-like manner. A much more pure product can be obtained by heating tartaric acid at 150–160°C in a vacuum. The polymerisation between the tartaric acid molecules is *via* an esterification reaction between an acid function of one molecule and a secondary alcohol function of another molecule. In practice, not all of the acid functions react, but, legally, in order to qualify for the name 'MTA' there must be at least 40% esterification during the heating process. There are many MTA preparations with different anti-crystallising properties, depending on the average esterification rate of their acid functions. This so-called 'esterification number' may be determined by acidimetric assay before and after saponification. The main impurity in MTA is pyruvic acid, which can be present at levels of 1–6% by weight, according to the method of manufacture. Oxaloacetic acid may also be a trace impurity. The formation of these two keto-acids results from the intramolecular dehydration of a tartaric acid molecule and subsequent decarboxylation. MTA, which is therefore not really a 'pure' substance, acts by forming a sheath around microscopic crystals of KHT and CaT, thus preventing them from further growth and deposition (and visibility in the bottle). It should be used at rates of around 10 g hL^{-1} wine (the limit imposed by the EU is 100 mg L^{-1}, and there is little point using less than this). During wine storage, MTA slowly decomposes to tartaric acid and the protective effect is lost, so the addition is only really for use in wines that are not going to be stored for very long. Decomposition rates are temperature dependent, and wines so treated should be stored below 20°C. To emphasise the importance of this, total hydrolysis of a 2% solution of MTA takes 10 months at 5°C, and 3 months at 23°C. Typically, protection of wine with MTA lasts *ca.* 12–18 months. Because of their instability, MTA solutions for wine treatment should always be freshly prepared and should never be considered for wines that are meant to be stored for an extended period. Having said all this, MTA is very useful in wineries where cold stabilisation is not possible (*i.e.* small concerns where refrigeration equipment is not available). A new generation of MTA products is being developed, which are a blend of MTA and soluble gum arabic. Such a combination gives a much longer period of protection.

It has been appreciated for many years that white wines stored *sur lie* for several months are highly unlikely to deposit tartrate crystals and do not require cold stabilisation. The fact that such wines become enriched with yeast mannoproteins, which are secreted during lees storage, has also been long recognised, but it was not until 1993 that Lubbers *et al.*[19] connected the two phenomena. They showed that yeast mannoproteins have an inhibitory effect on tartrate crystallisation, *albeit* in model wine solutions. Critics have rightly pointed out that the mannoproteins used in these experiments were extracted by heat in alkaline solution, circumstances that are far removed from the enzymatically released compounds found after lees storage. The following year, Dubourdieu and Moine-Ledoux[20] demonstrated the crystal-inhibiting effect of mannoproteins extracted by enzymic treatment of yeast cell walls. These mannoprotein preparations were prepared by digesting yeast cell walls with Glucanex™, an industrial preparation of β-(1–3)- and β-(1–6)-glucanases, a product permitted in winemaking. These mannoprotein preparations inhibit tartrate crystallisation in white, red and rosé wines at a concentration of 25 g hL^{-1}. The mannoproteins responsible have an average weight of 40 kDa. An industrial preparation Mannostab™ has been produced by Lafforte

Oenologie from purified yeast cell wall mannoproteins and comes as a soluble, odourless, tasteless, white powder. It will inhibit KHT crystallisation at doses between 15 and 25 g L^{-1}. Larger doses reduce the stabilising effect. Unlike MTA, mannoprotein preparations are stable and have a long-lasting effect on tartrate instability. According to the makers, Mannostab treatment does not affect the sensory qualities of wine. Sensory analysis of both white and red wines has shown no statistically significant differences in triangle tests with Mannostab-treated wine and untreated wine.

6.2.6 Protein Instability

The grape is the major source of protein in wine. After véraison, protein synthesis in the grape proceeds rapidly and parallels the rapid accumulation of sugars at this time. Some of these proteins are rather insoluble and are precipitated during normal winemaking processes; thus, they do not reach the bottle. Others, however, are more soluble and do persist in the wine up to the point where it is bottled. Some proteins in the latter category are unstable when subjected to even mild heat, and throw a troublesome haze in the bottle, the 'protein haze' or the 'heat haze'. The phenomenon occurs most frequently in white wines or wines of low polyphenol content. It is rarely met with in wines with relatively high levels of phenolics, especially tannins that complex with and precipitate proteins. It can occur in red wines if they have been over-fined with, say, gelatin. Heat haze is the result of a wine containing thermolabile grape proteins, which slowly denature and precipitate as the wine warms up. The phenomenon is obviously more of a problem in warmer winemaking areas, but this is compounded by the fact that grapes grown in hotter climes synthesise more of these heat-sensitive proteins. In addition, some cultivars are more likely to produce such proteins than others, the most notorious being Muscat, Semillon and Traminer. Under most conditions of growth, grapes will synthesise more proteins as they mature on the vine. The proteins identified as being involved in heat haze have molecular weights in the range 40–200 kDa and a pI varying between 4.8 and 5.7. Other smaller, grape-derived nitrogenous compounds such as amino acids and peptides are not contributors to heat haze.

The exact nature of protein instability in wine is difficult to discern because many factors are involved. In particular, differences in protein nature occur according to cultivar, maturity of the grape at harvest and climate, and all of these factors will determine how they react in the winery (*i.e.* their molecular size and electrical charge will control how they react with other components). In the context of the last point, some proteins appear to act as nucleation centres around which iron, copper and other metals may deposit. Protein haze, for example, is linked to 'copper haze', which involves the formation of a cuprous sulfide–protein complex. It has been shown, however, that protein haze is due to not only the precipitation of thermo-labile molecules, but also the formation of tannin–protein complexes. It has been estimated that around one half of total wine protein is bound to grape phenolics.[21]

Ewart[22] maintained that the so-called 'protein haze' is a complex of proteins, polysaccharides and polyphenols, with a small amount of inorganic ash. Indeed, unstable polysaccharide and polyphenol complexes may go a long way towards determining why many of the tests used to assess protein instability are not effective predictors of the problem. Certainly, the total protein content of a wine sample should not be used to gauge its protein instability potential, nor should total nitrogen determinations, because according to Gorinstein *et al.*[23] the proteins that result from yeast autolysis do not normally contribute to protein haze. Some authorities argue that reliable methods for assaying soluble proteins in wine have yet to be devised.

Protocol during harvest and in the winery may determine what happens to the soluble protein fraction, since the solubility of wine proteins is dependent upon temperature, pH, alcohol level and ionic composition. Changes in any of these parameters may influence the potential for protein precipitation in bottle. It has been shown that the extractability of proteins from the grape is influenced by harvesting methodology. For example, juice obtained from the pressing of whole-cluster grapes will be expected to have lower protein levels than that obtained from de-stemmed fruit. The inference here is that stems must play an important role in limiting protein diffusion, which has implications for wineries that use mechanically harvested grapes. Settling and racking white grape juice before fermentation reduces total nitrogen by up to 15%, and a 50% reduction may be achieved by using bentonite, although the latter can also remove essential amino acids if care is not taken. It has also been shown that practising skin contact increases the protein content of the juice of certain cultivars, such as Sauvignon Blanc and Semillon.

Wines fermented and/or aged in oak containers have lower levels of unstable protein and are much less likely to throw a haze than those that have only ever seen stainless steel. This, of course, is a result of the interaction between proteins and wood tannins, and champagne producers often take advantage of this fact by adding a small amount of tannic acid to their wine in order to bind potentially unstable proteins, formed from both primary and secondary fermentations, before bottling. The champagne producer has to be careful, however, because proteins are the most important foam-active components in his/her product, with hydrophobic forms contributing more than the hydrophilic variety. Thus, there has to be a balance between sufficient protein for the 'sparkle' and minimum protein for precipitation (and haze). In another context, the act of fortification initiates large quantities of proteinaceous material to be precipitated.

6.2.7 Assessment of Heat (Protein) Stability

Even as we embark upon the 21st century, there are no unequivocally accepted industry standards for measuring protein instability. The most widely used methods available involve subjecting a wine sample to heat or to a chemical oxidant such as trichloroacetic acid (TCA), and then examining it for haze development. When chemicals are used, one usually requires a nephelometer to measure haze. Various combinations of heat and time have been used to assess heat stability, probably as good as any being to micro-filter the wine sample, heat it to 80°C for 6 h and then examine it after cooling. If no haze is evident, the wine is heat stable. This protocol emerged from work carried out in Australia by Pocock and Rankine,[24] who evaluated treatment temperatures and times over a range varying from 50 to 90 °C. They found that about 40% of wine protein was precipitated when a sample was held at 40°C for 24 h, and that 95–100% was precipitated by holding at 60°C for the same period of time. They also found that the time necessary for haze formation decreased with increasing temperature. In their previous work, Ribéreau-Gayon and Peynaud[25] had considered that a wine sample heated to 80°C for 10 minutes should be considered stable if no haze is developed on cooling. Chilling wine samples after heat treatment may increase haze formation.

In addition to the array of heat tests, a number of chemical tests have been devised to assess the heat stability of wine proteins. These include the use of ethanol, ammonium sulfate, trichloroacetic (trichloroethanoic) acid, phosphomolybdic acid, phosphotungstic acid and tannic acid. Most of these will denature and precipitate all protein fractions. Probably the most expeditious and stringent of all these is the 'Bentotest' first developed by Jakob,[26] and one that does not require the use of heat. The method uses a mixture of phosphomolybdic acid and hydrochloric acid to denature and precipitate protein by forming cross-linkages with the molybdenum ion. The subsequent development of haze is proportional to protein content and may also be used to assess bentonite addition levels. Rankine and Pocock[27] demonstrated that the Bentotest is more sensitive than heat tests using 70°C for 15 min. Evaluation of haze can be *via* nephelometry or by the naked eye. Perhaps the simplest visual test for heat-labile proteins is to heat 95 ml of wine with 5 ml of saturated ammonium sulfate to 55°C for 7 h, and then chill it in an ice bath for 15 min. Precipitation indicates the presence of thermo-labile proteins. An equally simple test involves mixing equal volumes of absolute ethanol and wine sample and observing haze formation. The oft-used TCA test[28] is a little more elaborate, and is fully described, as are other methods, by Zoecklein *et al.*[29]

6.3 PRESERVATION

While ethanol and tannins serve as natural preservatives in wine, other precautions have to be taken to ensure that the product reaches the consumer in the best possible condition. This entails the addition of preservatives, the most frequently used of which are sulfur dioxide, potassium sorbate and ascorbic acid. Because we are dealing with a food and because of the modern antipathy towards the use of 'additives', it is normal to talk in terms of 'biological stabilisation' rather than 'preservation', the latter insinuating that a 'preservative' has been added.

6.3.1 Sulfur Dioxide

The use of SO_2 gas as a fumigant in wineries dates back to Roman times,[30] when sulfur was burnt inside wine casks in order to 'freshen' them. There are various references to its occasional use in winemaking from the 15th century onwards and there are indications as to the more general use of SO_2 in this capacity towards the end of the 18th century, but the widespread deliberate use of the gas in viniculture is very much a 20th century pheno-menon. Pre-occupation with food additives throughout the Western world during the last decades of the 20th century has, to some extent, cast SO_2 in a villainous role on the world stage, and the *raison d'être* for its use in the food industry is now coming under increasing scrutiny. Regardless of potential drawbacks, its many useful properties make it an almost indispensable aid to wine-making and to other food manufacturing processes.[31] Most nota-bly, SO_2 is both a germicide and an anti-oxidant at levels that are non-toxic to all but a minute fraction of the population and in instances where it is used over-enthusiastically , the fact is easily betrayed by its odour. At present, the inhibition of wild, spoilage yeasts prior to the onset of fermentation is achieved by the addition of SO_2 to freshly prepared must. The substance is then re-added at the end of fermentation when its anti-oxidant qualities are sought. Research is now underway (see Section 9.1 on 'killer yeasts' in Chapter 9) to utilise the lethal properties of some wild yeasts during the pre-fermentation stage, which would limit SO_2 use to only the post-fermentative stage, thus reducing the total level of this anti-microbial in the final product.

Most of the SO_2 found in wine is deliberately added at some stage during processing, but small amounts are contributed by the fermenting yeast.[32] In general, yeast-generated levels are no more than 10 mg L^{-1}, but higher concentrations (up to 100 mg L^{-1}) have been recorded, and so nowadays the total absence of SO_2 in wine is very rare. The production of SO_2 by yeasts tends to be higher in musts with a low level of suspended solids. The levels of SO_2 now used in viniculture are greatly reduced when compared to the high levels (>200 ppm) that were commonplace during the latter dec-ades of the 20th century. Such excesses used to strip wine of colour, in particular, conferring a seemingly permanent greenish-yellow tinge to white wines, and prevented or at best retarded malo-lactic fermentation. Wines made under these conditions were invested with what has been called a 'Peter Pan' quality, which meant that they often took decades to reach their peak. Many wines from that era possessed a characteristic 'sulfurous prickle', which only receded upon prolonged ageing. Nowadays, most commercial wines contain < 10 mg L^{-1} free SO_2, and very few quality dry table wines will have a total (free and bound) SO_2 content in excess of 100 mg L^{-1}.

Under normal conditions of temperature and pressure, SO_2 is a colourless, pungent gas with a molecular weight of 64.06, a melting point of $-72.7°C$ and a boiling point of $-10°C$. It is very soluble in water, its solubility varying from 228.3 g L^{-1} at 0°C to 78.1 g L^{-1} at 30°C. In aqueous solutions, the pH at saturation is *ca.* 0.8 at 20°C, and the saturated solution at STP contains 53.5% by weight of the gas.

When dissolved in water it behaves as a fairly strong acid, commonly known as 'sulfurous acid' (H_2SO_3). It has been usually assumed that the dissolution of SO_2 in water is accompanied by the reaction

$$H_2O + SO_2 \leftrightarrow H_2SO_3$$

although no evidence for the existence of the free acid has been forthcoming. Raman and infrared spectra of SO_2 dissolved in water indicate solutions of molecular SO_2, often denoted as $SO_2 \cdot H_2O$, with a limited amount (*ca.* 3%) of H_2SO_3 present. The relatively strong acid behaviour of dissolved SO_2 and the high solubility of the gas have led to the suggestion[33] that there might be tautomeric equilibria between 'sulfurous' species, that is

$$SO_2 \cdot H_2O \leftrightarrow SO(OH)_2 \leftrightarrow HSO_2(OH)$$

the latter being strongly acidic.

Assuming the presence of H_2SO_3, the pK_a for the first dissociation is 1.86 (although with the effects of ethanol and solutes characteristic of the wine environment, it is nearer to 2) and governs the reaction that liberates the bisulfite or monohydrogen sulfite ion:

$$H_2SO_3 \leftrightarrow H^+ + HSO_3^- \text{ (bisulfite)}$$

The pK_a for the second dissociation is 7.18 and is responsible for the production of sulfite, the second ionised form:

$$HSO_3^- \leftrightarrow H^+ + SO_3^{2-} \text{ (sulfite)}$$

Both dissociations are almost instantaneous. Thus, dissolution of SO_2 results in three entities: undissociated molecular form,

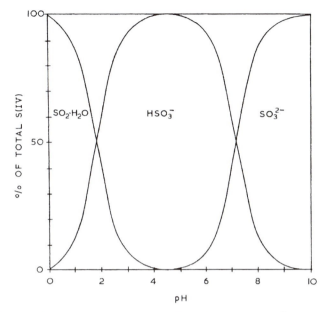

Figure 2 *Distribution of the species $SO_2 \cdot H_2O$, HSO_3^- and SO_3^{2-} as a function of pH in dilute solution (after Wedzicha[31])*
(Reproduced by kind permission of Reed Elsevier)

hydrated SO_2 and two ionised forms, bisulfite and sulfite, and in dilute solution the distribution of these three is dependent on pH, as can be seen from Figure 2. At pH values below 1.86, hydrated SO_2 (H_2SO_3; $SO_2 \cdot H_2O$) predominate, and it is this that has been traditionally referred to as sulfurous acid. The first ionised form, HSO_3^-, is the major species at pH values between 1.86 and 7.18, which effectively means grape juices and wines, while the second ionised form, SO_3^{2-}, is the dominant form above pH 7.18.

Molecular (or active) SO_2 exists as a gas or as single molecules in juice and wine, and is the most important form viniculturally. It is responsible for the observed anti-microbial activity, which, in respect of some bacteria, is some 500 times greater than the bisulfite ion. It also possesses anti-oxidant activity, and, being volatile, is responsible for the 'sulfury' odour and taste. As intimated, bisulfite is the predominant form of free SO_2 in juice and wine, and while it is odourless, it has a bitter, slightly salty taste. It causes inactivation of polyphenol oxidase (PPO) enzymes (which cause oxidative browning of juice) and the binding and/or reduction of brown quinones in juice. Bisulfite will aid extraction of anthocyanins, although it will also cause them some bleaching and slow down the anthocyanin polymerisation reactions with other phenols. Bisulfite has a very low anti-fungal activity. At the pH values prevalent in wines, the amount of sulfite present is negligible. It is the only form of SO_2 that will react directly, albeit slowly, with oxygen. In the presence of suitable catalysts, the sulfite ion is oxidised to sulfuric acid as follows:

$$SO_3^{2-} + H_2O \rightarrow SO_4^{2-} + 2H^+ + 2\,\text{electrons}$$

This reaction results in an increase in acidity and makes the sulfite ion a powerful reducing agent. At concentrations normally encountered in wine, sulfite is tasteless and odourless.

From a winemaking point of view, it is important to distinguish between 'free' and 'bound' SO_2 since a proportion of that added to wine will bind with other compounds present, and consequently not be readily available for desired reactions. This is the 'bound', 'combined' or 'fixed' SO_2, and is the result of activity of the bisulfite ion. The remainder is called 'free' SO_2 and is often simply referred to as 'sulfur'. Free 'sulfur' exists in two forms, the most abundant and efficacious of which is the molecular or unionised form. Bound SO_2 compounds are often called 'bisulfite addition products' or 'hydroxy-sulfonates', and they serve to provide a reservoir of SO_2 and feed it into the wine as it disappears through oxidation or evaporation. Having said that, the bisulfite ion is considered most undesirable to the winemaker because of its capacity to bind to the carbonyl oxygen atoms of a number of useful compounds, including aldehydes (notably acetaldehyde), keto-acids (*e.g.* pyruvic, ketoglutaric), glucose, quinones and monomeric anthocyanins of red wines (rendering them colourless). In some instances, the bonding is not very strong and in some cases reversible, and so some of the 'bound' form is available to react with enzymes or metabolites of bacterial biochemistry. It is the free form that is responsible for anti-oxidant activity and most of the anti-microbial activity, although fractions of SO_2 bound to acetaldehyde and pyruvic acid exhibit slight anti-bacterial activity (5–10 times weaker than free SO_2, even though they may be present in 5–10 times the concentration). There is no known anti-fungal activity attributable to bound SO_2.

By far the most significant carbonyl compound involved in SO_2 binding is acetaldehyde (ethanal), a natural fermentation intermediate and the result of oxidising ethanol. At elevated levels, the presence of acetaldehyde is not desirable in table wines giving them an over-oxidised or 'tired' character, although it is an essential component of sherries, where it imparts the typical oxidised aroma. Normal acetaldehyde concentrations in newly fermented table wines are <75 mg L^{-1}, with levels in mature products being anywhere in the range 20–400 mg L^{-1}, according to Rankine.[16] Berg *et al.*[34] report the sensory threshold of the substance as being 100–125 mg L^{-1}. Removal of excessive acetaldehyde from wines is said to give them a 'fresher' flavour.

The reaction for the binding of acetaldehyde and SO_2 is

$$CH_3-CHO + HSO_3^- \leftrightarrow CH_3-CHOH-SO_3^-$$

The acetaldehyde–bisulfite compound is technically acetaldehyde-α-hydroxysulfonate, and each milligram of acetaldehyde will bind with 1.45 mg of SO_2. It is reckoned that 60–100% of all wine, acetaldehyde becomes bound with SO_2, and that most bound SO_2 is in the form of acetaldehyde-α-hydroxysulfonate. In heavily sulfited wines, where all acetaldehyde has been sequestered, the 'excess' SO_2 appears to be preferentially bound to sugars. From a practical point of view, it is inadvisable to make SO_2 additions during fermentation because it will immediately be 'mopped up' by acetaldehyde and will not be able to fulfil its intended purpose. In addition, bisulfite binds to yeast, bacteria and other cellular material, especially proteins that will be maximised during fermentation, and as is expected sulfiting is most effective in clarified juice/wine.

In terms of sugar binding, glucose is known to form a bisulfite addition product and the binding is particularly important in grape juice, where some 50% of added SO_2 (at levels of 50–100 mg L^{-1}) may become bound, even though glucose has a low binding rate with bisulfite. Other sugars known to bind with bisulfite are mannose, galactose and arabinose, but binding with fructose is very low, and it is thought that no addition product is formed. A few polysaccharides are known to form bisulfite addition products.

Bisulfite binding can also occur with the oxidation products of phenols, and with ascorbic acid and phenolic compounds, particularly caffeic acid and *p*-coumaric acid, which bind reversibly with

SO_2. Such binding is pH- and temperature-dependent. SO_2 binds specifically with the four position carbons of the monomeric anthocyanins of red wines, resulting in a loss of colour.

Where bisulfite binding occurs, it is not instantaneous, although it is usually fastest within the first 24 h of contact. Full binding then takes place over the next few days, 4–5 days being required before binding ceases. Over this period, there will be a gradual decrease in free SO_2 levels until an equilibrium is reached. The rate of binding is dependent upon the dissociation constant, K, for each individual reaction; the lower the value of K, the more is the likelihood of a bisulfite addition product being formed. In the acetaldehyde–bisulfite reaction, for example, K is low and the binding between the two is rapid and strong. At pH 3.3, 98% of acetaldehyde is bound within 90 min of SO_2 addition, and the reaction is complete within 5 h. The value of K increases with increasing temperature, as can be illustrated by that for acetaldehyde, which shows a five-fold increase as the temperature rises from 25 to 37.5°C. The relationship between the amount of SO_2 added to a wine and the amount remaining free is an important and complex one, and one that will vary from wine to wine. It is largely governed by the total SO_2 content of the wine, and as can be seen from Figure 3 the rate of binding decreases as the free SO_2 content increases. There is a laborious method of measuring free SO_2, but most winemakers tend to make educated guesses as to how much will become bound and employ an empirical law of binding. Many winemakers make the assumption that at levels of 30–60 mg L^{-1}, around 50% of their SO_2 addition will become bound, while Peynaud[35] quotes that around one-third of all added SO_2 becomes bound. The degree of formation of bisulfite addition products can be minimised by keeping the fermentation strictly anaerobic, fermenting at lower temperatures, using certain strains of yeast and adding ammonium ions to the must/juice. White winemaking generally requires a more liberal use of SO_2 than does the red winemaking process, and, as a rule of thumb, the use of sound grapes (free from mould) can minimise its use completely.

In the winemaking context, there are three principal properties exhibited by SO_2:

   (i) Anti-microbial; both anti-bacterial and anti-fungal activity
      has been demonstrated by molecular SO_2, with the greater

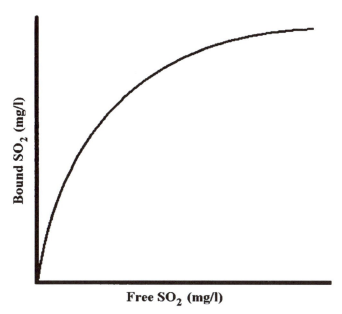

Figure 3   *Relationship between free and bound SO$_2$*

activity being shown against bacteria. At low concentrations it is bacteriostatic, but can be bactericidal at higher levels. Generally bound SO$_2$ is not anti-bacterial, but on binding to some compounds, such as acetaldehyde and pyruvate, does produce active addition products. Against a variety of yeast species, molecular SO$_2$ acts as a fungistat rather than being fungicidal. One of the known effects on yeasts is that in the presence of SO$_2$, the lag phase is extended and growth is slowed down. At the end of the lag phase, growth often returns to normal. Some yeasts (including those used for vinification) are remarkably resistant to free SO$_2$ and can tolerate levels up to 700 mg L^{-1} or more, and fermentation has been demonstrated in musts with as much as 2000 mg L^{-1}. Evidence suggests that yeasts can become resistant to prolonged exposure to SO$_2$, and it is therefore important to avoid adding successive doses to a ferment. To some extent, it is possible to use SO$_2$ as a selective agent in a mixed yeast population since it will retard the growth of 'wild' forms (*Candida, Torulopsis, etc.*) and leave desirable strains of

Saccharomyces relatively unaffected. In an extensive study, Beech *et al.*[36] found that within 24 h, there was a thousand-fold reduction in the number of viable cells of a variety of wine spoilage organisms at a molecular SO_2 concentration of 0.8 mg L^{-1}, and that this would appear to be the sort of level required for the routine suppression of the majority of unwanted yeasts and bacteria. SO_2 is especially effective during wine storage and controls the growth of a variety of bacteria, thus preventing hazes.

(ii) Anti-oxidant; SO_2 will protect must, juice and wine from excessive oxidation, although the mechanisms are imprecisely understood. It is known that in the presence of suitable catalysts, SO_2 will bind with dissolved oxygen, thus,

$$SO_2 + \tfrac{1}{2}O_2 \rightarrow SO_3$$

The reaction is very slow, but it may well help to protect wine from chemical oxidation (but not enzymic oxidation). The anti-oxidant capability of SO_2 is primarily due to its ability to react with oxidants, and/or to become preferentially oxidised over other compounds, which, when oxidised, present deleterious flavour and aroma problems. In particular, SO_2 protects phenolic compounds, the primary 'oxidisable' components in wine, from excessive oxidation and prevents the formation of other oxidised wine components, as can be illustrated by the following, which documents the oxidation of a vicinal dihydroxyphenol:

o-dihydroxyphenol   o-quinone

In this instance, the phenol is oxidised to its quinone and the strong oxidising agent hydrogen peroxide, which is very active and, if left unattended, will oxidise other wine compounds. With SO_2 present, the peroxide is 'mopped up' and oxidises it (as sulfite) to sulfate as follows:

$$SO_3^{2-} + H_2O_2 \rightarrow SO_4^{2-} + 2H_2O$$

It is thought that SO_2 can also reduce compounds that have already been oxidised in wine, such as quinones:

| o-quinone | | o-dihydroxyphenol |

SO_2 aids the formation of a low oxidation–reduction potential, thus aiding wine aroma and taste development during ageing.

(iii) Anti-enzymatic; it instantly inhibits the activity of oxidative enzymes such as PPO and laccase, and in time actually destroys them. Enzymes such as PPO catalyse oxidative reactions in juice and must, and in doing so reduce the amount of dissolved oxygen available to yeast cells during their vegetative growth phase. When added to juice/must, therefore, SO_2 has a protective effect against oxidations before fermentation proper ensues. PPO enzymes also cause browning reactions in juice according to the following reaction:

(colourless)　　　　　　　　　(browned)

PPO activity has been shown to be reduced by more than 90% by the addition of 50 mg L^{-1} SO_2, and according to Sayavedra-Soto and Montgomery,[37] it seems as though the bisulfite form might be responsible for this, and that it results *via* an irreversible structural modification rather than a binding product. PPO (also known as tyrosinase and catecholase) activity is at its height in newly prepared musts/juices, but it decreases rapidly with time and is usually inactive after fermentation. Laccases (*p*-phenoloxidases) are another important group of oxidative enzymes originating from grapes infected with *Botrytis cinerea*. They will oxidise 1,2-, 1,3- and 1,4-diphenols and anthocyanins, and are particularly active at wine pH. Laccases are more difficult to inactivate than PPOs (only a 20% reduction with a free SO_2 concentration of 150 mg L^{-1}) and they are thus able to consume oxygen over longer periods of time.

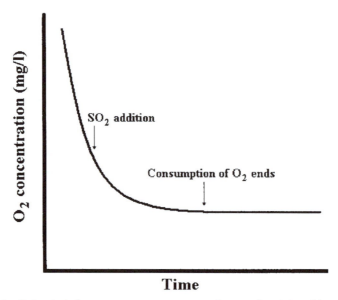

Figure 4   *Delay in halting oxygen consumption of must after SO₂ addition*

Laccase activity generally persists into the final wine, and in cases where it is impractical to add SO_2, ascorbic acid may have to be used as an anti-oxidant.

Upon the addition of SO_2 to a must, oxygen consumption by oxidase enzymes will start to decline in a matter of minutes until it ceases completely, and the dissolved oxygen content stabilises (Figure 4).

Sulfur dioxide has known effects on humans, and inhalation can result in serious respiratory illnesses such as lung oedema, bronchial inflammation and bronchospasm, making it particularly problematic for asthmatics. A small percentage of the population, mainly asthmatics, have an allergy to the gas. It is the molecular form that is responsible for the sensory effects of SO_2, and so the sensory threshold will be temperature- and pH-dependent. There is an enormous variation within a population, but the threshold is generally within the range 0.8–2.0 mg L^{-1} molecular SO_2 and most people will certainly be able to detect the 'burnt match' odour at around 15 mg L^{-1}.

There is a WHO/FAO recommendation on the maximum daily intake from all sources of 0.7 mg kg^{-1} of body weight. American labelling law requires that the words 'contains sulfites' must appear on labels, while in Australia and New Zealand wine containing SO_2 would be labelled as 'contains preservative 220' or 'sulfur dioxide added'. In the EU, wine producers put 'E220' on bottle labels of wine containing SO_2.

The need for the use of SO_2, and preservatives generally, can be minimised by paying attention to methodology, and the following can be recommended: (1) careful harvesting of grapes to avoid damaging the fruit; (2) rapid transition of grapes from vineyard to winery; (3) sanitary conditions in winery, especially in the fermentation area; (4) use of refrigeration where applicable; (5) regular topping up of barrels to exclude air; and (6) use of filtration to remove microbes, especially prior to bottling.

No exposé on SO_2 would be complete without some reference to that perennial stand-by of the home winemaker, the Campden tablet. Each tablet consists of a mixture of potassium metabisulfite ($K_2S_2O_5$) and sodium metabisulfite ($Na_2S_2O_5$), together with inert 'bulking' material. The 'active' mass of the two salts in each tablet is designed to be 0.44 g.

In water, the potassium salt dissociates as follows:

$$K_2S_2O_5 + H_2O \rightarrow 2K^+ + 2HSO_3^-$$

The sodium salt behaves in a similar way. In the acidic wine environment, the bisulfite ion unites with protons to form SO_2 as follows:

$$HSO_3^- + H^+ \rightarrow H_2O + SO_2$$

The molecular weight of sodium metabisulfite is 190.2, and that of potassium metabisulfite is 222.4, and, in effect, both compounds dissociate to give two moles of SO_2 (molecular weight 64.1) for each mole of salt. This means that the theoretical SO_2 content of the sodium salt is $(2 \times 64.1)/190.2 = 67.4\%$ and that of the potassium salt is $(2 \times 64.1)/222.4 = 57.6\%$

If given a choice, winemakers generally prefer to use the potassium salt since this increases the level of that element in wine, which may later help in the precipitation of tartrates during cold stabilisation. In addition, it is sometimes claimed that sodium metabisulfite can impart a 'salty' flavour to wine.

6.3.2 Dimethyldicarbonate

In the never-ending search for feasible alternatives to SO_2 as a cold sterilant in winemaking, the approved use of diethyl carbonate (DEDC) in wine by the US government in the 1960s appeared to signal the end of the quest. For here was a compound with broad-spectrum anti-microbial activity that decomposed into harmless products after addition to wine (ethanol and CO_2). In addition, DEDC actually killed its target microbes. Everything was fine until it was found that under certain conditions, ethyl carbamate (see page 231) may also be produced when DEDC decomposes in wine. Thus, the compound that was originally approved by the Bureau of Alcohol, Tobacco and Firearms (BATF) in 1963, was deemed unfit for use in 1972. The following year, an analogue, dimethyldicarbonate (DMDC), was put forward as a suitable alternative. It has similar anti-microbial properties and breaks down principally to methanol and CO_2, but does not liberate ethyl carbamate. Dimethyl carbonate and methyl ethyl carbonate are formed in minor quantities on DMDC breakdown, besides traces of carbo-methoxy adducts of amines, sugars and fruit acids. In the presence of ammonia and ammonium ions (*i.e.* wines), DMDC forms trace quantities of methyl carbamate. It is now a permitted wine and soft drinks additive in some countries. In 1988, the American FDA approved DMDC as a cold sterilant in wine at bottling, in amounts not exceeding 200 mg L^{-1}, which seems to be a generally accepted level. DMDC is a colourless, volatile liquid with a specific gravity of 1.26. It is highly heat-labile and is best stored frozen. In aqueous solution it rapidly hydrolyses to methanol and CO_2, the rate of hydrolysis being dependent on wine temperature (it is completely hydrolysed after 5 h at 10°C, 2 h at 25°C and 1 h at 30°C). The actual dosage used is determined by the number of microbes present, the ethanol content and the pH. The compound is most effective at lower levels of contamination, higher ethanol concentration and low pH. The effects of ethanol and temperature on the growth of wine yeasts in the presence of DMDC was reported by Porter and Ough,[38] who found that at wine pH, temperature and ethanol work synergistically with the sterilant. The 'normal' rate of addition is 50–100 mg L^{-1} wine, which is dosed in just prior to bottling. The mode of action of DMDC involves the hydrolysis of the important yeast enzymes, glyceraldehyde-3-phosphate dehydrogenase and alcohol dehydrogenase. The actual site of inactivation involves methoxycarbonylation of the imidazole ring of the histidine residues in those enzymes. DMDC is extremely lethal towards many wine spoilage yeasts, but it is far less effective against spoilage bacteria (at permissible usage levels), and so there is still no absolute substitute for SO_2 and membrane filtration. Although the FDA classifies DMDC as a 'direct secondary food additive' and therefore does not have to appear on the label, there is still the stigma of a 'chemical additive' for many winemakers. A proprietary brand of DMDC, Velcorin®, is produced by the German chemical company, Lanxess. A review of the use of DMDC in wine, and foods generally, was written by Ough in 1983.[39] There is no report of any 'off' flavour or aroma being imparted to wine at the permitted levels of addition.

6.3.3 Sorbic Acid

This short-chained, unsaturated, *trans–trans* fatty acid (2,4-hexa-dienoic acid) was first discovered by A.W. Hoffman, who reacted rowanberry oil with strong alkali. Its chemical structure was determined between 1870 and 1890, and it was first synthesised by Doebner in 1900.[40] Sorbic acid is fungistatic and in the context of winemaking, it is used an as additive just prior to bottling to prevent 're-fermentation' in the bottle. Its sole function is to prevent unwanted yeast growth in wine; it has no fungicidal activity. Its anti-fungal properties were first reported by Gooding in 1945,[41] and it was legalised for use in France in 1959 and in Germany in 1971. It is now a permitted additive in most wine-making countries. It is not a natural constituent of wine and has no bactericidal properties whatsoever. It is best used in wines that are especially prone to yeast spoilage, such as sweet white table wines. The solid form of sorbic acid is sparsely soluble, and it is the potassium salt, which is readily soluble, that is the usual form of application. Once added to wine, potassium sorbate is readily converted to sorbic acid, which is important because it is this undissociated sorbic acid that actually possesses anti-fungal activity, and not the sorbate anion. There will be a greater proportion of the undissociated form in wines of lower pH. One commercial preparation commonly in use is Sorbistat K, which contains the equivalent of 75% sorbic acid.

Conveniently for the winemaker, the efficacy of sorbic acid is largely dependent on the combined presence of ethanol, low pH and SO_2, and it is a fact that the compound is less anti-fungal at lower ethanol levels and in the absence of SO_2. Sorbic acid efficacy is also reduced if a wine contains many viable cells, and so it is important to filter before adding the fungistat. Peynaud[35] recommended that the yeast cell concentration should be reduced to around 100 cfu ml^{-1} before sorbate addition. Before it can exert its effect, sorbic acid has to enter the yeast cell and so has to circumvent the cell membrane and the cell wall. The former carries a net negative charge, and so the relative concentrations of the unionised (sorbic acid) and ionised (sorbate) forms become important. The pK_a of sorbic acid is 4.7, and so pH values lower than this diminish the concentration of the ionised form and proportionally increase the concentration of the desired, undissociated species. At pH 3.0, there is 98.4% of the undissociated form, while at pH 3.7, the level reduces to 92.6%. Once incorporated into the cell, sorbic acid appears to exert its action by inhibiting the dehydrogenase system of the yeast, interfering with the oxidative metabolism of carbon. The most obvious manifestation of yeast growth inhibition is the cessation of budding. Some yeasts, such as *Saccharomyces bailii*, are naturally resistant to sorbic acid and will actively grow in wine to which it has been added. Yeast species differ widely in their resistance to the substance, with *Kloeckera* spp. being among the least resistant.

In the EU, the limit of use is 200 mg L^{-1}, the same as is prescribed for Australia and New Zealand, while in the USA, BATF has defined its limit in wines as 300 mg L^{-1}. The 'normal' rate of addition seems to be 150 mg L^{-1}, which is fine in a 12% volume wine, but provides little protective effect in a 10.5% volume product. This latter level is fairly close to the taste threshold for sensitive subjects (130 mg L^{-1}). Sorbic acid should be added to wine immediately prior to bottling, preferably in conjunction with a low level of free SO_2 (20–30 mg L^{-1}) to act as an anti-oxidant and a bactericide. Sorbic acid should never be seen as a replacement for SO_2.

If sorbic acid is added to red table wines intended for malo-lactic fermentation (*i.e.* low in SO_2), serious tainting can occur when certain malo-lactic bacteria metabolise the additive and produce a taint reminiscent of crushed geranium leaves, 'geranium tone', which is quite undesirable. The compound responsible is 2-ethoxy-hexa-3:5-diene and the mechanism of its generation was studied by Wurdig *et al.*,[42] who concluded that it resulted from the formation of 2,4-hexadien-1-ol and its lactate esters and acetates. It now appears that most of the off-character results from the formation of the ether 2-ethoxyhexa-3,5-diene, which is formed by rearrangement of hexadienol.[43] Of the malo-lactic bacteria present in wine, Edinger and Splittstoesser[44] report only strains of *Oenococcus oeni* carry out the conversion; neither *Pediococcus* nor *Lactobacillus* spp can convert sorbic acid into 'geranium tone'. In an ideal world, sorbic acid should not be used in red table wines.

Sorbic acid is generally considered to be non-toxic and is metabolised by the human body. The acute level of toxicity for the acid is *ca.* 10 g kg^{-1} body weight, which is double that for benzoic acid. The WHO has set the highest acceptable daily intake at 25 mg kg^{-1} body weight. The use of sorbic acid in winemaking is diminishing, not for health reasons but because of the general antipathy to the use of additives. In reality, its use should not be necessary because good hygiene and effective filtration are all that are required for aseptic bottling. Indeed, according to many authorities, the need for sorbic acid is an admission of a lack of confidence in the bottling process. Sorbic acid and its sorbates are labelled E200–E203 for use in food.

6.3.4 Benzoic Acid

The use of benzoic acid (phenylformic acid or benzenecarboxylic acid) as its potassium or sodium salt is commonplace in food preservation, with little apparent adverse effect on the human body; indeed, the acid is a natural constituent of cranberries. Its use is relatively restricted in the USA, especially in drinks where it is BATF-approved for coolers but not for table wines. The legal limit of addition is 1000 mg L^{-1}, but the likelihood of unpleasant sensory effects demand levels lower than this. Like sorbic acid, the anti-microbial activity of benzoic acid is linked to its unionised form. Thus, at pH values below its pK_a of 4.2, the percentage of the active form increases significantly, and the levels needed for microbial growth inhibition decrease concomitantly. The maximum pH for anti-microbial activity is 4.5, while it is most effective

in the range 2.5–4.0. At pH 6.0, its activity is 100 times less than at
4.0. In wine coolers, benzoates are commonly used in conjunction
with potassium sorbate and SO_2. The combination of sorbate and
benzoate provides the desired level of anti-microbial activity at a
dosage that is not regarded as sensorially unpleasant. The sodium
salt is often used to arrest the fermentation of sweet wines. According to Chichester and Tanner,[45] benzoate does not accumulate
in the human body since it becomes conjugated with glycine to
produce hippuric acid, which is excreted. Benzoic acid and its salts
are labelled E210–E213 for food usage.

REFERENCES

1. M. Rüdinger and E. Mayr, *Zeits. Ange. Chemie*, 1928, **29**, 809.
2. M. Rüdinger and E. Mayr, *Kolloid-Zeitschrift*, 1928, **46**, 81.
3. M. Rüdinger and E. Mayr, *Kolloid-Zeitschrift*, 1929, **47**, 141.
4. P. Ribéreau-Gayon, Y. Glories, A. Maujean and D. Dubourdieu (eds), *Handbook of Enology, Vol II, The Chemistry of Wine Stabilisation and Treatments*, Wiley, Chichester, 2000.
5. J.L.P. Rodriguez, A. Weiss and G. Lagaly, *Clay Clay Miner.*, 1977, **25**, 243.
6. L.G. Saywell, *Ind. Eng. Chem.*, 1934, **26**, 379.
7. A.J. McKinnon, R.W. Cattrall and G.R. Scollary, *Am. J. Enol. Vitic.*, 1992, **43**, 166.
8. H.W. Blade and R.B. Boulton, *Am. J. Enol. Vitic.*, 1988, **39**, 193.
9. H.W. Berg and R.M. Keefer, *Am. J. Enol. Vitic.*, 1958, **9**, 180.
10. H.W. Berg and R.M. Keefer, *Am. J. Enol. Vitic.*, 1959, **10**, 105.
11. J. Perin, *Le Vigneron Champenois*, 1977, **98**, 97.
12. T. McKinnon, *Aust. Grapegrower Winemaker*, 1993, **352**, 89.
13. R.B. Boulton, V.L. Singleton, L.F. Bisson and R.E. Kunkee, *Principles and Practices of Winemaking*, Chapman & Hall, New York, 1996.
14. O. Abgueguen and R.B. Boulton, *Am. J. Enol. Vitic.*, 1993, **44**, 65.
15. O. Rhein and F. Neradt, *Am. J. Enol. Vitic.*, 1979, **30**, 265.
16. B.C. Rankine, *Making Good Wine*, Pan Macmillan, Sydney, 2004.
17. R.T. DeSoto and H. Yamada, *Am. J. Enol. Vitic.*, 1963, **14**, 43.
18. T. Muller, G. Wurdig, G. Sholten and G. Friedrich, *Mitteilung Klosterneuburg*, 1990, **40**, 158.

19. S. Lubbers, B. Léger, C. Charpentier and M. Feuillat, *J Int. Sci. Vigne et Vin*, 1993, **1**, 13.
20. D. Dubourdieu and V. Moine-Ledoux, *Brevet d'Invention Français*, 2726284, 1994.
21. T.C. Somers and G. Ziemelis, *Am J. Enol. Vitic.*, 1973, **24**, 47.
22. A.J.W. Ewart, Polysaccharide instability in wine, in T.H. Lee (ed), *Physical Instability of Wine*, Australian Society of Viticulture and Enology, Glen Osmond, South Australia, 1986, 99–108.
23. S. Gorinstein, A. Goldblum, S. Kitov and J. Deutsch, *J. Food Sci.*, 1984, **49**, 251.
24. K.F. Pocock and B.C. Rankine, *Aust. Wine Brew. Spirits Rev.*, 1973, **91**, 42.
25. J. Ribéreau-Gayon and E. Peynaud, *Traité d'Oenologie*, Libraire Polytechnique, Paris, 1961.
26. L. Jakob, *Das Weinblatt*, 1962, **57**, 805.
27. B.C. Rankine and K.F. Pocock, *Aust. Wine Brew. Spirits Rev.*, 1971, **89**, 61.
28. H.W. Berg and M. Akiyoshi, *Am. J. Enol. Vitic.*, 1961, **12**, 107.
29. B.W. Zoecklein, K.C. Fugelsang, B.H. Gump and F.S. Nury, *Wine Analysis and Production*, Chapman & Hall, New York, 1995.
30. F.T. Bioletti, *8th Int. Congress Appl. Chem.*, 1912, **14**, 31.
31. B.L. Wedzicha, *Chemistry of Sulphur Dioxide in Foods*, Elsevier Applied Science, Amsterdam, 1984.
32. C. Weeks, *Am. J. Enol. Vitic.*, 1969, **20**, 31.
33. P.J. Guthrie, *Can. J. Chem.*, 1979, **57**, 454.
34. H.W. Berg, F. Filipello, E. Hinreiner and A.D. Webb, *Food Technol.*, 1955, **9**, 23.
35. E. Peynaud, *Knowing and Making Wine*, Wiley, New York, 1984.
36. F.W. Beech, L.F. Burroughs, C.F. Timberlake and G.C. Whiting, *Bull. OIV*, 1979, **52**(586), 1001.
37. L.A. Sayavedra-Soto and M.W. Montgomery, *J. Food Sci.*, 1986, **51**, 1531.
38. L.J. Porter and C.S. Ough, *Am. J. Enol. Vitic.*, 1982, **33**, 222.
39. C.S. Ough, Dimethyl carbonate and diethyl carbonate, in A.L. Branen and P.M. Davidson (eds), *Antimicrobials in Foods*, Marcel Dekker, New York, 1983, 299–325.
40. O. Doebner, *Berichte der Deutschen Chemischen Gesellschaft*, 1900, **33**, 2140.

41. C.M. Gooding, *Process of Inhibiting Growth of Molds*, U.S. Patent 2 379 294, 1945.
42. G. Wurdig, H.A. Schlotter and E. Klein, *Allg. Deut. Weinfachztg.*, 1974, **110**, 578.
43. E.A. Crowell and J.F. Guymon, *Am. J. Enol. Vitic.*, 1975, **26**, 96.
44. W.D. Edinger and D.F. Splittstoesser, *Am. J. Enol. Vitic.*, 1986, **37**, 34.
45. D.F. Chichester and F.W. Tanner, Antimicrobial food additives, in T.E. Furia (ed.), *Handbook of Food Additives*, 2nd edn, CRC Press, Cleveland, 1972, 115–184.

CHAPTER 7

Maturation and Ageing

The ancients, notably the Romans, stored and transported their wine in large, stoppered earthenware vessels called amphorae, while other early cultures made use of inverted animal skins to achieve the same aims. The invention of glass bottles in the early 17th century permitted wine to be stored and transported in smaller containers, and once cork was used to seal bottles, it became feasible to age wine in bottle at an even temperature. The importance of the glass bottle to the wine industry can be ascertained from an old French maxim, which states, 'There are no great wines; only great bottles of wine'.

In its broadest sense, wine ageing encompasses all the reactions and changes that occur after the first racking, post-fermentation, which lead to the improvement of a wine, rather than its deterioration. There has always been some debate as to the difference between 'maturation' and 'ageing', but for convenience it is permissible to define a bulk storage period as 'maturation', and storage in a smaller container as 'ageing'. The latter, of course, very often takes place in a bottle. In another age, malo-lactic fermentation (MLF), tartrate stabilisation and clarification would have been considered as being part of the maturation/ageing processes, but they are now regarded in a different light and are considered in this work in their own right. After years of study, it is now evident that maturation and ageing should be thought of as a family of interlinked, chemical changes. One should think in terms of a 'mature' wine being 'ripe' to bottle, in which it will 'age', as appropriate. All of the processes, including blending, that bring a wine to the optimum condition for bottling, should be complete by the time that a wine is considered mature. Ageing will then proceed *via* a series of very slow reactions, which will occur under a vastly different set of conditions.

Given that there is a semantic difference between maturation and ageing, it should be stressed that there is a considerable difference in the chemistry of the two processes. Maturation, whether in vat or barrel, is essentially all about oxidative reactions, although it also entails the extraction of wood compounds. 'In-bottle' ageing does not involve oxygen or wood extractives, although many important chemical changes are still undertaken. Such processes are generally considered to improve the quality of a wine, especially of red wines made from the better quality grapes, such as Cabernet Sauvignon, Tempranillo and Nebbiolo. The same thing can be said about white wines made from, for example, Chardonnay and Riesling grapes. As a general rule, only those wines with a high phenolic content and/or acidity will improve dramatically with age. According to Robinson,[1] only some 10% of red wines are actually matured in cask. It is a fact of life that, with many wines, the modifications made during ageing are synonymous with an increase price, and while the results of the slow chemical modifications may be attractive to the consumer, the price increase may not!

In modern wineries, wines can be bulk stored in large stainless steel tanks, which are far removed from the large wooden vats of yesteryear. The flavours that develop in steel tanks are, of course, different from those arising from wine stored in wood. Indeed, maturation in wood is more efficient if it is carried out in smaller (say, 225 L) barrels, even though this is inevitably more expensive (cost of barrels and labour). Maturation in wood can vary, time-wise, from 3 months to 3 years or even longer, and wine should preferably be kept at a constant temperature, somewhere around 15°C, a temperature not uncommon in underground cellars. Some wines, such as Spanish Riojas, require some degree of 'oakiness' in order to satisfy the connoisseur. As Boulton *et al.*[2] observed, the methods of wine maturation and ageing considered traditional today were not deliberately developed, but arose out of concerted efforts to extend the availability of wines, *i.e.* to aid their preservation. Procedures such as the use of wooden casks and corks were simply the result of utilising natural materials that were at hand.

The general objectives of maturation and ageing fall into four categories: subtraction, addition, carry over and multiplication. To these can be added the fact that these objectives must be integrated, and the mode of integration will be determined by the desired style of the wine. Subtraction involves the removal of unwanted characters, such as too much CO_2 or a discernible sulfidic taint. In addition, young, 'green' wines are often excessively 'tart' or astringent (tannic). All of these defects can be ameliorated during maturation/ageing. Some imperfections, such as H_2S, are best avoided in the first place, but if they do arise, they should be eradicated as soon as possible, else they interact and give rise to modified compounds/aromas (*e.g.* 'skunky', in the case of H_2S), which are even more difficult to remove. Examples of addition would be the extraction of oak volatiles in red wines and the imparted oxidised flavour of sherries. Such additions should complement the underlying flavours of the particular wine in question and not be allowed to override them. The development of colour in some red wines also comes into this category of objective. For most wines, carry over concerns the survival and retention of the grape characters after maturation/ageing, and is especially important when varietal characters are required. When multiplication is considered, we are talking about the increase in complexity of a wine, which should not be to the detriment of its fundamental nature. Aficionados maintain that this facet of winemaking can be likened to the individual instruments in an orchestra, all of which are playing to the same score. Boulton *et al.*[2] give an excellent, albeit hypothetical, example of the possible consequences of multiplication, whereby they consider a wine in which, at the end of fermentation, there are four alcohols present: ethanol and three amyl alcohol isomers. If each alcohol was partially oxidised to an aldehyde, and these in turn were oxidised to carboxy-acids, then the 4 original compounds would have become 12. If each alcohol esterified with each acid, 16 additional compounds would result, giving a total of 28. The aldehydes might form acetals and hemiacetals with the alcohols, and, theoretically, this could yield 64 acetals and 16 hemiacetals to give a possible 108 compounds from the original 4! All this would certainly add to the complexity of the wine. Under another set of circumstances, if the four alcohols had been completely oxidised to acids, the wine would have contained just four additional (uninteresting) compounds.

Most white wines and many reds are sold ready to drink, and they do not require any further ageing in bottle, but it should be realised that there is often an inadvertent delay between bottling and the sale/consumption of a wine. Many bottled wines are classified according to their 'drinkability', with terms such as 'not ready', 'at peak' and 'fading' becoming part of everyday parlance.

7.1   *SUR LIE* STORAGE OF WINE

As a rule of thumb, once a microbial population has fulfilled its technological role in the preparation of an alcoholic beverage, the cells should be removed. The practice of storing wine *sur lie* rather goes against this concept, for it describes the practice of leaving newly fermented wine on the gross yeast sediment (lees) for an extended period of time, usually for a period of 6–9 months. The most renowned instances of such a method relate to the great white wines of Burgundy, such as Chardonnay, but there is now a widespread interest in lees utilisation, and *sur lie* storage is becoming a more important facet of red wine manufacture.

In an extensive paper outlining the practical aspects of lees utilisation in the making of red, white and rosé wines, Delteil[3] maintains that distinction should be made between 'light' and 'heavy' lees, especially in view of the fact that he maintains that there is no advantage to be had from storing wine on the latter. Heavy lees are the particles that are deposited from a pectin-free wine within 24 h of the cessation of fermentation. Such material is composed of large particles (> 100 µm), consisting of plant debris, bitartrate crystals and protein–polysaccharide–tannin complexes. The exact composition of these lees will vary between red wines, and white and rosé wines, and they are continually being slowly formed throughout wine maturation. As Delteil says 'heavy lees are never very interesting to the winemaker, and should be removed whenever necessary'. Many risks are associated with leaving wine on heavy lees, particularly in respect of bad odours and 'grassy' flavours. They also bind with SO_2, and hence increase the likelihood of microbial contamination; the higher the pH of the wine, the higher the risk. Light lees, on the other hand, can be defined as those smaller (1–25 µm) particles that precipitate from the wine more than 24 h post-fermentation and/or any movement of the wine (*e.g.* racking or stirring), and consist of yeasts, bacteria, tartaric acid, protein–tannin complexes and some polysaccharides. Light lees are highly regarded by some winemakers, and have a similar composition irrespective of whether they arise from red, white or rosé wines. Indeed, lees are so highly prized in Burgundy that those from a great wine, such as Montrachet, may be transferred to a lesser wine to ennoble it. In addition, traditional lees usage in this region of France can be somewhat intricate; for instance, wine may be aerobically racked off heavy lees in March, when MLF has been completed, and placed in wooden barrels on light lees (to nourish the wine), in the presence of SO_2. A further anaerobic racking onto more light lees often occurs in July.

Storage on light lees can give a wine a definite advantage in structural balance, complexity and stability (particularly protein stability), and being strongly anti-oxidant, the sediment plays a significant role in keeping a wine fresh and crisp. Contact with lees contributes to wine complexity by integrating yeast characters with fruit and wood flavours, and lees management is therefore an important flavour and aroma consideration. The technique also reduces the 'yellowness' of white wines and is said to impart a creamy texture to their mouthfeel, besides introducing 'smoky', 'toasty' and 'cheesy' flavours. During lees contact, the yeast cells undergo enzymatic hydrolysis of some of their intracellular constituents, perhaps the most important for the winemaker being proteolysis, which liberates peptides and amino acids into the wine. It is known that the latter can act as flavour precursors, thus enhancing wine complexity.

The nitrogenous compounds nourish a wine during ageing and serve to support the microbes responsible for the MLF, where this occurs, and they appear to be actively released from the yeast cell, rather than merely being set free after the cell disintegrates. LeRoy *et al.*[4] have shown that, during the ageing of champagne, maximum production of these nitrogenous compounds occurs after 5 years contact with lees. Cell wall polysaccharides, especially mannoproteins, are also released into the wine, and these provide a degree of 'sweetness' as a result of their binding with wood-derived phenols and organic acids. This promotes the harmony of some of the wine's structural elements, mainly by softening tannins. The release of these macromolecules arises as a result of the activity of β-glucanases present in the yeast cell wall. The enzymes remain active for several months after the death of the cell, thus slowly releasing mannoproteins into the wine. Mannoprotein levels in wine can increase by up to 30% over a 3-month period on the lees. Yeast-derived mannoproteins can act as inhibitors of the formation of crystalline deposits in wine; certainly the longer the period of lees contact, the greater the degree of potassium bitartrate stability. Storage on lees also modifies 'oaky' aromas, mainly due to their ability to bind with wood-derived compounds, such as vanillin.

Leaving aside the role of yeast sediment in champagne production, lees contact exerts its maximum effect during the making of muscadet. The best examples of this style are invariably bottled straight from unracked barrels, without any resort to filtration. Because of this, muscadets begin their maturation in a reducing environment, and if care is not taken, malodorous notes can easily be imparted. Of notable importance here, is the formation of H_2S from autolysed yeast cells. *Sur lie* wine storage should ideally be carried out in small barrels, so that the lees can settle around the walls and not form thick sediments in the bottom, which are conducive to the anaerobic formation of H_2S. In small casks, wine can be kept for up to a year on lees before sulfidic taints become detectable, whereas in large tanks, where the yeast settles out at the base, the unwanted aroma can manifest itself within a fortnight.

One of the ways to prevent the build-up of anaerobic conditions in and around the yeast sediment, is by stirring, a process called *battonage*. The frequency of stirring is important, since it generates oxidative changes and alters the sensory balance between fruit, yeast and wood by enhancing the 'yeast' component at the expense of the 'fruit' component and, to a lesser extent, the 'wood' component.

So far, none of the volatile compounds resulting from yeast autolysis have been unequivocally identified, although increase in ethyl lactate and diethyl succinate have been reported in wine stored on lees for 18 months. Some clue as to the nature of these volatiles might come from the discernible aroma/flavour changes in wine that is stored *sur lie*. In making Chardonnay, for example, once the wine has been stored on lees, it will be less 'buttery' and more 'toasty' than if there has been little or no lees contact. In a slightly different vein, some champagnes stored for 18 months on lees develop far stronger 'vanilla' and 'butter' aromas than do the base wines from which they are made.

7.2 OAK AND WINE

The organoleptic complexity of many beverages, including wines, brandies and whiskies, can be increased by ageing in wooden cask; certainly, most quality red wines are matured for some months, or even years, in oak cooperage, before they are aged in bottles. The relationship between wine and wood was appreciated by the Romans, who realised that it gave many of their wines an added dimension. Over the centuries, various kinds of wood have been used to aid winemaking, and some, such as chestnut, walnut and cherry, still have uses in certain European countries. The oak tree, however, is king, its wood having certain physical features that make it highly beneficial in the winery; it is light, it has high tensile strength, it is malleable and it is relatively impermeable to water; all these characters, in fact, have made it suitable for building ships! In addition, oak wood contains a variety of beneficial phenolic compounds that can be extracted into wine. Apart from the contribution of the wood itself, wine produced in traditional small cooperage will be subjected to routine handling procedures, which will result in intermittent aeration.

The principal species of oak used for making wine barrels are *Quercus alba* (native to central and eastern North America), *Q. petraea* and *Q. robur*, which are indigenous to both Europe (including the UK) and western Asia. Within Europe, there are oak industries in France, Germany, Portugal and the Balkans, with the former being predominant. Growth of the tree varies markedly in different geographical regions, and it is from areas where oak growth is slowest that the best wood for winemaking barrels emanates. Such wood has a tighter grain, which permits more 'oakiness' to be extracted and imparted into the wine. Certainly, more desirable phenols can be extracted from wood produced by slower growing trees. The extraction of phenols and other substances from oak depends upon a number of factors, including the ratio of oak surface to wine volume, humidity, temperature, ethanol level and the number of times that the barrel has been filled.

In addition to the origin of the wood, the coopering method also affects barrel character, particularly in respect of whether the individual staves have been bent by use of direct heat. This browns the wood and modifies the surface extractables – a technique called 'toasting', and has a major effect on final wine flavour. Effects can be undesirable if heating is overdone, and as a consequence, the wine picks up 'smoky' or 'bacon-like' flavours. As a result of toasting, a number of new compounds are introduced onto the wood, and this contribute to the 'oakiness'. They include some 18 phenolics, 7 aldehydes and the same number of lactones. Singleton[5] has provided an extensive resumé of the contribution made by wood to the maturation of wine.

As has been said, it was in ancient times that mankind began to appreciate the ameliorating effect of oak on wine, and that more interesting (complex) beverages resulted from fermentation/storage in oak barrels. It was much later that it was realised that such modifications to the wine were due to slow oxidation reactions resulting from the ingress of oxygen through the wood itself, and through the working apertures (bung-hole, *etc.*). It is now generally accepted that during barrel ageing, the winemaker is looking for a slow, controlled oxygenation, and at the same time the avoidance of over-oxidation of his product. As an aid to this aim, wine lees conveniently encourage oxygenation and deter oxidation. Experience has shown that, if barrels are stored in cellars with low relative humidity, then a considerable amount of wine can be lost by evaporation, something that serves to hasten the barrel-ageing reactions and concentrate the wine and enhance its flavour. It has also been noted that the larger the barrel, the slower the ageing processes, particularly oxidation.

Winemakers soon discovered that the age of a cask played an important role, there being a world of difference between 'new' and 'old' containers. It is generally agreed that, for making decent quality wines, in which oak aroma and flavour are important, the useful life of a barrel is between 4 and 6 years. Rous and Alderson[6] showed that extraction of phenols from new casks indicates diffusion control: the extraction from the second fill of the barrel being much less than the first fill, and by the third fill, the extraction curves were linear, indicating that hydrolysis of phenols becomes important for extraction. Oak barrels that have been used repeatedly impart little or no 'oakiness' to wine, this being partly due to the accumulation of tartrate crystals on their inner surface. Such deposits increase the impermeability of the wood, thus slowing ageing, but, on the credit side, serve as sites for the enhanced precipitation of more tartrate from newly introduced wine. In general, the ageing of wine in oak aims to impart flavours, either by the extraction of volatile oak compounds from new wood or through slow oxidative changes from older wood.

During oak ageing, wine flavour is changed *via* a number of mechanisms: slow oxidation; extraction of hydrolysed or ethanolysed wood components; evaporation of ethanol, water and volatiles; and a variety of chemical reactions, resulting in the formation of entirely new compounds. Perhaps the most important 'wood compounds' during maturation are the oak tannins, which, because of their very low redox potential, increase the capacity of the wine to consume oxygen and lead to the production of peroxides and acetaldehyde by oxidation.[7] The oak tannins are mainly ellagitannins – hydrolysable glucose esters of ellagic and gallic acids. The main ones identified are vescalagin and castalagin, and others include roburins A–E, grandinin, pedunculagin, castalin, vescalin and pentagalloyl glucose. Hydrolysis of oak tannins yields glucose, ellagic acid and gallic acid. At high levels, ellagic acid can form a yellow precipitate in wine, but gallic acid always remains soluble. Each ellagitannin has 15 highly reactive phenolic hydroxyl groups per molecule, which render these compounds more readily oxidised than flavonoid phenols and produce hydrogen peroxide and further reactive quinones and polymers by coupled oxidation. Hydrogen peroxide leads to acetaldehyde production in wines, which in turn has an influence on phenol and anthocyanin polymerisation reactions. According to Vivas and Glories,[7] the liberation of acetaldehyde by this means is responsible for the improved colour and colour stability in wines that have been matured in oak, and it also helps to protect wines against oxidation. Klumpers *et al.*[8] have reported that the heartwood of the oak *Q. robur* contains *ca.* 10% by weight of C-glucosidic ellagitannins.

Directly extractable compounds in oak include vanillin, hydroxymethylfurfural, maltol and ethylmaltol. The 'vanilla' flavour, characteristic of most oak-aged alcoholic beverages, arises from the oxidation of coniferyl and sinapic alcohols, which produce vanillin and syringaldehyde, respectively. The two alcohols are, themselves, produced by the ethanolysis of oak lignins. The 'spicy' aroma imparted to wines that have been aged in new oak is mainly due to eugenol and a few other volatile phenolics. An extensive review of the contribution made by wood to the flavour of alcoholic beverages has been provided by Maga.[9]

Another means of introducing desirable oak character to a wine is by using wood chips or finely ground shavings, especially during fermentation. Singleton and Draper[10] reported that during the ageing of wines in metal containers, up to 90% of the total potential extraction from wood could be achieved within 1 week by using oak chips. It is generally agreed that it is very difficult to totally mimic the maturation of wine in barrel. Irrespective of how much care is taken, most products eventually deteriorate (usually due to excessive oxidation) if ageing is too lengthy.

7.3 MATURATION REACTIONS IN RED WINE

During ageing in wood, red wines undergo a number of oxidative and non-oxidative polymerisation reactions, which involve phenols and anthocyanins. The colour of the wine moves towards browner hues, the colour density decreases and the astringency decreases, producing a more 'supple' mouthfeel. The colour change is measured as a decrease in absorbance at 520 nm in the red range, and an increase in absorbance around 420 nm in a more orange/brown range.[11] In a young red wine, most of the colour is due to monomeric antho-cyanins, but after a year, at least 50% of the colour can be attributed to polymeric anthocyanin-derived pigments. Compared to mono-meric anthocyanins, these pigments are less affected by changes in pH, temperature and SO_2, and so they give greater colour stability.[12] The incorporation of anthocyanins in tannin polymers helps them to remain in solution. Over a period of time, as polymerisation increases, aggregates form, which eventually become too large or too insoluble to remain in solution, and a coloured precipitate forms, along with a reduction in wine colour density.[13] In synthetic wine systems, Timber-lake and Bridle[14] reported that they could demonstrate at least six important categories of reaction involving anthocyanins and phenols:

  (i) reactions between anthocyanins and phenolics;
 (ii) transformations of phenols alone;
(iii) degradations of anthocyanins alone;
 (iv) reactions between anthocyanins and acetaldehyde;
  (v) reactions between anthocyanins, acetaldehyde and phenolics;
 (vi) reactions between phenolics and acetaldehyde.

The presence of acetaldehyde in wines is due to either microbial action, especially during fermentation, or slow oxidation of ethanol by coupled oxidation of *o*-hydroxy phenols.[15] The effect of acetal-dehyde addition to red wines (or to port wines) is an immediate increase in colour density and polymeric pigments and a shift towards a more violet hue. This is followed by a decrease in colour, and the formation of a coloured precipitate.[16]

7.4 MICRO-OXYGENATION (MOx)

This is a relatively new, now widely popular, technique that involves oxygenation of fully fermented wine. The technique has been employed commercially in France since 1991, when Patrick DuCournau and Thierry Lemaire of the French company Oenodev began experimenting on the wines of Madiran in south-western France. A period of successful research and development resulted in the sale of the first commercial equipment in 1996, and the technology is widely used in France and the USA. The theory behind the process is quite simple and is based on the notion that if the expensive process of wine maturation in wood is at least partially attributable to the slow ingress of oxygen, then why not pass a steady stream of the gas through the wine to achieve the same effect? The dose of oxygen required is minute, something in the order of 1 cm^3 oxygen per litre of wine per month, which has to be supplied as microscopic bubbles. The simplest means of effecting this is to force the gas through an unglazed, cylindrical, porous pot, something like a filter candle. Equipment designed to deliver the controlled mini-doses of oxygen (in this case, 0.5–60 cm^3 oxygen per litre of wine per month) has been produced by Oenodev and is illustrated in Figure 1.

The role of oxygen in the evolution of wines, especially red wines, has long been a subject of study. In tank, procedures such as racking at intervals during maturation result in the exposure of that wine to large volumes of air for short periods. On the other hand, wines stored in barrel are constantly exposed to small quantities of air that enters through the bung, during topping-up and, arguably, *via* the wooden staves. Most oenologists would agree that sensory assessment of the same wine matured in barrel is superior in terms of structure and flavour, and, although the ingress of oxygen by either means of storage is a matter of conjecture, it is considered to be a major factor in the preference of winemakers for ageing wine in oak. Analysis of wines stored in tank and wood has shown that their dissolved oxygen levels are comparable, typically 20 ppb.

The action of oxygen on wine is very complicated and depends on factors such as temperature, level of SO_2 and phenolic composition, and the introduction of regular, small amounts of the gas has been shown to enhance some organoleptic properties, such as

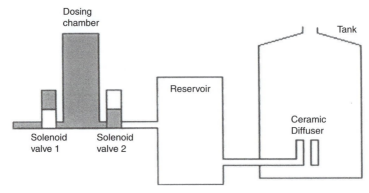

Figure 1   *Schematic diagram of micro-oxygenation (MOx) equipment*
(Reproduced by kind permission of Dr Thierry Lemaire)

aroma and astringency. The main effect of oxygen is to modify the wines' polyphenolic structure, with the result that it will be 'smoother' and 'softer', with better tannin integration.

According to its inventors, micro-oxygenation (MOx) treatment of a wine should begin with an assessment of that wine and a review of the winemaker's objectives for it. From this knowledge, an MOx programme can be instigated. The process of MOx can be started at any stage during winemaking, but, in practice, it is most effectively begun at the end of ethanolic fermentation and prior to MLF. As intimated, MOx is responsible for significant changes in the organoleptic qualities of a wine, and these occur in two main phases, which, chronologically, are the structuring phase and the harmonisation phase.

The former takes place at the end of primary fermentation and normally before MLF, and involves relatively high doses of oxygen (10–60 cm^3 per litre of wine per month) over a period lasting from 1 to 6 months. If treatment is commenced after the completion of MLF, slightly lower charges of oxygen can be employed. During this period, oxidative changes occur to the structure of anthocyanins, making them more stable, and the taste of the wine is characterised by an increase in the aggressiveness and intensity of tannins on the palate. Any 'vegetal' character is diminished; in short, the wine becomes totally 'out of harmony'. The more the tannins and phenolic compounds in the original wine, the longer the structuring phase will take, and the greater the structuring effect. Any practice such as early fining, which reduces phenolic content, will diminish the structuring effect. For red wines meant to be left on lees, it is essential to allow the sediment to settle before oxygenating, because lees will avidly take up oxygen and make it unavailable for structuring. Towards the end of this phase, the organoleptic changes start to reverse and the wine begins to 'soften'. One important occurrence is that any reducing character in the wine (*e.g.* sulfides) is ameliorated.

The harmonisation phase, which would normally take place after any MLF, is characterised by an increase in tannin softness and general wine complexity. Harmonisation employs much lower doses of oxygen (0.1–10 cm^3 per litre of wine per month) over a longer period. The length of this phase is determined by the structuring phase, and, as a rule of thumb, harmonisation should be twice as long as structuring. As one would expect, harmonisation presents the winemaker with more problems, because the wine at this time is at its most sensitive and changes are irreversible. The main risk involved is to produce wine with an over-dry palate and a lack of freshness. Such problems are more likely to occur in wines of lower alcoholic strength. A schematic representation of the organoleptic phases encountered in wine during MOx is given in Figure 2.

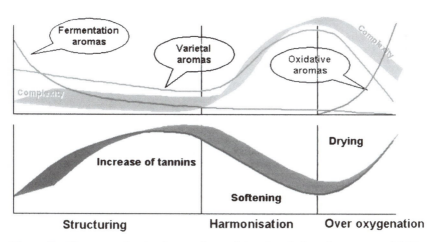

Figure 2 *The organoleptic phases observed in wine during the process of MOx (after Lemaire[17])*
(Reproduced by kind permission of Dr Thierry Lemaire)

The main criteria influencing MOx are the original composition of the wine, temperature, pH and, of course, volume of the oxygen added. Needless to say, the delivery of more oxygen than is necessary to effect the desired changes, results in the oxidation of other wine substrates, a process detrimental to wine quality. Devotees of this technology are adamant that its benefits are primarily an increase in overall wine quality, rather than any economic factor (*e.g.* a saving on the expense of barrels). An erudite mini-review of MOx in modern winemaking has recently been penned by Cottrell.[18]

7.5 CORKS

Closures are now a hotly debated topic, with the issue of cork 'taint' being seen as a serious and expensive drawback; annual damage to the wine industry being estimated at $10 billion. The 'mouldy', 'musty', 'earthy' or even 'medicinal' smell that masks or dominates the fruit aroma of a wine, and generally impairs its quality, is responsible for that wine being described as 'corked'. This may be an over-simplification, and cork manufacturers are vehement and quick to point out that 'cork taint' is a misnomer – the cork itself does not affect the wine, but the cork may have become contaminated at some stage. Oenological studies indicate that from 2 to 5% of all wines are affected by some sort of taint, of which cork taint is one contributor. The first identified, and probably the primary cause of the problem, is 2,4,6-trichloroanisole (TCA), whose aroma detection threshold in humans seems to range from 1.4 to 4 ng L^{-1}, depending on wine type. Land[19] reported that 50% of the population can detect 1 mg of TCA in 5 million gallons of water, but that human sensitivity is decreased by the presence of ethanol. Other chloroanisole derivatives identified and implicated in taint are, 2,3,4,6-tetrachloroanisole (TeCA), pentachloroanisole and 2,4,6-tribromoanisole (TBA). The second most regularly detected cause of taint, TeCA, has a slightly higher detection threshold, being recorded as 14 ng L^{-1} in Pinot Noir wine. There are undoubtedly many more contributory compounds, but the above-mentioned are the most notorious.

Most would agree that during the winemaking process, the most common causes of 'corking' are the cork closure and contaminated oak products, such as casks or shavings. Experience has shown that if the cork alone is implicated, then TCA will be responsible, whereas if oak is the culprit, TCA and TeCA can be identified.

Other likely sources of contamination are unhygienic machinery (especially bottling equipment) and airborne moulds (particularly in the cellar). Some authorities maintain that the historical extensive use of pesticides, such as pentachlorophenol, in cork forests may have exacerbated the TCA problem, which is not confined to wine. TCA is a worldwide pollutant affecting many foods and beverages. A reliable method has now been developed to detect and remove TCA from wine before it is bottled. Analysis involves a technique known as 'headspace-solid phase micro-extraction' (HS-SPME), which was developed by the Australian Wine Research Institute.

Cork is a unique substance and a long-proven closure for wine. No other stopper combines cork's inert nature, impermeability to liquids, flexibility, sealing ability and resilience. A cork usually gives an airtight seal, especially if the bottle is stored lying down at a cool temperature (*ca.* 16°C). Being a natural product, it is renewable, recyclable and biodegradable; *i.e.* environmentally friendly. Wine corks are graded by quality into seven categories, there being different types of cork for fortified, sparkling and still wines. To harvest the cork, bark is stripped from the tree once every 9 years, thus allowing time for regeneration. Natural corks are punched whole from the best quality bark. The Twin Top® cork, known as the 'technical cork', is assembled from granulated material, with a disk of natural cork at either end. The granulated cork body is made from reconstructed high-grade natural cork left over from cork punching. Champagne corks are made from a single moulding of high-quality, natural granulated cork, giving it uniform physical and mechanical characteristics. The end of the cork which will be in contact with the wine will consist of two or three discs of fine natural cork. 'Colmated' corks are made from lower grade natural cork, with larger lenticels (structural imperfections). After punching, the corks are coated with a mixture of fine cork particles and natural latex to seal the surface and improve its performance as a stopper. Artificial 'corks' are used, but can sometimes be very difficult to extract. Screw caps are very efficient closures and protect the wine well, but they do not really suit the image of the wine bottle – or wine consumers therefrom!

REFERENCES

1. J. Robinson, *Jancis Robinson's Wine Course*, BBC Books, London. 1995.

2. R.B. Boulton, V.L. Singleton, L.F. Bisson and R.E. Kunkee, *Principle and Practices of Winemaking*, Chapman and Hall, London. 1996.
3. D. Delteil, *Aust. N Z Grapegrower Winemaker*, 2002, Annual Technical Issue No. **461**a, 104.
4. M. LeRoy, M. Charpontier, B. Duteurtre, M. Feuillat and C. Charpentier, *Am. J. Enol. Vitic.*, 1990, **41**, 21.
5. V.L. Singleton, *Adv. Chem.*, 1974, **137**, 254.
6. C. Rous and B. Alderson, *Am. J. Enol. Vitic.*, 1983, **34**, 211.
7. N. Vivas and Y. Glories, *Am. J. Enol. Vitic.*, 1996, **47**, 103.
8. J. Klumpers, A. Scalbert and G. Janin, *Phytochemistry*, 1994, **36**, 1249.
9. J.A. Maga, *Food Rev. Int.*, 1989, **5**, 39.
10. V.L. Singleton and D.E. Draper, *Am. J. Enol. Vitic.*, 1961, **12**, 152.
11. T.C. Somers and E. Verette, Phenolic composition of natural wine types, in H.F. Linskens and J.F. Jackson (eds), *Modern Methods of Plant Analysis*, Springer-Verlag, Berlin, 1988, 219–257.
12. T.C. Somers, *Phytochemistry*, 1971, **10**, 2175.
13. C. Saucier, G. Bougeois, C. Vitry, D. Roux and Y. Glories, *J. Agric. Food. Chem.*, 1997, **45**, 1045.
14. C.F. Timberlake and P. Bridle, *Am. J. Enol. Vitic.*, 1976, **27**, 97.
15. H.L. Wildenradt and V.L. Singleton, *Am. J. Enol. Vitic.*, 1974, **25**, 119.
16. P. Ribéreau-Gayon, P. Pontallier and Y. Glories, *J. Sci. Food. Agric.*, 1983, **34**, 505.
17. T. Lemaire, *La micro-oxygénation des vins*, École Nationale Supérieur Agronomique, Montpellier, 2004.
18. T.H.E. Cottrell, *Wine Bus. Mon.*, 2004, **11**(2), 12.
19. D.G. Land, Taints – Causes and prevention, in J.R. Piggott and A. Paterson (eds), *Distilled Beverage Flavour*, Ellis Horwood, Chichester, 1989, 17–32.

CHAPTER 8

Fortified Wines

Fortified wines are characterised by their high concentrations of ethanol and sugar. Often called 'dessert wines', they may be defined as beverages in which some of the ethanol is derived from the fermentation of grape juice, and some from the addition of a distilled spirit, usually something akin to brandy. Such wines are all aged in a manner that would not be applicable to table wines, and involves much blending and oxidative storage. With its usual wisdom, the EU has decreed that these wines should now be described as 'liqueur wines'. According to Goswell and Kunkee,[1] such wines may be placed into two categories, namely, those which consist of, and owe their characteristic flavour to, wholly or partially fermented grapes and distilled spirits; and those which in addition contain, and owe part of their flavour to, non-grape products. In the first category are placed the products most usually sold under such names as 'port', 'sherry' and 'madeira'. In the second class we would find vermouths, aperitif wines and some other wines flavoured with fruits other than the grape. Wines in the latter category are often referred to as 'flavoured fortified wines'. The *OIV* defines fortified wines as: 'Special wines having a total alcohol content (both potential and actual) between 15 and 22% by volume'. They then recognise two types: 'spiritous wines', which receive only brandy, or rectified food quality alcohol during fermentation, or 'syrupy sweet wines', which can receive concentrated must, or *mistelle*, in addition to brandy or alcohol. In both instances, the natural alcohol potential of the grape juice must be 'at least 12%', and 'at least 4% of the alcohol in the final product must originate from fermentation'.

When compared to still table wines, whose manufacture goes back to antiquity, fortified wines have a relatively short history, and only emerged after methods for the distillation of alcohol had become commercially available. Thanks to the Dutch, and their work on distilled spirits, fortification became a commercial proposition in Europe in the 17th century, and since that time the production of wines with added alcohol has become widespread in warm winemaking areas, particularly in the Mediterranean basin. A number of regions in southern Europe, notably Jerez de la Frontera in Spain, the Douro valley in Portugal and the island of Madeira, specialised in the manufacture of such wines, most of their production being exported to cooler regions of the world. This is significant because these areas enjoy a very warm climate, and it is thought that fortification evolved to take advantage of the preservative properties of ethanol, rather than for any specialised flavouring character. Since those early days, the sterilising and stabilising effect of adding several degrees of alcohol to table wines has often been used to protect wines shipped in barrel. There are those, however, who feel that fortification was originally practised to conserve wine sweetness, albeit while lessening the likelihood of microbial contamination. Ribéreau-Gayon *et al.*[2] state quite categorically that these wines were created in the past as a response to technical problems that were frequently encountered in warm regions. Sugar-rich grapes and elevated temperatures resulted in explosive fermentations, which often became 'stuck'. This partially fermented wine was very unstable, particularly to bacterial attack, and the addition of alcohol was a simple means of stabilising it. The endproduct was sweet and alcoholic, with an agreeable taste.

Characteristic production methods have evolved in each of the three above-mentioned regions, and this has resulted in the very different styles of wine that we now know definitively as sherry, port and madeira. Some products, such as marsala from Sicily, are made from, a hybridisation of the three main processes, in the case of marsala a combination of the sherry and madeira processes. Theoretically, fortified wines can be prepared by the addition of any sort of potable spirit to any type of table wine, but experience has shown that, in reality, many of the resultant concoctions are not always organoleptically very attractive. Fortified wines normally have an alcohol content of between 15 and 20% vol (EU regulations define liqueur wines, generally, as 'those having an acquired ABV of between 15 and 22%'), but specifications vary. The minimum alcohol level of fortified wine for sale in Australia, under the State Food & Drug Acts, is 17% vol, and the maximum is 23% vol. The high levels of alcohol in these wines normally dictate that their consumption is either as an aperitif, or as a dessert wine at the end of a meal.

Australia's most successful answers to the great fortified wines of Europe are its creamy, long-matured, sweet 'Muscats'. Brown Brothers' vineyards in north-east Victoria provide the grapes for one of the most renowned of these luscious 'liqueur-wines'.

The apparent climatic requirement for producing fortified wines is a hot dry summer, which yields grapes with high sugar and low acid. The higher the must sugar content, the less alcohol needs to be added, thus reducing production costs. Grape varietal characters are less important than for table wines, although port, sherry and madeira makers do have their own traditional varieties. Most flavour and aroma characteristics are imparted during the processing of these wines, but, ultimately, the three major groups of fortified wine owe much of their distinctive character to their ageing and blending regimes. Traditional fortified wines have evolved within regions where the climate and soil conditions are not conducive to the cultivation of grape varieties suitable for high-quality table wines. The size of the crop, and the sweetness of the grapes (often in excess of 14° Baumé) are important criteria.

The amount of spirit added varies considerably with wine style, as does the timing of the addition, and whether it is affected in one step, or in several. The possibilities are:

(i) *Before fermentation.* Some wines are fortified before the onset of fermentation, and consequently never ferment out at all. A few table wines are made in this way (called *vins de liqueur* in France). As would be expected, they lack complexity and mature very little in bottle.

(ii) *During fermentation.* All vintage and tawny ports, and many muscats are fortified thus, which effectively arrests fermentation at the point of addition. The quantity of spirit added, and its timing determine the ultimate style of the wine. Some wines are fortified in this way in order to purposely make them sweet, the *vins doux naturels* from southern France (as exemplified by Muscat de Beaumes de Venise) perhaps being

the best example. Sweetness (residual sugar) can be controlled by the timing of the spirit addition, which is usually
added in a quantity sufficient to bring the alcohol content
up to 15–18% ABV. Subsequent treatments then influence
the nature of the final product. Muscat wines, for example,
are stored under anaerobic conditions, while rancio wines
are deliberately oxidised by being stored in large wooden
vats for a period of time.

(iii) *After fermentation.* The best example here is sherry. The
degree of fortification required for these wines is less than
that for port, say, because fermentation has proceeded
further, and there will have been more 'natural' alcohol
present. Also, the desired final ABV of a sherry is normally
slightly lower than that required in a port.

8.1   FORTIFICATION

Two types of grape spirit are used to fortify wine: highly rectified
fortifying spirit with an ethanol content of 94–96% by volume
(SVR, or *Spiritus Vinum Rectificatum*); or 'brandy spirit', distilled
to a strength of anything between 57 and 83% by vol. The latter,
which is used to fortify port, is often erroneously referred to as
brandy, and contains a number of congeneric compounds, such as
higher alcohols, aldehydes and esters, which add to the complexity
of the wine. Fortifying spirit, on the other hand, which is used in
the manufacture of sherry and madeira, is essentially high-proof
alcohol. It is repeatedly distilled from wine, or wine-derived products, such as pomace, and is a neutral spirit, containing no other
constituents apart from water. Since ethanol and water form a
constant boiling mixture at 96.4% ethanol by volume, the two
cannot be separated by distillation, even in the most efficient stills.
Once fortifying spirit has been added to a fermentation, it should
be thoroughly mixed in, otherwise it will float on top and allow
fermentation to continue underneath.

The calculation of the amount of fortifying spirit to be used is
complex, and dependent on several factors: the original sugar
content of the must; how much sugar at the point of fortification;
how much sugar is wanted in the final product and what degree of
alcohol is required. Fortification is not necessarily a very precise
operation, because, as alcohol is added, the sugar level is diluted.

There can be difficulty in accurately measuring the Baumé in fermenting must with a hydrometer, for, as soon as ethanol (SG 0.789) is added, the instrument is encouraged to sink. Winemakers either use their intuition, or a complicated set of tables, or a chart to calculate exactly how much spirit must be added. When spirit is added to a fermenting must, heat is evolved as a result of the reaction between ethanol and water. The source of the fortifying agent is not necessarily identical to the region of wine production, and the nature of the spirit is not always as refined as the final product might suggest. Until recently, the supply of 'brandy spirit' to port houses was monopolistically controlled by the Junta Nacional de Aguardiente, who would often use distillates of poor-quality wines from Portugal's southern vineyards. The resultant spirit could often be raw and fiery (*aguardiente* = burning water)! The volume of spirit added to fortified wines is quite substantial, and is sufficient to dilute the remaining natural sugar content of the wine. This invariably necessitates later adjustments to the sweetness level.

8.2 PORT

Is probably the best known, and mostly highly regarded example of a fortified wine, and its manufacture took on its present form during the third quarter of the 18th century, although it was not until around 1850 that fortification of the ferment became widespread. Its very method of production, involving an incursion into the vinification fermentation, brought consternation to the oenophiles of yesteryear, who found the production methodology very perplexing. As late as 1754, British merchants in Oporto complained about 'the habit of checking the fermentation of the wines too soon, by putting brandy into them while still fermenting, a practice which must be considered diabolical, for after this the wines will not remain quiet, but are continually tending to ferment and to become ropey and acid'. Thus, interrupting the fermentation was a risky business, and there was always a chance that the residual sugar (mainly glucose and fructose) would re-ferment too vigorously. Port wine is named after Oporto, a large coastal town in northern Portugal. Port wine grapes are, therefore, grown in a warm climate, and the young wines tend to have high phenolic contents, making them astringent, but highly suitable for ageing.

Ports are made from grapes grown in a specific area in the upper valley of the river Douro. The agronomic conditions in the area are quite unique, the soils are mainly schistous, the relief is rugged and there is a high seasonal temperature variation, with low average rainfall and intense sunlight. Such conditions yield highly pigmented, aromatic grapes, with elevated levels of sugars and phenolics. The best wines are made from very sweet grapes (14° Baumé, or more), and care is taken to prevent them raisining. Worldwide, numerous grape varieties (around 50 red and 35 white) can be used to make port, principally 15 red and 6 white in the Douro, but only half a dozen, or so, will be used in the best vintage ports. The most highly recommended varieties seem to be Touriga Nacional, Touriga Franca, Tinto Roriz, Tinto Cão and Tinto Barroca. Many vines are grown on hillside terraces so characteristic of this part of the Douro. The *terroirs* are classified in a decreasing scale from 'A' to 'F', according to soil type, grape variety, vine age, altitude, *etc.* The higher graded vineyards (*i.e.* A and B) are authorised to convert a greater percentage of their fruit into port, and the grading letter generally determines the price paid for the grapes. The area in which grapes for port must be produced and vinified constitutes the world's oldest demarcated wine region, the original definition and marking dating from 1761.[3] Within the EU, the name 'port' can only be used for wines produced in this designated region, although many countries now produce port-style wines. Australia and South Africa, for example, produce sweet red wines of considerable quality, although the grapes varieties used are different from those grown in the upper Douro. In its country of origin, the regulation of the production and sale of port is controlled by a professional body consisting of growers and shippers, and the *Instituto do Vinho do Porto* (IVP), a government body responsible, among other things, for research and development, maintenance of standards (both analytical and organoleptic) and the quality of fortification spirit. They also issue certificates of authenticity for the drink. Although port wines are made in the upper Douro, they are usually matured in port lodges downstream in Vila Nova de Gaia.

With the exception of fortification, many of the steps involved in the production of standard port wines are similar to those employed for red table wine vinifications. Grapes are either grown by local farmers, or by companies who actually make the wine. Since many such companies were also responsible for exporting the wine, they have tended to be called 'shippers'. Some farmers still make port from their own grapes, but most of them sell their produce to shippers, who vinify in their own wineries. Traditionally, all grapes for making port are very ripe and harvested by hand (automation on the hillside terraces being difficult, in any case), and taken to small farms, or *quintas*, where they are trodden by gangs of enthusiastic bare-legged workers. Grapes have to be sorted carefully to remove spoiled bunches. Treading is carried out in large, shallow stone (usually granite) baths, called *lagares* (Figure 1), which are filled with grapes to about knee height. Typical dimensions for a *lagar*, which can hold around 7 tonnes of must, would be 4 m × 4 m, and about 1 m deep, and treading occurs several hours each day, until the third day of maceration, and is often accompanied by music and general merriment (especially when visitors are present). The finest wines are still crushed by this method, since the shallow nature of the *lagar* encourages maximum contact between skins and juice (giving maximum extraction of pigments), and allows oxygen to dissolve readily, which helps to stabilise essential anthocyanins. Fermentation often commences during the period that treading is being carried out.

Modern *lagares* may be epoxy-lined, or even made of stainless steel, but most port is now fermented in stainless steel tanks. Nowadays, it is probable that the majority of port grapes are not trodden, but, to compensate for this, modern vessels are designed to achieve maximum extraction of pigments At least one

Figure 1   *Stone lagares*
           (By kind permission of Symington Family Estates)

manufacturer employs thermovinification, whereby must is heated to 70–75°C for about 15 min before pressing and inoculation with yeast. This means that extraction is effected before fermentation, which can take place in any vessel capable of being attemperated. The process yields dark, fruity wines. A mechanical version of the *lagar* exists, which closely mimics the treading action. The robotic device, which is peculiar to the Douro, consists of a stainless steel trough situated underneath the machine that contains the automatic 'feet', which are composed of rectangular pistons fitted with silicone rubber pads. The machine runs across the trough on rails, the feet working vertically in piston-like fashion. Such machines are now used for making high-quality port. To date, there have been few meaningful comparisons between the various port vinification methods, and producers tend to adhere to the system that has been proven for them. Despite ever-increasing mechanisation of vinification processes, there is still no substitute for the skill of the winemaker in the timing of the fortification, and in the assessment of how to blend port wines.

In the smaller *quintas*, fermentation takes place in the same *lagar* as was used for treading, so movement of must is unnecessary, but in larger wineries crushed grapes are pumped into a fermentation tank, often with an SO_2 addition. The large surface area of the *lagar* encourages air circulation for cooling, but, if tank fermentation is used then attemperation systems must be fitted. In this warm climate, the acidity of the grapes may be too low, and so the pH often has to be adjusted down to *ca.* 3.6 with tartaric acid. Similarly, in a hot year stems are not removed, their bitter tannins being seen as a bonus, whereas in cooler years most of them are removed, because these tannins conflict with the fruit character of the grape. Fermentation takes place mostly with the aid of naturally occurring yeasts, and the temperature is maintained at 26–28°C. Residence time in fermentation vessel is usually around 72 h. Thus, unlike red table wines, red ports are macerated for a short period of time, and so it is important to keep the fermenting must agitated to maximise extraction from the skins. Artisanally, this was accomplished by treading, and/or by paddling with wooden oars, but pumping-over is now more common, especially where electricity is readily available. Pumping-over, whereby must is sprayed over the skin cap in an open vessel, is considered the most gentle way of achieving turbulence in the ferment. Ports from some of the larger shippers will have been fermented in an autovinifier, or Ducellier system, where turbulence is driven by the CO_2 produced during the fermentation, but results for the better quality port wines have not apparently been spectacular. Irrespective of the precise vinification method, approximately half-way through the fermentation (*i.e. ca.* 4–5% ethanol by volume, or *ca.* 9° Baumé sugar remaining), juice is run off the pomace, which is then pressed and the released free-run juice and first pressing is run into a tank and fortified. Pressing can be effected in several ways (horizontal piston; pneumatic), and can often be via a continuous (screw) process in the larger houses. The precise point of fortification is dependent upon the composition of the must, and the eventual wine style intended. Because of the colour and tannin remaining in the partly fermented skins, pressing is a thorough process – designed to extract every last drop of juice. Some of the more tannin-rich pressings are often separated, and allowed to ferment out to dryness, the resultant 'wine' being destined for distillation. A few makers fortify in the presence of grape skins (*i.e.* before pressing), in which case, the ethanol intensifies the extraction of pigments.

Bakker *et al.*[4] compared two current methods for making port; a traditional stone *lagar*, with crushing by treading, and a tank fermentation, where crushing was mechanical. Yeast growth, sugar depletion, ethanol formation and changes in amino acid composition were monitored. No marked differences in these fermentation characteristics were found between the *lagar* and tank systems. The effect on the quality of the end product was assessed by colour analysis and sensory assessments of the wine during maturation. The results showed initial differences in pigment extraction and significant differences in sensory attributes; but as the wine matured over 3 years, this difference in sensory quality became insignificant, suggesting that the sensory quality of the finished wines did not depend on production method.

As a wine, port is unique in offering a range of styles (ranging from ruby through to tawny) that have all come from a common starting point, and have followed a similar vinification pathway. They are all basically 'sweet' wines, and will all have been stored in the presence of sugars. Admittedly, certain grape varieties from selected vineyards tend to go towards the same end product every year, but the destiny of each batch of wine is ultimately decided by exhaustive tastings over the first 18 months of its life. The first decisions, taste-wise, will be made after the first racking, which normally occurs during the winter following the harvest. Then, in an exceptional year, the very best samples will be put aside and will be destined for bottling as vintage port. For any 1 year's growth, vintage port is seen as the premium port wine, and, after an early aeration to stabilise colour, it will be matured in barrels, normally for *ca.* 2–3 years, and bottled. All further transformations to such wine will occur in the bottle, and, in the absence of air, vintage ports have an enormous ageing potential, thanks to their high polyphenol content. During ageing in bottle, a reduced environment develops, which yields a characteristic bouquet, and any iron present will be in the Fe^{2+} state, thus, the characters developed are entirely different from those imparted to wines matured solely in vats. Bottle-aged port should be cold stabilised (cooled to $-10°C$) prior to bottling in order to remove excess tartaric acid, which might cause problems during maturation. When young, these wines are very robust, but, once bottled and matured, they are sensitive to oxygen; an opened bottle of vintage port rapidly loses its quality.

The remaining wines (most of them) will be blended, and will be destined for either of the two main port wine categories, ruby and tawny, which are both blends of grape varieties/cultivars, and of wines made in different years. This allows the shipper to obtain a consistent product, despite crop variations, and any indication of age on the bottle will be a 'typical average age' of that wine. Following the appropriate blending, wines will be aged in oak barrels, called for a number of years, under oxidative conditions (high redox potential), whence harshness due to tannins softens, and colouring matter precipitates (causing a change in colour). Ruby ports will normally be bottled after 3–5 years ageing in oak, and, being only moderately oxidised, have a fruity character, a deep red colour and plenty of phenolics to provide a 'full' body. By contrast, tawny ports, which are actively manipulated by shippers for several years, until the desired end-product is attained, are aged for much longer periods in old oak casks (10 years, or even up to 30 years), and are consequently much more oxidised. Once its destiny as a tawny port has been confirmed, a wine will be transferred to elongated oak casks, called pipes (*pipas*), with a capacity of around 105 gallons. For the first 10 years, or so, the wine will be racked annually, and, apart from clarifying, this manipulation helps to keep it oxidised. With time, the deposits get lighter, and racking is reduced to once every 18 months. The nature of the deposit changes over time, initially consisting mainly of moribund yeast cells, but latterly being composed almost entirely of polymerised anthocyanins. The deposition of the latter is responsible for the gradual change in colour of the wine, from red to reddish-brown, to light brown, and, finally, to the almost golden hue of a very old tawny. These colour changes, and the chemical reactions that they indicate, are closely tied to oxidations, and the production of aldehydes that gives these wines their *rancio* character. The harsh phenolics tend to soften during the extended maturation period, and it is claimed that such wines take on 'nutty', 'raisin-like' or even an 'oaky' character. The latter should not be too pronounced, because shippers go to considerable lengths to ensure that barrels are well seasoned before use. Top-quality tawny ports will have some indication of age on the bottle, even though they might contain wines from a number of vintages. Younger, more commercial products will have been matured for shorter periods, and to compensate, will have been made from lighter-coloured base wines. Such tawnies will lack the complexity of truly aged samples. At the time of their bottling, ruby and tawny ports are stable in the presence of air, and will improve imperceptibly during ageing in bottle. Some shippers produce blends of real wood-aged tawny ports from a single vintage, and these are called 'vintage character' ports, or *Porto Colheita*. Ruby and tawny ports, which are aged under oxidised conditions, will have their iron content manifested in the Fe^{3+} state, and the prolonged oxidation, and intense esterification give these wines a rich and complex bouquet. Many wood-aged ports may be lightly fined before being bottled, gelatine being the most commonly used agent. This would not normally be necessary with a very old tawny, because it would have been racked numerous times during its years of maturation. Many shippers use purpose-made sweetening wines, called *geropiga*, to adjust the sweetness of their final blends before bottling, thus allowing themselves to attain a consistent product. Occasionally, an additional volume of spirit may have to be added. The final sweetness of a wine depends upon the style demanded in any one location; Portuguese port wines normally ending up in the 3–4° Baumé range, while Australian products are much sweeter at between 3.5 and 6° Baumé.[5]

White ports represent only a small fraction of total production of the Douro. They too are blended products, but made only from white grapes, such as Viosinho, Malvasia Fina and Verdelho. As with other ports, grapes are pressed, allowed to partially ferment, then fortified. Skin contact is reduced to a minimum in order to produce a bright, clear wine, and fermentation is carried out at lower temperatures (18–20°C) than are used for ports made with red grapes. This allows white port to retain its natural fruit flavour. Some shippers ferment clarified juice, in order to obtain a very light colour. These wines will be matured in a similar way to the ruby and tawny ports, and are mostly aged in wood, whereby they assume a golden colour, and a rich, nutty complexity. Shippers place slightly less emphasis on oxidation during maturation, so that they retain their fruity aroma and pale colour. Some non-wood-aged wine is often blended in to add 'freshness' and a fruit character.

Apart from their obvious sweetness, and elevated ethanol level, port wines differ from table wines in several ways, most of which result from their elongated ageing period. There are two main reasons for these differences. Firstly, the residual sugars in port break down slowly during maturation to give flavour-active compounds not found in table wines. Secondly, port wines tend to contain higher levels of acetaldehyde than red table wines, and this greatly influences both their flavour development and their colour. The 'acetaldehyde factor' is seen as a key factor in port ageing, and the substance can be derived from fermentation, from the fortifying spirit, and can be generated during maturation. The role of acetaldehyde in winemaking generally has been expertly reviewed.[6]

One of the main reasons why port wines have elevated levels of acetaldehyde is, that during fermentation maximal amounts are formed when yeast activity is at its most vigorous, and that, just after this phase, fermentation is arrested by fortification, which effectively 'preserves' the compound. In table wines, which are fermented out further, acetaldehyde levels are reduced by a variety of means. Also, the capacity of a yeast to liberate acetaldehyde during fermentation, is strain-dependent; some strains being able to produce uncomfortably high concentrations (> 600 mg L^{-1}). On top of all this, fortifying spirit invariably contains appreciable levels, and even more is formed by the oxidation of ethanol under the oxidative conditions prevalent during ageing in wood. As Bakker and Timberlake state,[7] the discernible level of acetaldehyde in port helps to explain some of the observed colour changes during ageing. As is documented elsewhere, acetaldehyde binds reversibly, but strongly, in equimolar amounts, with SO_2, forming an acetaldehyde–bisulfite complex (see page 279). In table wines most acetaldehyde tends to get bound to SO_2, but in young port wines it is usually present in excess of the preservative gas, giving rise to both 'free' and 'bound' acetaldehyde. Only the former can participate in any maturation reactions. Bakker and Timberlake also found that, in 55 port wines analysed, the average 'total' acetaldehyde content was 127.3 ± 25.6 mg L^{-1}, while the average 'free' level was 65.1 ± 24.2 mg L^{-1}.

To make a port wine that is capable of being blended, and matured for several years, it is essential that the base products contain sufficient colour to start with. The colour of fresh red port wines is mostly due to the monomeric anthocyanins, which are extracted from grape berries during vinification, and, during maturation, there is a gradual 'loss' of these compounds, resulting in a change in wine colour. With a relatively short maceration period, the chemistry of grape anthocyanins becomes an important subject, especially since varieties vary greatly in their pigmentation. In grapes, anthocyanins are linked to a sugar molecule (forming anthocyanin monoglucosides), which confer on them much greater chemical stability. All of these compounds are chemically related to each other, and their colours are determined by a number of factors, notably pH. Some coloured tannins are also present in port wines, and their concentration and molecular size increase as the wine ages, until, in aged wines, they become so large that they settle out as bottle crust.

As Brouillard *et al.*[8] remark in their fascinating paper, the anthocyanins of the red berries of *Vitis vinifera* are among the simplest encountered in higher plants, and that among beverages, such as tea, beer, coffee and juices, red wine is the only one whose organoleptic properties improve with time. The grape/fresh red wine pigments disappear from the wine, after a few months, giving rise to new pigments which result from some spontaneous wine chemistry, thus allowing the colour to remain stable for many years. As they say: 'The structural simplicity of grape anthocyanins and the long-lasting colour of red wine is another French paradox; we call it French paradox II'.

The general structure of *V. vinifera* anthocyanin monoglucosides is shown in Figure 2. These compounds are all in the 3-glucosidic

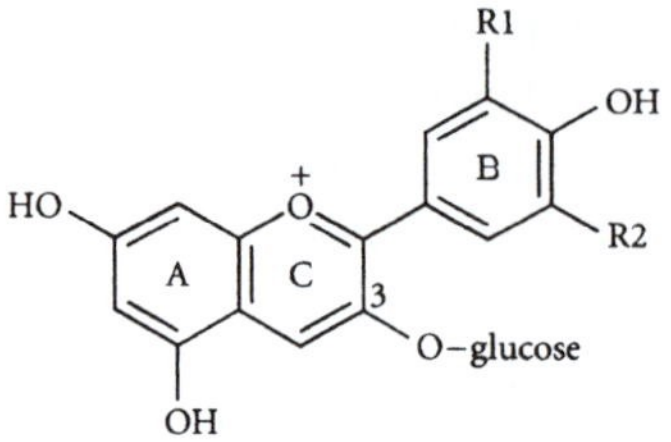

Anthocyanidin	R1	R2
Delphinidin	OH	OH
Cyanidin	OH	H
Petunidin	OCH₃	OH
Peonidin	OCH₃	H
Malvidin	OCH₃	OCH₃

Figure 2   *Structure of* V. vinifera *anthocyanidin monoglucosides (flavylium form)*

form, and differ in their hydroxylation and methoxylation patterns of ring B, yielding a wide range of colours, from orange-red to violet (at high acidity). The glucosyl moiety linked at the 3-O position of ring C may also be acylated with acetic acid, coumaric acid or caffeic acid. In a study of 16 *V. vinifera* varieties/cultivars used to make port,[9] it was found that all but three of the antho-cyanins encountered were located in the skins, and that, in general those based on the malvidin molecule predominated. Malvidin 3-glucoside was the major pigment overall, with a constitution varying from 33 to 94%, and, in order of abundance, this was followed by malvidin 3-*p*-coumarylglucoside (1–51%), and malvidin 3-acetylglucoside (1–18%). Peonidin 3-glucoside (1–39%) was prominent in four varieties/cultivars, and delphinidin 3-glucoside (1–13%), petunidin 3-glucoside (2–12%) and cyanidin 3-glucoside (trace–6%) were only detected in low concentrations. These seven anthocyanins accounted for some 90% of the total anthocyanin content of the grapes studied. Although the same anthocyanins could be found in all varieties/cultivars studied, the authors found that the ratio of malvidin 3-acetylglucoside: total malvidin gluco-sides[*], was characteristic of variety, independent of location and a useful aid to varietal identification.

Anthocyanins undergo chemical transformations during ageing, yielding new pigments that become responsible for the changing colour, and longevity of colour in the wine. These new pigments were first thought to result mainly from condensation reactions between anthocyanins and flavanols, either directly, or mediated by acetaldehyde. Polymerisation reactions between anthocyanins and other phenolic compounds leads to the formation of larger

---

[*]That is the sum of the percentages of malvidin 3-glucoside, malvidin 3-acetylglucoside, and malvidin 3-*p*-coumarylglucoside.

oligomeric or polymeric molecules, in which anthocyanins are essentially acting as building blocks. In port wines, two different polymerisation reactions are thought to occur concurrently, although the kinetics of each reaction depends upon the concentrations of certain compounds, such as acetaldehyde. These polymerisations are known to coincide with the observed decline in anthocyanin content of the wine. During the first months in maturation, a port typically becomes more intensely coloured, a phenomenon known as 'closing-up'. The increase in colour density can be up to 80%, depending on the acetaldehyde content, and the increase is attributable to formation of acetaldehyde-containing oligomeric pigments that are more highly coloured, at the pH of port, than the anthocyanins from which they originated. Other phenolic compounds (mostly flavan-3-ols) are also involved in these reactions, leading to the formation of acetaldehyde-bridged polymers of anthocyanins and other phenolics. While individual anthocyanins react strongly with acetaldehyde, the oligomers become increasingly less reactive as their size increases. At, or just after, 'closing-up', the larger oligmers (*i.e.* polymers) become insoluble and start to precipitate out. For a period of time, the formation of new oligomers and the precipitation of polymers counteract each other, and the wine colour remains stable, but, gradually, the precipitation process exceeds the formation of new oligomers, and colour intensity is reduced. The rate of acetaldehyde-induced polymerisation is governed by the concentration of 'free' acetaldehyde. During ageing of red wine, the free form is liberated from the acetaldehyde–bisulfite complex by the oxidation of SO_2.

Until relatively recently, the main alterations in anthocyanin composition, and the resultant colour changes in the 'redness' of the wine, were believed to be solely due to the polymerisation reactions mentioned above, but recent work has discovered the presence of new malvidin-derived pigments, indicating that anthocyanins undergo even more complex changes during ageing. During routine research into the colour stability of ageing red wines, Sarni-Manchando *et al.*[10] reported the presence of novel anthocyanin-derived pigments, which appeared to involve a new ring structure resulting from the direct reaction of an anthocyanin with another, small, molecule. The new pigments were very stable, even at high pH, and proposed structures were reported by the same team later that year.[11] This group proposed that their new pigments, which they called pyranoanthocyanins, were formed from direct covalent reactions between anthocyanins and vinyl phenols, the latter being wine fermentation products resulting from enzymatic decarboxylation of phenolic acids. They suggested two possible pathways for the formation of these pigments: a cyclo-addition, or an electrophilic addition of a vinylphenol double bond to an anthocyanin nucleus, followed by an oxidation step. The latter was favoured, because of the electrophilic site at C-4, and the nucleophilic site at the C-5 hydroxyl group of the anthocyanin. The reaction mechanism of pyranoanthocyanin formation is still not certain, and it is likely that several different pathways exist. Bakker *et al.*[12] reported the isolation and identification of an anthocyanin-based pigment, which they named vitisin A, from aged red wines. The compound was more intensely coloured in the wine than the anthocyanins from which it derives, and was later shown to be resistant to SO_2 bleaching.[13] Vitisin A is formed by a reaction of malvidin 3-glucoside with pyruvic acid, a natural component of wine. In 1998, Fulcrand *et al.*,[14] working independently, recovered this novel pigment from grape pomace, and reported further upon its structure. They identified it as a malvidin 3-monoglucoside pyruvic acid adduct, and found its formation resulted from cyclisation between C-4 and the hydroxyl group at C-5 of the original flavylium moiety with the double bond of the enolic form of pyruvic acid, followed by dehydration and rearomatisation steps. The new pigment significantly increased colour stability, and its discovery stimulated much international research into the field of red wine colour chemistry. Using model solutions, Romero and Bakker[15] studied the effects of storage temperature and pyruvate on the kinetics of anthocyanin degradation, and the formation of vitisin A, and some of its derivatives. They found that acylated forms of vitisin A, having the 6-position of the sugar acylated with acetic acid (3-acetylvitisin A) and *p*-coumaric acid (3-*p*-coumaryl-vitisin A), were also formed through the interaction between pyruvic acid and malvidin 3-acetylglucoside and malvidin 3-*p*-coumarylglucoside, respectively. Maximum levels of degradation of the anthocyanins were observed at the higher temperatures used (20 and 32°C), the reaction following first-order kinetics, both with and without pyruvic acid. At lower temperatures (10 and 15°C), the presence of pyruvic acid accelerated the kinetic reaction, but at the higher temperatures, it decreased it. The activation energy values for the degradation of the three anthocyanins in model solutions with and without pyruvic acid were not significantly different from each other, and it was at the lower temperatures that the highest concentrations of the vitisin A compounds were obtained.

In a later piece of work, the same workers[16] investigated the formation of vitisin A-type compounds in four maturing port wines, stored for 29 weeks at 15°C, and the effect of pyruvic acid addition on such storage. Anthocyanin concentrations were monitored by HPLC, and colour changes were assessed spectrophotometrically. The anthocyanin losses obeyed first-order kinetics, and the formation of polymeric pigments was found to be concomitant, albeit being present in very low concentration. The addition of pyruvic acid to the wines led to the production of much higher levels of vitisin A-type compounds, and up to 23 mg L^{-1} of vitisin A, and its acylated forms, acetylvitisin A and *p*-coumarylvitisin A, could be determined. Owing to their having greater colour expression and greater stability than malvidin 3-glucoside, these new pigments played an important role in the colour quality of the ageing wines. To illustrate the significance of the pigments, a subsequent analysis of 32 red port wines, that had been matured for between 2 and 6 years, showed that vitisin A-type compounds were the main, and sometimes the only, anthocyanins present. The structures of vitisin A, and vitisin B, the latter being the smallest of the pyranoanthocyanins, are shown in Figure 3.

Results obtained over the last decade have shown that anthocyanins, in fact, react with a number of different compounds, including: pyruvic acid, vinyl phenol, vinylcatechol, α-ketoglutaric acid, acetone and 4-vinylguiacol, yielding pigments which mostly have spectroscopic features that contribute to an orange-red colour. This new group of pigments are now much studied, and several unique examples have been identified from aged red port wines. For example, Mateus *et al.*,[17] using ultra-modern techniques, detected and identified three new pigments from a 2-year-old port wine. The structure of the pigments was found to correspond to the vinyl cycloadducts of malvidin-3-coumarylglucoside, bearing either a procyanidin dimer or a flavanol monomer ((+) -catechin, or (–)-epicatechin)). The same research group[18] reported two new, blue anthocyanidin-derived pigments whereby the pyranoanthocyanin moiety is linked to a flavanol by a vinyl

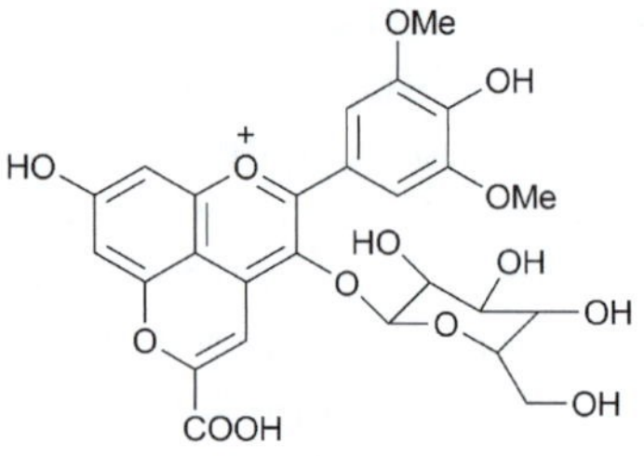

Vitisin A (malvidin-3-glucoside pyruvic acid adduct)

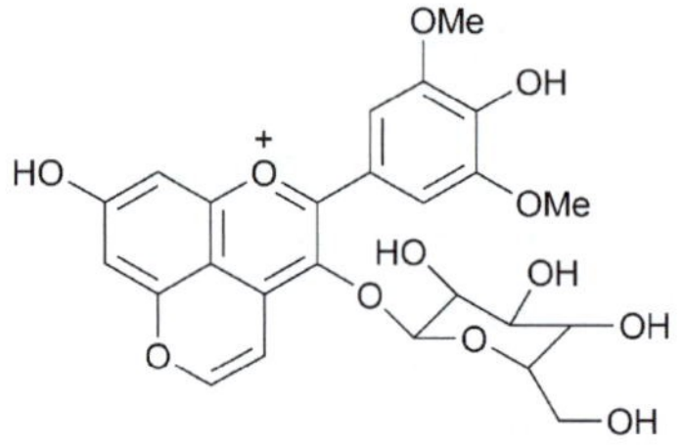

Vitisin B (pyranomalvidin-3-glucoside)

Figure 3 *Structure of vitisins A and B*
(By kind permission of Dr Nuno Mateus)

bridge. These compounds, which are essentially vinyl-pyranoanthocyanins, showed maximum absorption in the visible region at 583 nm, and were named 'portisins'. According to the authors, this new class of anthocyanin-derived pigments may be obtained through a reaction between anthocyanin–pyruvic acid adducts and other compounds, such as flavanols (*e.g.* catechins, procyanidins) or phloroglucinol in the presence of acetaldehyde, or directly by reaction with *p*-vinylphenol. The last step of their formation is thought to include decarboxylation, dehydration and oxidation, yielding a structure with extended conjugation of the π-electrons, which is likely to confer a higher stability on the molecule, and is probably the origin of the intense blue colour.

Using a slightly different approach, Morata *et al.*[19] investigated the production of pyruvic acid and acetaldehyde by ten different strains of *Saccharomyces cerevisiae* which they used to ferment samples of must of the red grape variety Tempranillo. The idea was to investigate the effect of these two compounds on the pyranoanthocyanins, vitisin A and vitisin B (malvidin 3-*O*-glucoside-pyruvate, and malvidin 3-*O*-glucoside-4-vinyl respectively). Pyruvate production reached a maximum on day 4 of fermentation, while acetaldehyde production was at its peak in the final stages. The correlation between pyruvate production and vitisin A formation was especially strong on day 4, when the greatest quantity of pyruvate was found in the medium. On the other hand, the correlation between acetaldehyde production and the formation of vitisin B was strongest at the end of fermentation, when the acetaldehyde content in the medium was at its highest.

The original pyranoanthocyanins exhibited a more orange-red colour than their corresponding intact anthocyanins, and are thus considered to be in part responsible for the colour of aged red wines. A hypsochromic shift of the maximum absorbance wavelength (λ_{max}) was observed with malvidin 3-glucoside derivatives of around 18–19 nm for vitisin A, and 36–39 nm for vitisin B, depending on the solvent used.[13] Since pyranoanthocyanins confer a more stable colour and changed hue to a mature wine, it has been mooted that their effect on colour might be classified as an example of co-pigmentation. Interestingly, it was found[20] that the anthocyanin–catechol pigments showed the same λ_{max} as the pyruvate derivatives, which are hypsochromatically shifted from that of their parent anthocyanins.

A method to quantify anthocyanin-derived pigments, particularly vitisins A and B, has been developed by the Co-operative Research Centre for Viticulture in Australia. The technique employs HPLC-electrospray tandem mass spectrometry (HPLC–UV–MS/MS), and it was found that both vitisins were primarily formed during fermentation, and declined, slowly, thereafter. The acetaldehyde derivative, vitisin B was detected in significantly enhanced concentrations in wines fermented with *Saccharomyces bayanus*, as opposed to *S. cerevisiae*, but that it declined rapidly during ageing.

The volatile components of port wines derive from three sources: the grape, fermentation and the fortifying spirit. Of the 200, or more, volatiles so far detected in port wines, around one quarter remain unidentified.[21] Those compounds that have been unequivocally identified include: 81 esters; 14 alcohols; 9 dioxolanes; 6 hydrocarbons; 5 acids; 5 carbonyls; 4 nitrogen-containing compounds; 3 hydroxy carbonyls; 2 phenols; 2 alkoxy phenols; 2 alkoxy alcohols; 2 lactones; 2 oxygen heterocyclics; 2 halogen compounds; 1 diol and 1 sulfur-containing compound.

Recent reviews of the chemistry and production of port wines have been provided by Jackson,[22] Reader and Dominguez,[23] and Clarke and Bakker[24] and Bakker[25] has documented the chemical composition of these beverages. For an erudite exposition on port wine, the reader is directed to Mayson's latest offering.[26]

8.3 VINS DOUX NATURELS (VDN)

Some of the French wines classified thus are among the most celebrated of fortified wines, and originate from a dozen or so appellations across three regions in southern France, the best known appellations, perhaps, being Muscat and Rivesaltes. The fortified wines of the fiercely Catalan most southerly part of the Languedoc-Roussillon region are world famous. Their very name is somewhat misleading, because they are not naturally sweet wines, as sauternes are, for example, since the sweetness is introduced by intervention. The parameters for their production are quite exacting, with only certain grape varieties, such as Muscat, Grenache, Macabeu and Malvoisie (Pinot Gris), being permitted. Crop yield limits are set, and the grape juice must contain a minimum 252 g L^{-1} fermentable sugar, which approximates to a potential ethanol level of 14.5% by vol. The proportion of spirit used for fortification must comprise between 5 and 10% of the must volume, and the must is fortified when the fermentation has exhausted just over half of the grape sugars. The final product must contain between 15 and 18% actual alcohol by volume, and at least 21.5% total alcohol.[†] Maximum residual sugar concentrations may vary between 95 and 125 g L^{-1}, depending on appellation. Of the types of VDN that are made, the whites are usually made from white Grenache, or Macabeu grapes, which are macerated very lightly, if at all, and give light, fruity, non-oxidised wines suitable for early drinking. Red VDN, often made from Mourvèdre and Syrah grapes, are macerated for up to a fortnight, and the resultant wines are richly coloured and benefit greatly from ageing. The addition of spirit to these wines is known as *mutage*. Those VDN

[†] Total alcohol includes the actual alcohol by volume and the potential alcohol level, which is the amount of alcohol that might theoretically be produced from the residual sugar, according to the formula as follows: potential alcohol = residual sugar concentration (in g L^{-1})/17.

that are aged are done so under oxidative conditions in oak casks or barrels, but some are not deliberately oxidised, including Muscat de Rivesaltes, probably the most famous of this particular style.

8.4 MADEIRA

The volcanic island of Madeira, which is part of Portugal, lies some 1000 km from the Portuguese mainland, and some 750 km off the coast of North Africa. Madeiran wine, in its original form was probably first popularised by Venetian merchants during the 13th century, when there was a strong market for sweet, tasty wines, especially in Britain. The original vines on the island are thought to have come from Crete, and, in particular from the city of Candia, a fact that can be gleaned from one of the early names for Madeiran wine; *Candia Malvasia*. Madeira became Portuguese in 1419, and by the end of the 15th century, madeira had attained a distinguished reputation throughout the wine-drinking countries of mainland Europe, where, together with sherry and 'Canary wine' it constituted 'sack'. Although exports grew, the main source of cash on the island was sugar-cane, and it was not until the end of the 17th century that madeira became the major export commodity. In those days, madeira was a sweet table wine, but as a result of a consignment that became overheated while in equatorial waters during transit, and subsequently exhibited improved organoleptic characters, it became routine for barrels to be shipped back and forth over the equator until 1794, when a technique evolved for purposely heating the wine in large ovens, or hot houses, called *estufas*. The origins of the practice of maturing madeira by shipping it around the world as ballast, are lost in the mists of time, but the benefits of such prolonged heating were well appreciated by the mid-18th century, and the method actually continued until World War I. For many years, madeiras were considered as some of the greatest wines in the world, and during the mid-19th century there were some 70 British importers doing business on the island, shipping some two million gallons.

The overall wine style is determined by its unique maturation process, which involves heating at temperatures up to 50°C, but 'madeira' is not one generic kind of wine, but about half a dozen, some of which are named after the grapes with which they are made. This heating process, the *estufagem*, caramelises wine sugars, and confers a characteristic, caramel-like flavour and odour to the wine, while ageing imparts slightly oxidised notes. The wines are made from four, principal, traditional grape varieties (often called the 'noble' varieties), plus a few others, which are usually grown in small vineyards on terraces on steep slopes. Grapes for madeiras can, theoretically, be grown anywhere on the islands of Madeira, but, in practice they are grown on the main island, and the neighbouring Porto Santo. Most of the better quality grapes are grown west of the capital, Funchal. Grapes are picked, crushed and pressed, and fermentation is usually instigated by yeasts from the natural grape surface flora. Several different winemaking techniques are employed on the island, some of which involve terminating fermentation by fortification (thus preserving some grape sweetness, as in port), and then heated, while some wines are fermented to dryness, and fortified after the e*stufagem*. For most wines, however, certainly those destined for export, the *estufagem* commences immediately after fortification. The heating process is carried out in large (*ca.* 200–280 hL tanks), with an inert lining, such as cement, and, most commonly, wines are heated to *ca.* 45°C for a period of 3 months. Smaller volumes may be heated in wooden butts, which are placed in an *estufa*. After *estufagem*, the wine must be cooled slowly, a step that is more critical for the lesser-quality Sercials and Verdelhos, which, having not yet been fortified, are vulnerable after heating, especially to oxidation and acetification. Most madeiras are ultimately fortified to 17% alcohol by volume.

Nowadays, the largest single market for Madeira wine is in France, where most of it ends up in kitchens, for general cooking, or for making sauces (*sauce madère*)! Wine for this market is usually made from grapes of the, now widely planted, Tinta Negra Mole variety, which are harvested early (in August), and the must fermented out almost to dryness, before the wine is sweetened, and then fortified with an unfermented, alcoholised (20–23% by volume) must known as *vinho surdo*, before being subjected to 3 months in a heated vat at 50°C. This is madeira production at its most functional, and fine wines would be made by variations on this theme, most importantly involving grape selection, care and time. If possible, *vinho surdo* is prepared from Porto Santo grapes, which are very sweet and give a high density product (5–9° Baumé).

Madeiras can be both sweet and dry in style, and four types of madeira are sold under the name of the grape from which, in theory at least, they are made. In ascending order of richness and strength, these are: Sercial, Verdelho, Bual and Malvasia (Malmsey), with the last-named being the original grape variety on the island. Sercial and Verdelho are dry, while Bual and Malvasia are sweet, and they are traditionally consumed before and after dinner respectively. In practice, these varieties have been in short supply since oidium, and then phylloxera devastated the island's vines during the late 19th century, and so it is normal for a small quantity of Tinta (planted as a replacement in the late 19th century) to be used in all but the very finest products, so long as at least 85% of the grapes used are of the variety named on the bottle. Sercial and Verdelho grapes are pressed and the juice fermented, in the absence of skins, to near dryness, before being fortified and sweetened with *vinho surdo*. Some Sercial vines are grown some distance above sea level, and wines from these grapes require extra fortification to bring them into specification. The wines from Sercial and Verdelho grapes are the lightest (colour-wise) and driest of the madeiras, with final sugar levels of between 0.5 and 2.5° Baumé. Bual and Malvisia grapes are crushed, then fermented with their skins, as are red table wines. When a Bual ferment is half complete, and a Malvisia fermentation one-third over, both terminated by addition of spirit. This used to be via an indigenous sugar cane product, but a grape spirit from Portugal is now generally used. Buals and malmseys are both much darker products, some malmsey being almost black with age, and they also have higher levels of sweetness (between 2.5 and 6.5° Baumé). After its 'ordeal by heat' a madeira wine will be extremely stable, even though it will inevitably contain elevated levels of acetic acid (volatile acidity) and acetaldehyde. In a wood-aged, mature madeira, the acetic acid content may well be sufficiently high to contribute to both taste and smell. The greatest madeiras will have spent many years in cask, and will not generally be adversely affected by exposure to air.

As well as the four grape-determined varieties of madeira mentioned above, there are several categories of the wine that are determined by blending and ageing regimes, including 'Reserve', 'Special Reserve', 'Extra Reserve' and 'Vintage'. The latter must be made from one of the noble varieties, have spent 20 years in cask after *estufagem*, and an additional 2 years in bottle. It must also be made from 100% of its specified grape variety. The first grading of madeira occurs after the wine has been taken off its fermentation lees, each style being classified according to quality. Shipment of the cheaper wines (*e.g.* those destined for French cuisine), is permitted after 13 months, but superior quality products are matured for at least 5 years. Most wines are fined with isinglass and Spanish earth before shipment, but, owing to their age, there is little need for a cold treatment to stabilise against tartrate deposition. Madeiras are reckoned to be the longest lasting of all wines; the 'baking' process, their acidity and their alcoholic strength all contributing to their longevity. Relatively little is known about the chemical reactions involved in making madeira, but for highly readable accounts of this most characteristic of wines, one should read the books by Cossart[27] and Liddell.[28]

8.5. SHERRY

The legislation for making Spanish sherry are complex, and the drink is now only permitted to be made in a restricted area close to the town of Jerez de la Frontera, in the province of Cadiź, Andalusia, southern Spain. The area has a long history of wine-making, for it was the Phoenicians who first introduced the vine to the area around Cadiź. The phylloxera plague of 1894 almost totally destroyed the established vineyards around Jerez, which were then subsequently replanted with American and grafted varieties. English merchants have been involved in sherry production for several centuries, and several of today's major sherry companies, most of which have their own vineyards, date back to the 18th century. Until Spain's accession to the EU, some of the larger companies exported their sherry in bulk, and most of the blending and bottling was carried out overseas (*e.g.* in England). As a result of EU regulations, all Spanish sherries are now 'bottled at source', something that has necessitated the re-structuring of some of the large concerns, since everything from grape cultivation to bottling now has to be undertaken within the designated geographical region.

The materials and methods used in sherry manufacture, the vineyards where the grapes are grown, and the places where it may be fermented and matured, are strictly controlled by a committee known as the *Consejo Regulador de la Denomination de*

Origen 'Jerez-Xeres-Sherry', appointed under Government auspices. The area in which sherry grapes may be grown is carefully defined, but in years when the crop is depressed, the *Consejo* may give permission for new wines of a similar type to be brought into the approved sherry territory, from suitable places outside. The quantity of wine brought in must not exceed 10% of that produced in the sherry area in that year.

The climate around Jerez is generally warm, and rainfall is moderate, conditions that would be expected to yield white grapes capable only of producing rather uninteresting white table wines. Indeed, the main grape, the Palomino, gives rise to wine that is rather neutral, lacking in acidity and varietal character (often described as 'flabby'), and of little interest, but, as a result of a unique, complicated method of maturation and blending, sherries have evolved into the highly individual wines that we know today. The neutral base wine produced by the Palomino grape provides an excellent background for the delicate flavours produced as a result of the maturation and blending regimes. After fermentation, and the production of the base wine, sherries are differentiated into three broad categories: *raya, fino* and *oloroso*, the first-named being regarded as the most basic. From these the familiar wines marketed as 'Amontillado', 'Cream', 'Pale Cream', 'Medium' and 'Dry' are generated, mostly from the last two categories. After the initial classification of the base wine, batches will then be subjected to totally different maturation techniques, which enable the three main types of sherry, 'Fino', 'Oloroso' and 'Amontillado', to be made.

The sherry vineyards are situated in a diamond-shaped area, with Jerez being roughly in the centre. The best vineyards are in the hills, and are situated on a rather chalky soil type, known locally as *albariza*. It consists of 30–80% chalk, with the remainder being made up of clay, sand and humus. Such soil has the ability to retain moisture, from the winter rainfall, throughout the very dry summer. Even after many weeks without rainfall, it will retain moisture just below its surface. Other local soil types, on which many vineyards are situated, yield coarser, inferior wines. Irrigation in Jerez in forbidden.

Over the years, there has been a gradual diminution in the number of grape varieties used, and, presently, Palomino predominates in dry sherries. There are two 'strains' of this grape, Palomino

Fino, and Palomino de Jerez, the former yielding heavier crops. For some special sweetening wines, the Pedro Ximinez grape is used, although it grows best in the hotter Montilla district, where most sweetening wines are made. Bunches of grapes destined to be used for sweetening, or colouring, wines are left out to dry in the sun on esparto grass mats, in order to increase their sugar content, a process known locally as *soleo*. In Jerez, as in the Douro, and on Madeira, vines are almost invariably grafted, since the soil is likely to be contaminated with phylloxera. Root stocks are hybrids between European vines and American species, with the latter conferring some degree of resistance to phylloxera, and eelworm. In practice, it is found that a small crop of poor-quality grapes can be obtained 2 or 3 years after grafting, and after 6 years a commercial yield can be obtained, which gradually increases in size for the next 20, or so, years. From that point, the crop size diminishes, and, after 40 years, the vine is no longer economically viable (in the Douro, vines are often commercially useful for 40–50 years).

The vintage (*vendimian*) in Jerez normally begins between the 1st and 9th of September each year, and lasts for about 15 days. Such a regime presents the winemaker with logistical problems, especially during processing, and inevitably means that there is a high capital investment in plant that is only used once a year. Grapes are harvested very ripe, when their sugar content gives a reading of between 11.5 and 12.5° Baumé, and they have suitable acidity. Towards the end of vintage, measurements are taken every couple of days until it is considered that the grapes will not improve, whence they are removed with a knife and stored in wicker baskets. In very hot, dry years, the vintage can be a little earlier, and such conditions will favour the production of *oloroso*-style wines. Conversely, in a damper year, a late vintage may result, and this will favour *fino* production. Traditionally, grapes were/are crushed in a *lagar* by treading in nailed boots, and the resultant free-run juice removed. Subsequent pressing of the pomace would result in the first pressings being combined with the free-run juice. Second pressings would only be suitable for distillation purposes, or for making vinegar. Gypsum, or *yeso*, is invariably sprinkled onto the grapes, both to mechanically assist pressing, and to increase must acidity. This process, known as 'plastering', has been used since Roman times, and a typical rate of addition would be 1 kg gypsum to 700 kg grapes. The precise reactions emanating from *yeso* addition are still being studied, but the desirable effects for the winemaker are: moderation of fermentation, increase in acidity, and the production of aromatic esters, especially ethyl tartrate. One of the chemical reactions long known to be a result of gypsum addition is the reduction of must pH, due to precipitation of calcium tartrate:

$$2KH(C_4H_6O_6) + CaSO_4 = Ca(C_4H_4O_6) \downarrow + H_2(C_4H_6O_6)$$

potassium bitartrate   calcium tartrate   tartaric acid

This is obviously not the whole story regarding gypsum addition.

It is only really in the last few decades that treading in *lagares* has been largely abandoned in favour of more mechanical processes. Batch presses are generally preferred to continuous ones, with the pneumatic press being the most suitable. The rules of the *Consejo* prescribe that only 70% of the potential juice that can be squeezed from grapes can be made into sherry. Many grapes are crushed and pressed by the grower, and the musts run into casks before being transported to the *bodega*, where vinification and maturation take place. In many instances, juice will be differentiated at this point, that which is free-run, together with the early pressings, being most suitable for 'Fino', while juice containing more pressings will be earmarked for 'Olorosos' and 'Amontillados'.

Bodegas (Figure 4) are tall, well-ventilated buildings, preferably facing west to catch cooling winds, which are designed to avoid the extremes of temperature. Internal relative humidity is maintained at around 60% by frequent watering. At the *bodega*, must is left

Figure 4   *A sherry bodega*

standing for 24 h to allow gross solid matter to settle out, before being run/pumped into fermentation receptacles, where yeast activity commences within hours. Traditionally, fermentation took place in oak butts, often newly constructed, so that they became suitably seasoned for subsequent use during maturation (new wood introduces too much 'taint' during sherry ageing). The size of a butt means that, at the temperatures prevailing in most *bodegas*, fermentations can normally proceed without the need for any ancillary cooling device. The specific gravity of a must will typically be between 1.085 and 1.095 (11.5–12.5° Baumé), with a total acidity of 3.5–4.5 g L^{-1} (as tartaric acid), and a tannin content of 300–600 mg L^{-1}. If all grapes are sound, butts are treated with *ca.* 100 mg SO_2 per litre. Tartaric acid may be added if the must is of low acidity. Many of the more modern *bodegas* favour stainless steel fermentation vats.

When must arrives at the *bodega*, its temperature will be around 20°C, and within 24 h the first signs of fermentation can be observed, and, by 48 h, it will be in full spate. Most of the must sugar is removed within 1 week, and, during this time the temperature will have risen considerably. Ideally, the temperature should be kept below 25°C for *fino* wines, and below 30°C for *oloroso*. After the initial fermentation 'surge', metabolic activity slows down, and this is when many of the important flavour compounds are formed. At the very end of fermentation (late December–early January), the young wine falls 'bright' as suspended material settles out. Fermentation was always initiated by indigenous yeasts, but yeast cultures have found favour in the last few decades, as intimated by Goswell and Kunkee in 1977, who said: 'In Jerez, cultured yeast is rarely added to juice, as grapes have adequate amounts of suitable fermentative yeasts present on the skin. A few producers are beginning to experiment with cultured yeast, but it is too early to assess the results'. The principal yeasts[‡] involved in sherry fermentations in the Jerez area were studied by Inigo Leal *et al.*,[29] who found that the first phase was dominated by

---

[‡]The under-mentioned yeast names are given as per the nomenclature of the day. More modern taxonomy ('Register of Yeast Names', Barnett *et al.*[30]) is as follows: (i) now the anamorph of *Hanseniaspora uvarum*; (ii) now *Sacch. cerevisiae*; (iii) now *Sacch. cerevisiae*; (iv) now *Sacch. cerevisiae*; (v) now *Sacch. cerevisiae*. (NB. An anamorph is the asexual stage of a sexually reproducing species.); (vi) now *Torulaspora delbrueckii*; (vii) now *Torulaspora delbrueckii*; (viii) now the anamorph of *Hanseniaspora vineae*; (ix) now *Pichia anomala*; (x) now the anamorph of *Pichia guilliermondii*; (xi) now *Sacch. bayanus*.

Kloeckera apiculata[(i)], with *Metschnikowia pulcherrina* (and its ana-morph *Candida pulcherrina*), *S. cerevisiae* var. *ellipsoideus*[(ii)], *Sac-charomyces chevalieri*[(iii)], *Saccharomyces mangini*[(iv)], *Saccharomyces italicus*[(v)], *Saccharomyces rosei*[(vi)], *Torulaspora rosei*[(vii)], *Hansenia-spora valbyensis* and *H. guilliermondii* present in significant propor-tion. As the ethanol concentration increased, the population of *Metschnikowia* spp. declined in size, and there was some increase in the population of *S. chevalieri*[(iii)]. During the final phase of fermentation, *S. cerevisiae*, *S. chevalieri*[(iii)] and *S. italicus*[(v)] pre-dominated. The workers noted that there were differences in the yeast profiles of fermentations from disparate parts of the Jerez area, but that the final phase was always dominated by the three strains/species of *Saccharomyces* just mentioned. Other yeasts encountered in unfermented sherry musts included: *Hanseniaspora osmophila*, *Kloeckera africana*[(viii)], *Debaryomyces hansenii*, *Han-senula anomala* var *anomala*[(ix)], *Candida guilliermondii*[(x)], *Pichia strasburgensis*, *Saccharomyces uvarum*[(xi)] and *Kluyveromyces mar-xianus*. Some *Mucor* species were also encountered.

After grading, the base sherry will be racked off the lees, fortified (*encabezado*), and placed into seasoned oak butts within the *bode-ga*, before entering into a unique and elaborate fractional blending system known as the *solera*. Butts are stacked in long rows, up to five high, and, where possible, the 'Finos' are located in the cooler lower levels, and the 'Olorosos' in the upper tiers. The original method of maturing sherry, the *añada*[§] system, involved keeping wines from different vintages separate until they were ready for use. Such a system frequently resulted in end-product inconsistency, and so the dynamic *solera* system evolved. The *añada* now forms the first stage in sherry ageing, where unblended wines are allowed to mature for a short time (1 year in the case of *finos*). After a second selection process, the young sherry will then enter the *solera*, which, in larger companies, often means that *finos* will be totally separated from *olorosos*.

Fino wines are uniquely matured under a biological pellicle called *flor*, which develops naturally on the wine–air interface. *Flor*, often called the 'velum', or *velo de flor*, is a skin-like yeast growth which occurs naturally in the *bodegas* of the Jerez region, and imparts the characteristic aroma and flavour to a 'Fino' sherry. The formation

---

[§]The *añada* represents the year of the grape harvest in which a particular wine was produced.

of this floating film has been attributed to hydrophobic proteins on the cell surface of the yeasts involved.[31] A *flor* film will develop on any maturing 'Fino', as long as the base wine has suitable characteristics. Table 1 illustrates the composition of an *añada* wine that would be suitable for *flor* maturation. To encourage *flor* formation, butts of young *fino* wine are stored in the coolest part of the *bodega*, and are fortified to 15.5% v/v alcohol, with neutral grape spirit, which is not a sufficiently high level to inhibit *flor* yeast growth, but is high enough to inhibit deleterious acetic acid bacteria. The usual fortification spirit is what is known as *mitad y mitad*, or 'half-and-half', which is a clarified mixture of spirit and base wine. Such a preparation avoids the turbidity encountered if high strength alcohol is added directly to the wine. Butts containing fortified *fino*-category wine are filled to approximately 80% of capacity in order to ensure maximum surface area for the growth of *flor* yeasts. Young wines with a good *flor* film are all regarded as *finos* of varying quality, and will be consigned, unblended, to an *añada*, which is essentially a 'fresh wine' store, for a year, or until required for the *solera* system. The precise conditions that determine whether a good or poor yeast film develops, remain a mystery, but experience has shown that wines made from grapes grown in warmer vineyards, on soils with low chalk content, and containing a high percentage of pressed juice, are less likely to grow an extensive *flor*. Certainly, a vintage produced from grapes grown in lower than average ambient temperatures will produce more *flor* samples.

Table 1   *Suitable* añada *base wine characteristics for* flor *maturation (after Reader & Dominguez [23])*
(Reproduced by kind permission of Springer Science & Business Media)

Ethanol (% v/v)	14.8–15.3
Glycerol (g L^{-1})	6.7–7.2
Fermentable sugar (g L^{-1})	<1.5
Gluconic acid (g L^{-1})	<0.6
Acetic acid (g L^{-1})	<0.65
Malic acid (g L^{-1})	<0.15
Lactic acid (g L^{-1})	<1.15
Total phenolics (mg L^{-1})	<250
Total SO_2 (mg L^{-1})	<75
pH	3.00–3.25

Technically, as soon as the young *fino* wine becomes fortified it is described as *fino sobretablas*, the name emanating from the way that casks used to be stored on tables, before entering the *solera* system, and this is seen as the first stage of the ageing process (*i.e.* said to be in *añada*). Similarly, a young, fortified *oloroso* will be known as *oloroso sobretablas*. The act of adding spirit to a sherry is called 'heading' (*i.e.* sherry is a 'headed' wine). From this point, wine is gradually fed into the *solera*, where butts are stored at 15–20°C, and where most of the biological ageing occurs. The *solera* system consists of several stages of ageing wines called *criaderas*, each one consisting of *ca.* 100 butts. The oldest wines are to be found at floor level. 'Fino' sherry producers have anything from 3 to 14 *criaderas* per *solera*, more than are needed for other sherry styles. The way that the system operates is that wine for bottling is taken from the oldest *criadera*, and the volume removed is replaced by wine from the second oldest *criadera*. Wine from that *criadera* is replenished with some from the third oldest, and so on throughout the *solera* system. Wine from the youngest *criadera* is replaced with suitable wine from the *añada*. The wine drawn from the butts in each *criadera* is normally blended before being used to top up the butts in the next stage up. In effect, the young wine 'refreshes' older samples, as it is passed down the system, and just enough nutrient material is carried with it to keep the *flor* alive. Up to one-third of the wine in a 'Fino' *solera* can be withdrawn annually, without adversely affecting the wine. Indeed, it is essential for such wine to be removed regularly from the *solera*, so that fresh nutrients for the *flor* organisms can be introduced via fresh wine from the *añada*. It can be seen that the *solera* system is a means of fractional blending, that can incorporate wines of different ages and vintages. It aims to provide a constant supply of wine of consistent quality. This complex procedure involves many wine transfers, and is very labour intensive, and it is not easy to effect significant changes to the character of the wine. The success of the *solera* system depends on the fact that the younger wines rapidly take on the character of the older wines to which they are added. Because of the aerobic activities of the *flor* yeasts, the alcoholic strength of a *fino* is reduced slightly during its time in the *solera*. Having been originally fortified to *ca.* 15.5% vol., it can lose almost 1% of its alcoholic strength by the time it has completed its maturation phase (around four years in the *solera*), and it is usual for the wine to be rectified to 17% by vol. prior to bottling. Considering their alcoholic strength, dry 'Fino' sherries, especially the finest samples, are very delicate, and inherently unstable. If they are kept too long in butt (or in bottle) under less than ideal conditions, their fragrance and vitality disappears. A 'tired' 'Fino' may, however, be resuscitated in the *bodega*, and eventually appear on the market as an 'Amontillado'!

The integrity of the *flor* film is vital, a pellicle 3–6-mm thick being seen as ideal, and this discrete layer will persist as long as there is free access to oxygen. Much care is taken to minimise the disruption to the film, especially during the numerous transfers that are conducted in the *solera*. Elaborate transfer techniques have been evolved in the Jerez *bodegas*, all aimed at minimising disruption to the *flor* while wine movements are in progress. Many of these involve narrow-bore tubing, siphons and/or pumps, wine being transferred to a holding tank, where it can be homogenised before being carefully introduced to the next *criadera*. Even with extreme care, the *flor* film exhibits some degree of seasonal variation in respect of its biological activity. The *flor* organisms are most active between February and June, after which there is a steady decline until October, when activity increases again. The decrease in *flor* yeast activity in high summer is thought to be at least partly due to the formation of respiratory-deficient mutants at the elevated temperatures. *Flor* yeast development is temperature sensitive, cell growth only occurring between 13 and 25°C, with maximum cellular activity between 22 and 25°C. The optimum temperature range for *flor* characteristics is considered to be 17–20°C. Pantothenate is vitally important to film growth, and some amino acids are essential for its maintenance, once it has been established. A detailed and comprehensive study of sherry *flor* has been carried out for the Australian Wine Board.[32]

Maturation under *flor* results in numerous biochemical changes, which vastly influence wine flavour. One of the major changes is the gradual decrease in wine volatile acidity, and this is accompanied by a gradual reduction in glycerol content. In addition, ethanol is being used as a carbon source by *flor* yeasts, and this results in a perceptible rise in acetaldehyde content (normally in the region of 260–360 mg L^{-1}). This acetaldehyde is believed to originate exclusively from ethanol as a result of enzyme activity (alcohol dehydrogenase), and it acts as a precursor for a number of other flavour-active compounds, such as acetoin, diethyl acetal, and polyphenols-acetaldehyde complexes. In addition to the slight loss of ethanol by yeast activity, there is a loss by evaporation, estimated to be around 0.2–0.3% by volume, and, although this can be 'topped up', it will have some economic significance for the *bodega*. A list of the volatile constituents found in 'Fino' sherry is shown in Table 1, the data being drawn from the work of Begoña Cortes *et al.*[33]

In many modern *bodegas*, 'Fino' is generated in a bulk tank by inoculating juice with a *flor* yeast and bubbling air through it to supply the yeast with the oxygen necessary for the synthesis of aroma compounds. When the *flor* flora has developed sufficiently, the wine is put into a 'Fino' *criadera*.

A very small quantity of *fino* wine develops slightly differently, and becomes lighter in both body and colour. This is categorised as *palo cortado*, and is much prized locally, mainly because of its 'rarity' value.

Sometimes, a batch of wine in a *fino solera* will begin to lose its *flor*. This is usually because the wine is very old, and/or has not been regularly 'refreshed', for some reason. Such an event allows the ageing process to change course, and esterifications, and oxidations (especially of acetaldehyde) begin to dominate proceedings, resulting in the wine being classified as an *amontillado-fino*. This will still be straw-coloured, but much more full-bodied than a *fino*, and, if it is further aged, it will evolve into an *amontillado* proper. To effect this, the wine will be further alcoholised to 17–17.5% by volume (which totally inactivates any remaining *flor*), and taken to another *solera* system, where, with storage, its colour gradually deepens to amber/dark gold. The change in hue is associated with the development of a complex 'nuttiness'. The transformation of a *fino* into an *amontillado* takes at least 8 years. The finished wine can either be used for blending, or sold as a generic 'Amontillado'. As such, these wines can be sold in their naturally dry form, or they can be sweetened, and are usually re-fortified to a final alcohol level of 18–20% v/v.

Good quality wines, that do not develop much *flor*, are destined to become *olorosos*, and are matured as such. In crude terms, *oloroso* sherries originate from *añada* wines which have been decreed unsuitable for making *fino*, especially those made from later grape pressings. They are rectified with *mitad y mitad* to around 17–18% by volume, which inactivates any remnants of the *flor* organisms, and are then removed to an 'Oloroso *criadera*'. The wine in these butts is matured under oxidative conditions, which darkens colour and imparts robustness. Casks in an *oloroso solera* are normally kept 95% full and lightly stoppered, and require less topping up than 'Fino' casks. In addition, storage temperature for an 'Oloroso' sherry does not have to be as low, and as constant as is demanded by a 'Fino'. An 'Oloroso', is usually of a darker hue (due to its higher phenolic content), more full-bodied, and with a well-developed bouquet. The higher alcohol concentration, combined with an elevated storage temperature within the *bodega*, promotes extraction of wood phenolics during maturation. 'Oloroso' sherries actually increase in alcoholic strength while they are in the *solera*, since, without the *flor* covering they slowly evaporate, leaving the final ethanolic level at around 18–20% by volume. Evaporation occurs through the wood, and as much as 5% of the water content can be lost annually. The low humidity in the *bodega* means that water evaporates in preference to ethanol. This water loss, together with extraction of wood-derived compounds, and oxidative reactions, leads to an increase in levels of fuselols and some non-volatile compounds. Volatile acidity and ethyl acetate also increase, both by concentration and esterification. Although technically dry, most *olorosos* give an impression of sweetness because of their relatively high glycerol content (can be *ca.* 7–8 g L^{-1}). Glycerol originates from ethanolic fermentation, but is, again, concentrated in the *solera*. A decent quality 'Oloroso' requires around 7–8 years in the *solera*, in wood, and will pass through at least three *criaderas*, before it develops all of its characteristics.

Raya sherries are treated in the same way as olorosos, but are not stored for as long in an *añada*. Some bodegas store their raya wines outdoors in the sun, which is said to improve the maturation of this particular style.

Finished sherries have very complex sensory characteristics. 'Fino' wines are usually sold dry, and some of the very fine older wines may also be dry, but most sherries are sold after sweetening. 'Fino' has a pale, straw colour, and despite its dryness shows little acidity. It has a delicate, pungent bouquet, and an alcoholic strength of between 15.5 and 17% v/v 'Manzanilla' (Spanish diminutive of *manzana*, apple) is a regional variation of the 'Fino' style, and is produced in the coastal town of Sanlúcar de

Barrameda, which has a very specific microclimate. It is very dry, with a clean, somewhat bitter aftertaste, and has slightly less body than a 'Fino'. It is often described as tasting 'salty', and this would accord with the fact that it is manufactured on the estuary of the Guadalquivir river. The cooler, more humid climate on the coast is presumed to influence yeast metabolism, and thus cause the differences between 'Manzanilla' and 'Fino'. 'Manzanilla' has a pale, straw colour, and an alcoholic strength of between 15.5 and 16.5% v/v. Both of these wines should be drunk young, and preferably cold, and, once opened, bottles should be consumed within a short period, since the flavour deteriorates rapidly. 'Amontillado' is dry, clean, with a pungent aroma, and a discernible 'nuttiness'. It is more full-bodied than a 'Fino', with an alcohol content of 17–18% v/v. Its natural colour is amber, but this darkens with age. 'Olorosos' are the most full-bodied of all, with a strong bouquet, but less pungency than either 'Finos' or 'Amontillados'. They have the darkest colour of all sherries, and this intensifies with age. The colour of an *oloroso* is initially only slightly darker than a *fino*, but with the ingress of oxygen, the colour deepens, and passes through various shades of light brown. In practice, most *olorosos* will be destined for blending with sweetening wines to give 'Cream sherries', although some old, dry *olorosos* are highly regarded wines in their own right. Wines marketed as 'medium dry', 'medium', 'cream' and 'pale cream' will be based on 'Oloroso' wine, with a 'Fino' addition to lighten the colour and flavour, and a small amount of 'Amontillado'.

All sherries are dry after going through the *solera* system, and so adjustments have to be made if sweet styles are required. This is effected by blending with sweetening wines before bottling. Medium sherries are traditionally sweetened with *mistela*, the fortified juice from raisined Palomino grapes. For darker sherries, the sweet, dark, concentrated, fortified juice of the Pedro Ximenez grape, commonly known as 'PX', is used. As a general rule, grapes intended for sweetening or colouring purposes are left out in the sun to shrivel slightly, which tends to lower their acidity.

Other countries that produce this style of sherry have had to import these unique organisms in order to emulate Spanish version. The origin of *flor* yeasts, and their taxonomy, is still a matter of conjecture, but several are known to be strains/species of *Saccharomyces*. The organisms in the *flor* pellicle are respiratory (*i.e.* oxygen consumers) and so they protect the underlying wine from the influence of oxygen, and thus preventing the browning of phenolics. Thus, in effect, *fino* wines are aged under anaerobic conditions, and redox potentials may vary from 300–320 mV at the bottom of the butt, to 340–360 mV just below the *flor*. The environment within the *flor* pellicle is highly oxidative.

Both the origin and taxonomy of the yeasts growing in *flor* are still the subject of study. There has been much debate as to the origins of the *flor* yeasts, especially as to whether they are actually fermentation organisms that have become adapted to the ageing environment. The other likelihood is that they are different strains to those responsible for fermentation, but that they are present in small numbers during, or at the end of, fermentation. Their taxonomy has been equally questionable, with several new strains/species of *Saccharomyces* having been proposed over the years, including *Saccharomyces beticus*, *Saccharomyces cheresiensis*, *Saccharomyces montuliensis* and *Saccharomyces rouxii*. Of these, the first two are now considered to be synonymous with *S. cerevisiae*; *S. montuliensis* is now considered to be *Torulaspora delbrueckii*, and *S. rouxii* relates to *Zygosaccharomyces rouxii*. Other species that have been implicated are *Saccharomyces fermentati*, and *Saccharomyces capensis*. One thing is certain; *flor* yeasts are physiologically different from those responsible for alcoholic fermentation, principally because they exhibit oxidative rather than fermentative metabolism. Studies carried out thus far have shown that '*S. beticus*' is the dominant yeast in the youngest wines in the *solera*, and that their numbers decline with age, whence '*S. montuliensis*' becomes more conspicuous and dominant. '*S. cheresiensis*' is somewhat irregular in its frequency, and '*S. rouxii*' only sporadic.

With modern molecular methods of yeast taxonomy, it is now possible to assess far more precisely the contribution of individual species to wine fermentation and to subsequent biological ageing, and, in 2001, the results of an extensive study were reported by Esteve-Zarzoso *et al.*[34] The authors claimed that this was the first attempt to study yeast population dynamics during alcoholic fermentation, before and after fortification, and during biological ageing of a *fino*-type sherry. In their experiments, the early vinification protocol followed what they claimed to be common practice in Jerez, namely the progressive addition of must to the fermentation. Accordingly, they added a dried yeast sample to a volume of fresh must that was equivalent to one-third of the capacity of the fermenter. The same volume of must was added 4–5 days after fermentation had commenced (*i.e.* tank now two-thirds full), and after another 4–5 days, the fermenter was filled with must. This progressive addition of must helps to provide a uniform wine quality prior to maturation. Yeast strains were identified by restriction length polymorphism analysis. The work, which was based at the González-Byass wineries, showed conclusively that yeast population dynamics during biological ageing is a complex phenomenon, the intricacies of which vary from bodega to bodega. They found that there were four races of '*flor Saccharomyces*'; *beticus, cheresiensis, montuliensis* and *rouxii*, which all exhibited identical restriction patterns for the region spanning the internal transcribed spacers 1 and 2 (ITS-1 and ITS-2) and the 5.8s rRNA gene, but this pattern was different from those exhibited by non-*flor S. cerevisiae* strains. This *flor*-specific pattern was detected only after wines were fortified, never during alcoholic fermentation, and all the strains isolated from the velum exhibited the typical *flor* yeast pattern. By restriction fragment length polymorphism (RFLP) of mitochondial DNA and karyotyping, the group showed that: (1) the 'native' yeast strain was better adapted to fermentation conditions than commercial yeast strains; (2) two different populations of *S. cerevisiae* strains are involved in the production of *fino* sherry wine, one of which is responsible for must fermentation, and the other for wine ageing and (3) one strain was dominant in the *flor* population of sherries elaborated in González-Byass wineries. The last conclusion was interesting because it confirmed the notion that the composition of a *flor* is winery-specific. In 1997, for example, Martinez *et al.*[31] reported that they found an ecological succession of *S. cerevisiae* races during wine ageing, while, in the same year, Ibeas *et al.*[35] observed that a single strain dominated individual casks in the solera, and that the strain was stable for two consecutive years. The complexity of sherry ageing is undoubted.

As David Bird[36] pointed out, some of the romance of wine-making has been lost in this modern, technological age, and this is nowhere more apparent than in the production of sherry. No longer do we have to wait expectantly for butts of sherry to ferment mysteriously one way or the other towards *fino*, or *oloroso*.

Scientific knowledge has helped us to be able to guide the development of a wine whichever way is commercially desirable. When sherry is drawn from the *solera* it is slightly cloudy, and the traditional way to clarify it was via egg white and Spanish earth. In the larger *bodegas*, wine is now filtered before bottling. Because of the complex machinations of the *solera* system, the age of a sherry is almost impossible to ascertain.

For an erudite account of how sherry was made in the pre-technological era, the reader is recommended to Gonzalez-Gordon's little gem of 1972.[37] The account by Mey[38] also makes good reading, as does Jeffs.[39] Goswell[40] has reviewed the microbiology of sherry production, and Bakker[41] has summarised its chemical composition. The aroma of sherry wines has been the subject of an important work by Webb and Noble,[42] while flavour occurrence and formation was covered by Nykänen in 1986.[43]

8.6 COMMANDARIA

Although not a fortified wine, the little known commandaria has always fascinated me, and, because of its antiquity and unique mode of preparation, deserves a brief mention here. Commandaria is an amber-coloured, non-fortified, dessert wine, with a natural alcohol content of *ca.* 15% ABV, which is indigenous to the eastern Mediterranean island of Cyprus. No one is quite certain when this ancient wine style was first made, but it is certainly a lot older than its name, which dates from AD 1191, the year that Richard Coeur de Lion sold Cyprus to the Order of Knights of the Temple. This makes 'Commandaria' the world's oldest known wine appellation. The Knights Templar had settled on the island in territories which they called *Commanderies*, their main one being near Limassol, where they had their headquarters at the still extant Kolossi Castle. After a while, the Knights Templar sold much of their land to Guy de Lusignan, but kept a large estate, the 'Gran Commanderie' (Commanderia), which contained areas ideal for growing grapes. The Templars, and their successors, the Knights of St. John, who came to Cyprus in 1210, became expert winemakers, and exported much of their product back to Europe, where Commandaria became a vogue drink in European courts. Richard the Lionheart had apparently greatly enjoyed the wine at his wedding in Cyprus, and pronounced it: 'The wine of kings and the king of wines', while, in 1223, the King of France dubbed it the 'Apostle of Wines' after supposedly tasting over 100 wines from all over the known world. Much of the wine came over to England, where it was held in very high esteem by the Plantagenets.

The wine is still made in Cyprus, most notably in and around the village of Kalo Chorio ('good village'), near Limassol, where grapes still flourish in the gritty sand of the foothills of the Troodos range. Nowadays, it is made from equal proportions of two, high sugar, grape varieties, red (black) Mavro and white Xynisteri, which are harvested (Mavro at 12° Baumé; Xynisteri at 15–16° Baumé), spread out on mats in the sun, and left until evaporation raises the sugar content to 19–23° Baumé. In the hot September sun in Cyprus, a fortnight is sufficient to shrivel them and effect this. The grape-laden mats are then carried down to the press-house by donkey, where, after separation into red and white, the grapes are then crushed, pressed, and the juice placed in open containers, whence fermentation ensues. A few makers still ferment in rotund earthenware jars buried to their rims in the earth. Fermentation is quite slow, and proceeds until it finishes of its own accord. The young wine is then matured in oak, often in open air, for around 3 years. It does not need, or, indeed, benefit from ageing in bottle.

The wine style is said to date back to the ancient Greeks, when it was a popular drink at festivals celebrating the goddess Aphrodite. It was celebrated by Homer, and drunk by Egyptian pharaohs under the name 'mana', Greek for 'mother'. The drink was fermented/matured in large earthenware pots, and when the contents were poured out, some was always left in the bottom in order to aid fermentation/maturation of the next batch; the 'old' became the 'mother' of the new. Apparently, the method of its preparation has hardly changed since those far off days, for we have a short passage from Hesiod, that might be recognisable to the producers of today, Hesiod, the second great Greek epic poet, lived around 700 BC, and in his *Works and Days*[¶], he describes to his brother, Perses, how 'special sweet wine' is made:

But when Orion and Sirius are come into the mid-heaven, and rosy-fingered dawn sees Arcturus [September], then cut off all the

---

[¶] In effect, this was a dispute between Hesiod and his brother over the distribution of their father's land.

grape-clusters, Perses, and bring them home. Show them to the sun ten days and ten nights, then cover them over for five, and on the sixth day draw off into vessels the gifts of Dionysus.

REFERENCES

1. R.W. Goswell and R.E. Kunkee, Fortified wines, in A.M. Rose (ed), *Economic Microbiology, Volume 1, Alcoholic Beverages*, Academic Press, London, 1977.
2. P. Ribéreau-Gayon, D. Dubourdieu, Donèche and A. Lonvaud, *Handbook of Enology, Vol. 1. The Microbiology of Wine and Vinifications*, Wiley, Chichester, 2000.
3. A.M. Fonseca, A. da Galhano, E.S. Pimental and J.R.-P. Rosas, *Port Wine–Notes on its History, Production and Technology*, Instituto do Vinho do Porto, Porto, 1987.
4. J. Bakker, S.J. Bellworthy, T.A. Hogg, R.M. Kirby, H.P. Reader, F.S.S. Rogerson, S.J. Watkins and J.A. Barnett, *Am. J. Enol. Vitic.*, 1996, **47**, 37.
5. B.C. Rankine, *Making Good Wine*, Macmillan, Sydney, 2004.
6. S.-Q. Liu and G.J. Pilone, *Int. J. Food Sci. Technol.*, 2000, **35**, 49.
7. J. Bakker and C.F. Timberlake, *Am. J. Enol. Vitic.*, 1986, **37**, 288.
8. R. Brouillard, S. Chassaing and A. Fougerousse, *Phytochemistry*, 2003, **64**, 1179.
9. J. Bakker and C.F. Timberlake, *J. Sci. Food Agric.*, 1985, **36**, 1315.
10. P. Sarni-Manchado, H. Fulcrand, J. Souquet, V. Cheynier and M. Moutounet, *J. Food Sci.*, 1996, **61**, 938.
11. H. Fulcrand, P. Cameira dos Santos, P. Sarni-Manchando, V. Cheynier and J. Favre-Bonvin, *J. Chem. Soc. Perkins Trans.*, 1996, **1**, 735.
12. J. Bakker, P. Bridle, T. Honda, H. Kuwano, N. Saito, N. Terahara and C.F. Timberlake, *Phytochemistry*, 1997, **44**, 1375.
13. J. Bakker and C.F. Timberlake, *J. Agric. Food Chem.*, 1997, **45**, 35.
14. H. Fulcrand, C. Benebdeljalil, J. Rigaud, V. Cheynier and M. Moutounet, *Phytochemistry*, 1998, **47**, 1401.

15. C. Romero and J. Bakker, *J. Agric. Food Chem.*, 2000, **48**, 2135.
16. C. Romero and J. Bakker, *J. Sci. Food Agric.*, 2001, **81**, 252.
17. N. Mateus, E. Carvalho, A.R.F. Carvalho, A. Melo, A.M. González-Paramás, C. Santos-Buelga, A.M. Silva and V. de Freitas, *J. Agric. Food Chem.*, 2003, **51**, 277.
18. N. Mateus, A.M. Silva, J.C. Rivas-Gonzalo, C. Santos-Buelga and V. de Freitas, *J. Agric. Food Chem.*, 2003, **51**, 1919.
19. A. Morata, M.C. Gomez-Cordoves, B. Colomo and J.A. Suarez, *J. Agric. Food Chem.*, 2003, **51**, 7402.
20. N. Mateus, J. Oliviera, M. Haettich-Motta and V.J. de Freitas, *Biomed. Biotechnol.*, 2004, **5**, 299.
21. A.A. Williams, M.J. Lewis and H.V. May, *J. Sci. Food Agric.*, 1983, **34**, 311.
22. R.S. Jackson, *Wine Science: Principles and Applications*, Academic Press, London, 2000.
23. H.P. Reader and M. Dominguez, Fortified wines: Sherry, port and madeira, in A.G.H. Lea and J.R. Piggott (eds), *Fermented Beverage Production*, 2nd edn, Kluwer Academic, New York, 2003, 157–194.
24. R.J. Clarke and J. Bakker, *Wine Flavour Chemistry*, Blackwell, Oxford, 2004.
25. J. Bakker, Port, Chemical composition and analysis, in R. Macrae, R.K. Robinson and M.Y. Sadler (eds), *Encyclopaedia of Food Science, Food Technology and Nutrition*, Academic Press, London, 1993, 3658–3662.
26. R. Mayson, *Port and the Douro*, 2nd edn, Mitchell Beazley, London, 2005.
27. N. Cossart, *Madeira–the island vineyard*, Christie's Wine Publications, London, 1984.
28. A. Liddell, *Madeira*, 1st edn, Faber & Faber, London, 1998.
29. B. Inigo-Leal, D. Vasquez Martinez and V. Arroyo Varela, *Revista Ciencia Aplicada*, 1963, **93**, 317.
30. J.A. Barnett, R.W. Payne and D. Farrow, *Yeasts: Characteristics and Identification*, 3rd edn, Cambridge University Press, Cambridge, 2000.
31. P. Martinez, L. Pérez Rodriguez and T. Benitez, *Am. J. Enol. Vitic.*, 1997, **48**, 55.
32. J.C.M. Fornachon, *Studies on the Sherry Flor*, 2nd edn, Australian Wine Board, Adelaide, 1972.

33. M. Begoña Cortes, J.J. Moreno, L. Zea, L. Moyano and M. Medina, *J. Agric. Food Chem.*, 1999, **47**, 3297.
34. B. Esteve-Zarzoso, M.J. Peris-Torán, E. García-Maiquez, F. Uruburu and A. Querol, *Appl. Environ. Microbiol.*, 2001, **67**, 2056.
35. J.I. Ibeas, I. Lozano, F. Perdigones and J. Jimenez, *Am. J. Enol. Vitic*, 1997, **48**, 75.
36. D. Bird, *Understanding Wine Technology*, DBQA, Newark, Notts, 2005.
37. M. Gonzalez-Gordon, *Sherry, the Noble Wine*, Cassell, London, 1972.
38. W. Mey, *Sherry*, Asjoburo, The Netherlands, 1988.
39. J. Jeffs, *Sherry*, Mitchell Beazley, London, 2004.
40. R.W. Goswell, Microbiology of fortified wines, in R.K. Robinson (ed), *Developments in Food Microbiology, Volume*, Elsevier Applied Science, London, 1986.
41. J. Bakker, Sherry, Chemical composition and analysis, in R. Macrae, R.K. Robinson and M.Y. Sadler (eds), *Encyclopaedia of Food Science, Food Technology and Nutrition*, Academic Press, London, 1993b, 4122–4126.
42. A.D. Webb and A.C. Noble, *Biotechnol. Bioeng.*, 1976, **18**, 939.
43. L. Nykänen, *Am. J. Enol. Vitic.*, 1986, **37**, 84.

CHAPTER 9

Other Organisms Important in Oenology

9.1 KILLER YEASTS

Since its first report in *Saccharomyces cerevisiae* by Bevan and Makower in 1963,[1] killer activity has been demonstrated in a number of other yeast genera,[2] and is now found to be widely distributed among yeasts.[3] Killer yeasts have the intrinsic ability to extirpate sensitive yeast strains by secreting a proteinaceous toxin to which they themselves are immune. These toxins have no activity against microbes other than yeasts. Ecological studies indicate that killer activity is a mechanism of interference competition with the production of a toxic compound from one yeast strain excluding other, susceptible (killer-sensitive) strains from its habitat. Strains showing the killer phenotype (called K^+ strains) produce a protein toxin that kills sensitive (called K-sensitive) yeasts, but to which the K^+ strain is resistant. Neutral strains are resistant to killer toxin and do not themselves kill sensitive cells. As long ago as 1968, Woods and Bevan[4] showed that the death of sensitive cells is not coincident with absorption of the killer factor, but can be delayed or prevented by variation in environmental factors such as pH, temperature and aeration. They also showed that sensitive cells are most susceptible to the action of the killer factor when in the logarithmic phase of growth.

The first killer yeast was designated K1, when it was realised that another strain, called K2, showed no cross immunity to it. In fact, these two types killed each other. Owing to the discovery of similar cross-reactions, the first general attempts to classify killer yeasts emerged. One such scheme was reported by Rogers and Bevan,[5] who identified four groups. Some strains of killer yeasts of various genera were tested for their ability to kill each other, two neutral strains and four classes of mutants isolated as being resistant to the killer toxin of a stock killer strain. The strains tested (eight of *S. cerevisiae*, two of *Candida albicans* and one each of *S. drosophilarum* – now known as *Kluyveromyces lactis* – and *Torulopsis glabrata*) fell into four groups, designated as TOX1 to TOX4, according to their killing/immunity reaction with one another and their ability to kill resistant mutants. All the pH optima for killing fell in the range 4.0–4.7. In the same year, Young and Yagiu[6] working independently, but along the same lines, classified killer yeasts into 11 groups, K1 to K11, of which five have been identified from *S. cerevisiae*. After around three decades of research on killer yeasts, we now realise that the majority of killer toxins are acidic proteins with an isoelectric point *ca.* pH 4 and a molecular mass in the range of 10–20 kDa.

In a winemaking context, killer strains have been isolated from a variety of sources including grape skins, must fermentations and wine cellar equipment. Many surveys have been conducted to verify the evidence of killer yeasts in spontaneous fermentations and it has been shown that they are distributed differently in various wine-producing areas. In some regions, killer yeasts were not detected, while in others only killer yeasts were encountered. One of the most extensive surveys was carried out by Vagnoli and co-workers[7] from the University of Siena, who worked in the 'Brunello di Montalcino' area of production and reported on their findings in 1993. 'Brunello di Montalcino'[*] is a prestigious wine produced in a restricted area of Tuscany, a region in central Italy. Over the years, its production has been generally carried out by spontaneous fermentations, but with industrial winemaking in Italy inexorably changing over to the use of commercial starter strains, this study was conducted with a view to selecting a potential starter strain from the natural wine yeast population within the domain. The results showed that in the area where 'Brunello di

---

[*]Brunello is a variety of Sangiovese, the famous red grape of Tuscany and is the name for that grape in Montalcino. This aristocratic wine was the first in Italy to be awarded D.O.C.G. (denominazione di origine controllata e garantita) status, set up in 1980 by the Italian authorities.

Montalcino' is produced, killer yeasts are widespread, with almost all of the killer isolates belonging to class K2. K^+ isolates were found to be present in 88% of the spontaneous fermentations from the 18 wineries examined. The incidence of killers varied with respect to stage of fermentation and vintage period, increasing from the first vintage to successive ones and from the commencement to the end of fermentation. At the end of fermentation, the proportion of killer strains relative to the total yeast population was below 25% in 15 cases, above 75% in 6 cases, from 25 to 50% in 5 cases and from 50 to 75% in 3 cases. Thus, three different situations were encountered at fermentation end: (1) strong prevalence of K-sensitive yeasts; (2) strong prevalence of K^+ yeasts; (3) slight prevalence of K^+ yeasts. Since almost all of the isolates were phenotypically K2, it seemed as though the toxin exhibited variable degrees of toxicity against the K-sensitive strain. It was hypothesised that the K^+ isolates were producing either differing amounts of toxin, or toxins with different specific activities. This notion was further studied by the same research group,[8] which detected two different killer phenotypes, which could be distinguished by their degree of killer activity. The two phenotypes were designated as SK^+ (strong killer) and WK^+ (weak killer), and it was found that in a must environment the killer effect of both phenotypes was reduced, when compared to cells challenged in a synthetic medium. In mixed-culture must fermentations only the SK^+ phenotype was able to express itself.

A review of killer yeasts in the wine industry was given by Van Vuuren and Jacobs,[9] who evaluated the seemingly conflicting reports on the interactions of killer and sensitive yeasts during grape must fermentations. The paper emphasised the killer phenomenon in *S. cerevisiae*, particularly the significance of K2 killer yeasts that may dominate wine fermentations initially inoculated with sensitive, cultured wine yeast. Such fermentations can lead to the premature cessation of ethanol production, high volatile acidity, hydrogen sulfide production and off-flavours caused by fuselols, acetaldehyde and lactic acid. The K2 killer strains could prove to be important in vinification technology because the toxin is stable at normal must pH values. In their paper of 1987, Heard and Fleet[10] reported that some strains of *S. cerevisiae* isolated from well-established vineyards and wineries in European countries have exhibited the killer property. At that point, there were no reports on the occurrence of such killer strains in the newly developed winemaking regions of the USA and Australia, where a more technological approach to winemaking has been adopted.

The mechanisms of recognising and killing sensitive cells seem to differ for each toxin, although it has to be said that the K1 killer system of *S. cerevisiae* is by far the most extensively studied. The characteristics of the different killer toxins depend on the information stored in the nucleic acid and on post-translational modification, that is glycosylation. When the first studies on the nature of the killer phenomenon in *S. cerevisiae* showed the involvement of cytoplasmic, non-Mendelian genetic determinants, the occurrence in killer yeasts of double-stranded RNA (dsRNA) associated with viral-like particles was suggested. It is now known that in *S. cerevisiae* the killer phenotype is associated with the presence of two cytoplasmic genetic elements made of double-stranded, separately encapsidated RNA: mycovirus L-A and its satellite M.[11] The L-A genome encodes an RNA polymerase and capsid proteins, the functions of which are necessary for replication and maintenance of both elements, whereas the M genome encodes the killer toxin and immunity to its action. In addition to the presence of both elements, several host components are necessary for the expression of the killer phenotype. A single open reading frame encodes the toxin, which is synthesised as a single polypeptide pre-toxin. This pre-toxin, once synthesised, undergoes post-translational modifications *via* the endoplasmic reticulum, Golgi apparatus and secretory vesicles, resulting in the secretion of the mature, active toxin. While glycosylation is necessary for the efficient secretion of the toxin, the active K1 killer toxin is not glycosylated; thus, the glycosylated portion must be removed, somehow, during the toxin maturation process. The K1 toxin secreted by *S. cerevisiae* consists of a disulfide-linked α–β dimer (the subunits having a molecular mass of 9.5 and 9.0 kDa, respectively). Both subunits have a relatively high content of charged and hydrophobic amino acids. The toxin is capable of forming multimers from the basic dimer, but it is not clear how, or what, they contribute to toxin action. Taking into account the basic dimeric nature of killer toxin K1, several modes of action are possible, and what is probably the most widely accepted model is illustrated in Figure 1. In this model, the β-subunit is involved in cell wall receptor binding, and in the case of the K1 killer toxin, the cell wall receptor is a $(1 \rightarrow 6)$-β-D-glucan.

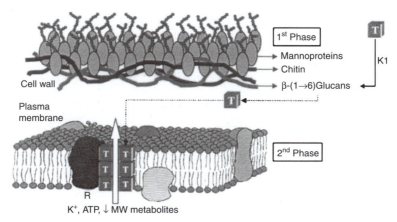

Figure 1 *Receptor-mediated killing of a sensitive yeast cell by the K1 killer toxin of S. cerevisiae. After binding to the cell wall, the K1 toxin is transferred to the cytoplasmic membrane and acts by forming voltage-independent cationic trans-membrane channels, which cause ion leakage and subsequent cell death. The existence of a receptor (R) in the membrane is postulated (after Marquina et al.[23])*
(Reproduced by kind permission of Springer Science & Business Media)

This reaction is strongly pH-dependent. The killer toxin then interacts with a receptor on the plasmalemma, which results in the membrane becoming permeable to protons and potassium ions. Later, the membrane becomes permeable to molecules of higher molecular mass, such as ATP. The K2 *S. cerevisiae* killer toxin has a similar mode of action, but it is a very different protein ($\alpha\beta = 21.5$ kDa). A review of the dsRNA viruses of *S. cerevisiae* has been prepared by Wickner.[12]

Although five *S. cerevisiae* killer types – K1, K2, K3, KT28 and K3GR1 – have been reported as being encoded by different M genomes, only three – K1, K2 and K28 – have been clearly defined and differentiated by genetic and molecular analyses of their dsRNA genomic determinants. It is now evident that K3 is not clearly distinct from K2, and that M_3 is in fact an M_2 mutant.

The effect of the K2 killer toxin from *S. cerevisiae* on the growth of a mesophilic wine yeast, as indicated by electron microscopy, was reported by Vadasz *et al.*,[13] who found that there was a change in integrity of the cell wall and cytoplasmic membrane of challenged cells. In particular, susceptible cell surfaces became 'rippled' and showed cracks and pores, whereas control cells, unaffected by the toxin, maintained a smooth surface.

The virally encoded K28 killer toxin of *S. cerevisiae* differs from K1 and K2 killer toxins in that it is bound to the mannoprotein component (the outer 1,3-α-linked mannose residues) of the sensitive yeast cell wall, a feature that has been used for its purification. In contrast, K28 has no ionophoretic effect, but rather inhibits nuclear DNA synthesis. DNA synthesis is rapidly inhibited, cell viability is lost more slowly and cells eventually arrest apparently in the S-phase of the cell cycle, with a medium-sized bud, a single nucleus in the mother cell and a pre-replicated (1*n*) DNA content. The action of this toxin is not immediate, and it must be present for more than 1 h before it can affect a significant proportion of the treated cells. Because K28 causes sensitive yeasts to arrest proliferation as unbudded cells, it is suggested that it blocks completion of the G_1 phase of the cell cycle. As is the case with K1 killer toxin, the mature and biologically active K28 killer toxin is a α/β-heterodimeric protein, with the subunits having a molecular mass of 10 and 11 kDa, respectively.

The dsRNAs associated with *Hanseniaspora uvarum* and *Zygosaccharomyces bailii* killer toxins are also encapsidated by protein capsomeres. The RNA of *Z. bailii* strain 412 contains three dsRNA plasmids, and has been extensively studied by Radler *et al.*[14] The killing spectrum of the toxin attracted interest because it is was somewhat limited, being restricted to strains of *S. cerevisiae*, *Candida glabrata* and most strains of *Zygosaccharomyces*. It was found by gel filtration and SDS-PAGE of the electrophoretically homogeneous killer protein that the apparent mass is 10 kDa. The protein does not appear to be glycosylated, and the primary target on sensitive strains was the mannan fraction of the cell wall, not the glucan (like the K28 killer toxin). Of the three plasmids (1.9, 2.9 and 4.0 kb), the smallest is thought to correspond to the killer-toxin-coding M-plasmids of *S. cerevisiae*.

The situation regarding the mode of action of the toxin from some strains of *Kluyveromyces lactis* is completely different. Killer strains secrete an exogenous, heterotrimeric (αβγ) zymocin complex, which in effect blocks the cell cycle. Sizes for the three subunits are 97, 31 and 28 kDa, respectively. Earlier reports that the zymocin functioned against sensitive strains of *S. cerevisiae* by inhibiting adenylate cyclase, and hence abolishing the roles of cAMP essential for mitotic growth and cell division, have been disproved, and, at present, its precise mode of action remains unclear. What is known is that zymocin causes sensitive yeast cells to arrest growth in the G_1 phase of their cell cycle. It appears that intracellular expression of its smallest subunit, the γ-toxin, is alone responsible for G_1 arrest, while the α- and β-subunits are required for docking to sensitive yeast cells by interacting with chitin (the α-subunit has a chitinase activity). Exogenously added chitin can compete for the zymocin receptor on the surface of sensitive yeast cells.

Klassen and Meinhardt[15] studied the cellular response of *S. cerevisiae* to a linear plasmid-encoded killer toxin from *Pichia acaciae*. This novel toxin was also shown to bind to chitin and probably acts by facilitating the import of a toxin subunit, as does zymocin; but unlike zymocin, it provokes S-phase arrest in the cell cycle and concomitant DNA damage checkpoint activation. It was found that death of sensitive cells was a two-step process, and that the chitin-binding and DNA-damaging toxin constitutes an apoptosis[†]-inducing protein. The *P. acaciae* toxin is also encoded for by double-stranded DNA, and consists of three subunits: α (110 kDa), β (39 kDa) and γ (38 kDa).

Over the years, killer yeasts have provided an interesting model for studying the mechanisms involved in the processing and secretion of extracellular proteins, and the identification of killer toxin receptors on the envelope of sensitive targets has helped in the elucidation of the structure and function of the yeast cell wall. The food and beverage industries were among the first to explore the ability of toxin-producing yeasts to kill other (unwanted) microbes.[16] Attention has been mainly focused on the toxins from *S. cerevisiae* and *K. lactis*, but more recent investigations have turned to yeasts such as *Z. bailii*, *H. uvarum*, *Pichia membranifaciens*, *Debaryomyces hansenii*, *Kluyveromyces phaffii* and *Schwanniomyces occidentalis*. Characteristics of the yeast killer toxins are outlined in Table 1.

One of the most topical subjects in winemaking is the reduction in the use of SO_2 and its partial or complete substitution with natural anti-microbials, which would be more compatible with the

---

[†]Apoptosis is one of the main types of programmed cell death (PCD). In multi-cellular organisms it is a process of deliberate cell relinquishment, in contrast to necrosis, which is a form of death that results from acute cellular injury.

Table 1 *Characteristics of the killer yeast toxins (after Marquina et al.[23])* (Reproduced by kind permission of Springer Science & Business Media)

Yeast species	Killer toxin	Subunits	Glycoprotein	Isoelectric point	Genetic basis	Primary receptor	Mechanism of killing	Application
Bullera sinensis	?	?	?	?	Chromosomal	?	?	?
Candida krusei	?	?	?	3.6–3.8	?	?	?	?
Candida glabrata	?	?	+	?	Chromosomal	?	Plasma membrane damage	?
Cryptococcus humicola	?	<1 kDa	?	?	Chromosomal	?	?	?
Debaryomyces hansenii	?	23 kDa	?	?	Chromosomal	β-(1 → 6)-Glucan	?	?
Hanseniaspora uvarum	?	18 kDa	–	3.7–3.9	dsRNA	β-(1 → 6)-Glucan	?	?
Kluyveromyces fragilis	K6	42 kDa	?	?	?	?	?	?
Kluyveromyces lactis	?	α (97 kDa) β (31 kDa) γ (28 kDa)	?	?	dsDNA (pGKL1)	Chitin <$\beta2$>	Inhibition of cell cycle, G_1 arrest	Avoid aerobic deterioration of silage
Kluyveromyces waltii	?	>10 kDa	?	?	?	?	?	Control of *S. pombe* in wine making
Pichia acaciae	?	α (110 kDa) β (39 kDa) γ (28 kDa)	?	?	dsDNA (pPac1–2)	Chitin	?	Cell cycle arrest in G_1, chitinase activity

Pichia anomala	?	83 kDa	?	?	?	?	?	Control of filamentous fungi in wood
Pichia farinosa	SMKT	α (6.3 kDa) β (7.7 kDa)	+	?	Chromosomal	?	Increase of membrane permeability to ions	?
Pichia fermentans	?	?	?	3.8–4.2	?	?	?	?
Pichia inositovora	?	>100 kDa	?	?	dsDNA (pPin 1–3)	?	?	?
Pichia kluyveri	?	19 kDa	+	4.3	Chromosomal	?	Formation of ion channel	?
Pichia membranifaciens	?	18 kDa[a]	–	3.9[a]	Chromosomal[a]	β-(1 → 6)-Glucan	Formation of ion channel[a]	?
Saccharomyces cerevisiae	K1	α (9.5 kDa); β (9 kDa)	–	4.5	M_1-dsRNA	β-(1 → 6)-Glucan	Formation of ion channels, activation of K^+ channel	Avoid undesired contaminants in wine, beer, sake, *etc.* Genetics
S. cerevisiae	K2	αβ (21.5 kDa)	+	4.2–4.3	M_2-dsRNA	β-(1 → 6)-Glucan	Increase of membrane permeability to ions	Wine fermentations
S. cerevisiae	KT28	α (10 kDa); β (11 kDa)	+	4.4	M_{28}-dsRNA	Manno-proteins	Entering into cell by endocytosis and inhibition	?

(Continued)

Table 1 *(Continued)*

Yeast species	Killer toxin	Subunits	Glycoprotein	Isoelectric point	Genetic basis	Primary receptor	Mechanism of killing	Application
							of cell cycle, G_2 arrest	
Schwanniomyces occidentalis	?	α (7.4 kDa); β (4.9 kDa)	—	?	Chromosomal	Mannoproteins	Plasma membrane damage	?
Tilletiopsis albescens	?	10 kDa	?	?	Chromosomal	?	?	?
Williopsis mrakii	HM-1	10.7 kDa	?	?	Chromosomal	β-(1 → 6), β-(1 → 3)-glucan	Inhibition of β-(1 → 3)-glucan synthesis	Silage, yoghurt, taxonomy of *Nocardia*, control of *C. albicans*
Williopsis saturnus	HYI	9.543 Da	—	5.8	Chromosomal	?	?	?
Zygosaccharomyces bailii	KT412	10 kDa	—	4.1	dsRNA	Mannoproteins	?	?

^a Unpublished results.

requests of consumers for safe and unspoiled food products. One of the most promising candidates, thus far, is the KpKt killer factor from the yeast *K. phaffii* (since classified as *Tetrapisispora phaffii*), which, according to Ciani and Fatichenti,[17] is active against wine spoilage yeasts under winemaking conditions and is therefore of particular interest for its potential application as an anti-microbial agent in the wine industry. KpKt exhibits broad-spectrum activity against potential spoilage yeasts such as *Saccharomycodes ludwigii*, *Z. bailli* and *Z. rouxii*, and has an extensive anti-*Hanseniaspora/Kloeckera* activity under winemaking conditions. The toxin is stable at wine pH, and partial biochemical characterisation of the purified form[18] showed that it is a glycosylated protein with a molecular mass of 33 kDa. In addition, it shows a 93% and 80% identity to a β-1,3-glucanase of *S. cerevisiae* and a β-1,3-glucan transferase of *C. albicans*, respectively, and is active on laminarin and glucan, thus showing β-glucanase activity. Competitive inhibition of killer activity by cell wall polysaccharides suggest that glucan (β-1,3, and β-1,6 branched glucans) represents the first receptor site of the toxin on the envelope of the sensitive target, and results indicated that the lethal activity of KpKt is mediated by a β-glucanase activity that results in cell wall permeabilisation and subsequent cell lysis. The killer toxins secreted by *S. cerevisiae* (K2 and K28), *K. lactis*, *Candida* sp. SW-55 and *Hansenula anomala* are also glycosylated proteins. The estimated mass of the carbohydrate fraction of KpKt is similar to that exhibited by the K28 killer toxin of *S. cerevisiae*, but its molecular mass and NH_2-terminal sequence do not show any similarities with those of other known killer toxins.

Also, with the oenologist in mind, considerable interest is now being shown in the killer toxin from *P. membranifaciens*, a common contaminant of fermenting olive brines, which shows a particularly strong, broad-spectrum zymocidal activity, besides being inhibitory to the growth of some filamentous fungi, such as *Botrytis cinerea*. The killer toxin from *P. membranifaciens* strain CYC 1106 was purified to electrophoretic homogeneity by Santos and Marquina,[19] who reported an apparent molecular mass of 18 kDa. Further investigations showed that at pH 4.0, optimal killer activity was displayed at temperatures up to 20°C, while at 25°C, the toxic effect decreased by 30%. Killer activity was higher in acidic media; it decreased markedly above pH 4.5, and at 6.0, it was barely noticeable. Interestingly, when the non-ionic detergent Brij-58 (polyoxyethylene 20 cetyl ether), known to be useful in the isolation of functional membrane complexes, was incorporated into the yeast growth medium, the culture supernatant exhibited the highest specific activity. The killer toxin from CYC 1106 is stable only within a narrow pH range (3.0–4.8). Experiments with healthy plants of *Vitis vinifera* showed that the presence of purified *P. membranifaciens* killer toxin prevents infection by grey mould (*B. cinerea*) in 80% of the plants challenged. Santos and Marquina found that the anti-fungal effect was even more marked (100% success) when actively growing cells of *P. membranifaciens* were used, as opposed to the toxin itself. With the evidence that this toxin offers protection to the vine against grey mould, further studies are underway in the hope of developing a biocontrol agent to combat *B. cinerea*.

Chemical control of *B. cinerea* has been partially successful, and fungicides are commonly used in the management of grey mould. The risk of the establishment of resistant *Botrytis* strains in the vineyard, however, is considerable. The evolution of strains of *B. cinerea* resistant to dicarboxymide and benzimidazole in some New Zealand vineyards was amply demonstrated by Beever *et al.*[20]

A number of review articles have appeared over the years, and the interested reader is directed to Young,[21] Shimizu[22] and Marquina *et al.*[23]

9.2   *BRETTANOMYCES*

The name was first given to a 'Torula-like yeast' identified by Claussen in 1904,[24] as being present in some faulty English beers. Yeasts of the genera *Brettanomyces* are asexual (imperfect) forms of the ascomycete genus *Dekkera* (which do produces sexual spores or ascospores), and are commonly found in fermenting musts and in wine, the organisms being commonly referred to as 'brett'. In mycological terms, *Dekkera* is the teleomorph of *Brettanomyces* (which, by definition, is an anamorph). Technically, one should think in terms of *Dekkera/Brettanomyces* species, but for the sake of brevity I shall refer only to the latter. Of the five species described by Barnett *et al.*,[25] only *B. bruxellensis* concerns the wine industry and is generally thought of as a 'wild yeast'. Both

Brettanomyces and *Dekkera* species are capable of causing alcoholic fermentation of grape juice, and the wines produced can be pleasant and fruity when consumed young.

The vegetative cells of *B. bruxellensis* are slightly smaller than those of *S. cerevisiae* and have a characteristic shape, which has been referred to as 'ogival' (reminiscent of a gothic arch or akin to the diagonal rib of a vault), and this is caused by repeated budding at the pole of each cell. Very occasionally, incomplete separation of daughter cells after budding may result in the formation of chains. Most typically, these yeasts grow after alcoholic and malolactic fermentations during storage in tank, barrel or bottle, and while they are not thought to be part of the grape surface flora, they are common components of the winery biosphere. They are the cause of the characteristic 'bretty' flavours, which are almost impossible to describe, but have been variously said to resemble 'phenolic', 'barnyard', 'smoky', 'plastic', burnt plastic', 'Band-Aid', 'sweaty horse' and 'creosote'. Wines infected with 'brett' also tend to take on a metallic bitterness. The yeasts contain cinnamate decarboxylase activity and can thus produce vinyl phenols from hydroxycinnamic acids, such as *p*-coumaric acid. Even worse, they contain vinyl phenol reductase, which converts vinyl phenols to ethyl phenols, thus giving wine a much stronger phenolic taint. Compounds known to be mainly responsible for 'bretty' taints are 4-ethyl phenol (4EP), 4-ethyl guaiacol (4EG) and *iso*-valeric acid. The compound that is responsible for the 'burnt plastic' aroma is unknown as yet. 4EP is used in some wineries as an indicator for the presence of 'brett', which is reasonable when one considers that this compound is formed at all stages of the *Brettanomyces* growth cycle. There is an anomaly, however, that some wines exhibiting strong 'brett' aromas and flavours contain little or no 4EP. 'Brett' can, and does occur in white wines, although it is predominantly a red wine problem because they are far higher in polyphenols and generally have a higher pH. The latter reduces the effectiveness of added SO_2, while polyphenols are precursors of the volatile phenols largely responsible for 'bretty' odours. The wild yeast, *Pichia guillermondii*, is also capable of producing 4EP in synthetic growth media, often at levels akin to *Brettanomyces*, but it has not yet been shown to do so in wine. Indeed, most available evidence suggests that *P. guilliermondii* is inhibited by the wine environment even without the addition of preservatives.

The *Brettanomyces* fermentation is rather unique since the relationship between yeast and oxygen is a rather complex one. In 1940, Custers[26] showed that oxygen actually stimulated fermentation in *B. claussenii*, rather than inhibiting it, which was a contradiction of Pasteur's findings with *S. cerevisiae*. Accordingly, the phenomenon was dubbed as the 'negative Pasteur effect'. A little later, Scheffers[27] showed that the Pasteur effect and the negative Pasteur effect are brought about by different mechanisms and can, in many yeasts, occur simultaneously during fermentation. To avoid any confusion, Scheffers proposed calling the negative Pasteur effect as the 'Custers effect'. He found that passage of cells from aerobiosis to anaerobiosis suddenly interrupted fermentation and growth in *B. claussenii*, but that this 'blockage' was only temporary and fermentation starts again after a period of latency, albeit much more slowly than when under aerobic conditions. It was also shown that the 'blockage' could be circumvented by supplying cells with oxygen or through addition of hydrogen acceptor compounds, such as carbonyl compounds (Scheffers used acetoin).

The end pathway of glucose fermentation to ethanol in *Brettanomyces* is outlined in Figure 2. Acetaldehyde coming from pyruvate is oxidised, at the expense of NAD^+, to acetic acid, rather than being reduced to ethanol (with the concomitant re-oxidation of NADH). The oxidation of aldehydes to organic acid end products may not be substrate specific, and it is this reaction that is

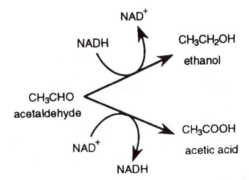

Figure 2 *The end products of glycolysis in* Brettanomyces, *showing the Custers effect*
(Reproduced by kind permission of Springer Science & Business Media)

probably the source of butyric acid and other medium-chain-length organic acids noticed in 'bretty' wines, as well as the increased volatile acidity often associated with *Brettanomyces* spoilage. The oxidation of aldehydes to organic acids also means that the energy economy of the cell is even more dependent on the re-oxidation of coenzymes than in the 'normal' fermentation of *S. cerevisiae*. This makes the control of redox potential, and air contact *via* 'headspace' in the bottle, of wines infected with *Brettanomyces* a matter of some importance.

Despite ever-improving standards of cleanliness, 'brett' is on the increase and has been since the 1990s. This has been partly attributed to the modern penchant for more 'natural' wines, in the making of which no SO_2 is added to the grapes during crushing. According to Peter Godden of the Australian Wine Research Institute, 'minimalist winemaking is a perfect recipe for bretty wine'. It is certainly a fact that the best way of preventing 'brett' is to maintain a suitable lethal level of free SO_2 throughout wine processing. It has also been claimed that the trend towards more 'international' styles of red wine made by extractions from super ripe grapes (which are high in polyphenols and have an elevated pH) also encourages 'brett'. From a purely chemical point of view, the increased use of diammonium phosphate (DAP) by winemakers as a supplementary nitrogen source (to prevent 'stuck' fermentations) is seen as being beneficial to 'bretts'. DAP, which is regarded by some as 'junk food' for yeasts, will often be preferentially used during fermentation, thus leaving some natural must amino acids in solution. These then become available for subsequent microbial growth – including that of *Brettanomyces*.

B. bruxellensis is a particularly difficult organism to control because its presence may go unnoticed until a wine is permanently tainted. It appears to spread from one winery to another through contaminated wine and/or equipment, and fruit flies seem to act as a vector once the fungus has become established. Management of *Brettanomyces* is fast becoming one of the more important oenological issues, and 'brett' is now a fashionable topic of discussion among wine enthusiasts. In the general absence of information from the winery, it is difficult to tell whether the 'earthy' or 'barnyard' characters of a red wine are attributable to terroir or to 'brett'. A classic, albeit rare, example of the resolution of this kind of situation has recently been forthcoming from some of the past vintages of Château de Beaucastel, the highly regarded Châteauneuf du Pape estate. Over recent decades, Beaucastel has been widely acknowledged as one of the great wines of the world, but some of the most highly acclaimed vintages have contained high levels of 4EP (above 900 mg L^{-1}) and have shown other evidence of 'brett' growth, suggesting that the organism should not necessarily be regarded as a spoilage yeast.

Traditional methods of detecting *B. bruxellensis* in wine involve lengthy enrichment techniques, which often overlook metabolically active but non-culturable (ABNC) populations, which can affect product quality. It is thus necessary to enumerate total *Brettanomyces* populations, and several methods exist, such as flow cytometry or microscopy, which use antibodies or nucleic acid-based probes, respectively, to quantify the organisms. Recently, Phister and Mills[28] have reported a quantitative real-time polymerase chain reaction (QPCR) method to directly enumerate this yeast in wine. Specific PCR primers to *B. bruxellensis* were designed to the 26s ribosomal RNA gene that did not amplify other non-target yeasts and bacteria commonly found in wine. The QPCR assay was linear over a range of cell concentrations (6 log units) and could detect as little as one cell per millilitre of wine. Addition of a large volume of non-target yeasts did not impair the efficiency of the assay. The whole assay procedure takes about 3 h and will allow the winemaker to make some important decisions quickly, thus reducing the threat of spoilage.

At the same symposium, Joseph and Bisson[29] presented evidence that highlighted the variation among *Brettanomyces* strains. They examined 35 isolates, which were isolated from wines of different type, vintage and geographic area. Strains were tested for growth on various carbon and nitrogen sources, vitamins and under other environmental conditions. The level of production of 4EP and 4EG in wine made from Grenache grapes was also determined. A number of important points emerged from their work, including the fact that all isolates grew well on arginine as the sole source of carbon and nitrogen, and that they had an absolute requirement for biotin and thiamine. They did not, however, show any such requirement for any specific amino acid. The strains differed widely in their ability to produce 4EP and 4EG, with 16% showing no evidence of the production of either compound. Most isolates grew well at pH 2.0 and could tolerate up to 10% v/v ethanol (with some strains being able to metabolise ethanol). Among them, 30% of strains studied could grow well at 10°C, and a few could grow at 37°C; 50% of the 'Brett' strains examined grew in 30 mg L^{-1} free sulfur dioxide at pH 3.4, and it is known that the organism generally demonstrates a large variation in sulfur dioxide tolerance – anything from 14 to 56 mg L^{-1} molecular free SO_2. This would suggest that it is much better to attempt control the organism with a few large doses of SO_2 rather than a number of smaller ones. Rather worryingly, it was found that 50% of the strains surveyed formed biofilms, which would indicate that cells, once present on the surface of a piece of equipment can slough off gradually into the winery environment. How well a sanitation procedure breaks down the biofilm will largely determine how effective it is.

Mansfield *et al.*[30] showed that, in model wine solutions, *B. bruxellensis* has the ability to break down phenolic glycosides like anthocyanins to liberate glucose. This breakdown can provide a carbon source for cell growth, and explains why 'Bretty' wines frequently lack a desirable colour. In a study on the growth of eight *B. bruxellensis* strains in simulated Pinor Noir wine, it was shown that there was a considerable variation in growth rate and population densities among strains, and that significant increases in the concentration of 4EP occurred after accumulated cell populations reached 2.5×10^5 cfu mL^{-1}.[31] A standard gas chromatographic method for the detection of 4EP is given by Zoecklein *et al.*[32] and Chatonnet *et al.*[33] have reviewed the origin and occurrence of ethyl phenols in wines.

9.3 *BOTRYTIS CINEREA*

A hyphomycete fungus (Fungi imperfecti) with a perfectly descriptive name, since 'botrys' means a bunch of grapes in Greek, which describes the arrangement of the asexual spores (conidia) on the conidiophore (Figure 3), and 'cinereus' equates to ashy-grey in Latin, which documents their colour. The fungus has been known since ancient times as one that grows on the skin of grapes, and the genus *Botrytis* is one of the first described genera of fungi, being erected by Micheli in 1729. It forms a blue–grey film on the grape surface and, according to climatic conditions, is responsible for two different effects: one undesirable and the other not so. Either the

Figure 3   Botrytis cinerea; *conidiophore bearing clustered conidia*

berries rot rapidly and completely or, in suitable conditions, they decay very slowly, permitting them to dry considerably. Overall, *B. cinerea* Pers. is a very important fungus, whose beneficial and deleterious effects on the world's grape and wine producers amounts to millions of pounds annually. Not only is grey rot a viticultural problem, since the mould is capable of infecting and spoiling a wide range of agricultural and horticultural crops, such as sunflowers, beans, rapeseeds, tomatoes, strawberries and cut flowers, but it also infects a number of wild plants and survives on these alternative hosts as a saprophyte (on dead tissue). Economic losses in the European Community alone have been estimated at between 50 and 100 million euros annually.

If the weather is consistently too wet and humid, growth of the mould gets out of control and the grapes are ruined. The grapes split, and a variety of other moulds and wild yeasts, together with, for example, acetic acid bacteria[‡], utilise the liberated juice. This destructive condition is called 'bunch rot' or 'grey rot' (*pourriture grise*; German *Graufäule*), and wine made from such grapes is of very poor quality. This particularly applies to red wines, which often have a sub-standard colour and texture and off-odours.

[‡] Normally, *Gluconobacter* spp. are found on the grape surface; but in rotted material, *Acetobacter* spp. dominate.

Unfortunately, the fungus is non-specific, and it will attack red/ black grapes as readily as it attacks white varieties. Under optimum conditions, the mould can infect and destroy a berry, and sporulate again in a matter of 2–3 days. The main drawback regarding red wines concerns colour, an enzyme produced by *Botrytis* (laccase) causing, among other things, decolouration and premature browning. In addition, white grapes can often tolerate a higher level of grey rot infection before their quality becomes compromised. *B. cinerea*, which appears to be ubiquitous in most geographic zones, can also attack the vine at flowering as long as there is a suitable host, leading to death of the bunch before the grapes have even formed. Under normal circumstances, however, the vine is resistant to *B. cinerea* during its purely vegetative phases of growth, and destructive rot rarely occurs between fruit set and véraison. Occasionally, vineyards can suffer early attacks (as happened in some northern European vineyards in 1983 and 1987), and this has been attributed to a loss of resistance within the plant.

Conversely, under the precise conditions of cold and dry nights (under 60% relative humidity to slow down the growth of the mould), and warm, humid, mid-day temperatures between 15 and 25°C (to promote mould growth), the fungus does not get out of control and magically becomes the adored 'noble rot' (*pourriture noble*; German *Edelfäule*). Noble rot appears to be encouraged by the dew, and frequent morning fogs that are characteristic of certain river valleys, where in the afternoons there is enough air movement to evaporate water. The best known example is probably that of Sauternais, where the small river Ciron creates an environment conducive to the production of Sauternes and Barsac. After a few weeks of this controlled growth, the fungal mycelium causes small apertures to occur in the grape skin, thus permitting water to escape from the grape flesh, causing the berry to shrivel. As water is lost, the concentrations of sugars and acids (and other solutes) rise, in some cases doubling their original concentrations. As long ago as 1888, Müller-Thurgau reported that grapes affected by noble rot lost about 25% of their weight mostly as water loss, while their relative sugar content increased from 11.7% in mid-October to 19.6% by the end of November. Over the same period, he found that their total acid content rose from 17.4% to 22.1%. Ribéreau-Gayon[34] found that in noble rot of Sémillon grapes, 22% of the sugars were utilised, together with 49% of the tartaric acid and 8% of the malic acid; in destructive rot of the same cultivar, 52% of the sugars, with 44% of the tartaric acid and 28% of the malic acid, were utilised. In the noble rot of Sauvignon grapes, the utilisation of sugars, tartaric acid and malic acid was 27%, 40% and 40%, respectively, compared with 36%, 27% and 24% for grey rot. Thus, in these varieties, it seems as though the metabolism of acids is relatively more important than that of sugars in noble rot, while the converse is true for grey rot. It is now known that *B. cinerea* can metabolise organic acids as an energy source. This causes an increase in pH, and imparts a 'sweeter' and 'softer' taste to juice and wine.

Exactly when and how (and by whom) noble rot was put to good use is shrouded in mystery, but over the years major claims have been made by the Hungarians (Tokaji Castle, 1650), the Germans (Schloss Johannisberg, 1775) and the French (Château d'Yquem, 1830s). All of these claims are likely to be spurious because the condition was known and described in Roman times, and in all probability goes back further still since most of the wines of the ancient Near East, ancient Egypt and classical Greece were seemingly made as sweet as possible. It is not without significance that the three claims mentioned above are representative of areas that are responsible for producing what are probably the three most famous sweet white table wines of the Old World; Tokay, Trockenbeeren auslese and Sauternes. Wines produced from *Botrytis*-infected grapes are the most complex and balanced (and sometimes most expensive) of the sweet white genre, and immense care is often taken in the harvesting of botrytised grapes. To illustrate this point, we can refer to the Auslese 'family' of German sweet white wines in which the *Auslesen* results from selected picking of infected bunches of grapes, *Beeren auslesen* comes from the selected picking of infected berries and *Trockenbeeren auslesen* emanates from the selected picking of dried, infected berries. Noble rot is often said to confer a 'honey' or 'roasted' component to the aromatic character of a wine.

When *Botrytis* occurs in a vineyard, not all of the grapes are infected equally, the most infected grapes being essential for the sweetest wines. In Sauternes, the great vintages coincide with a high incidence of *pourriture noble*. One of the mysteries of viticulture has been the reason why one vineyard may have noble rot, and its neighbour destructive rot. It was first thought that different races of the fungus with differing degrees of pathogenicity may be involved, but this has not been unequivocally proven. Ribéreau-Gayon[34,35] found that must from healthy grapes from a destructive rot vineyard had twice the total N and ammonium N content of healthy grapes from a neighbouring vineyard that harboured noble rot. In both samples of grapes, 75–90% of this N disappeared when they were parasitised by *B. cinerea*. He concluded that the N nutritional status of the vines had a profound effect on whether the fungus is noble, or otherwise. Noble rot vineyards tend to be on nutrient-poor, well-drained, limestone soils; the plants are deep rooted and have a constant water supply that tends to decrease at véraison. In contrast, vines on richer soils have more superficial roots and a widely fluctuating water supply; their fruits mature earlier and are more susceptible to cracking. Such conditions would seem to favour grey rot. Ribéreau-Gayon attributed the resistance of the St. Emilion cultivar to *B. cinerea* to its very low N content.

Whether one is talking about grey rot or noble rot, the fungus must first infect the grape plant before environmental conditions determine exactly which growth pattern is to be followed. *B. cinerea* is present in most vineyards as part of the natural microflora and survives on the vine throughout the year, its vegetative mycelium inhabiting dormant buds and crevices in bark, *etc*. It also produces highly resistant structures called sclerotia (singular, sclerotium), which are essentially knots of hyphae that have become surrounded by a thick, blackened, protective wall. Sclerotia, which have a diameter of approximately 3 mm, are the most important 'over-wintering' phases of the fungus, and under favourable conditions (normally in spring) germinate to produce hyphae, and, eventually, conidia. In culture, therefore, *B. cinerea* can be divided into three forms: mycelial, sporulating (conidia) and sclerotial.

Occasionally, the fungus will resort to sexual reproduction resulting in the production of cup-shaped fructifications (apothecia), which contain ascospores. This telomorphic stage, which is rarely observed, is known as the cup-fungus *Sclerotinia fuckeliana*. The normal mode of reproduction, however, is by asexual conidia, which are generated in large numbers at the distal ends of upright conidiophores (Figure 3). Vegetative mycelium and conidia are colourless, and the grey colour of mature fungal colonies is imparted by the colouration at the base of the conidiophores. The

relationship between *B. cinerea* and *S. fuckeliana* was first suspected in 1866 by de Bary,[36] who was convinced that the two were
closely related, and to most modern taxonomists, *B. cinerea* is to be
regarded as the conidial stage of the discomycete *S. fuckeliana*.
Although the nomen *B. cinerea* has remained unaltered over the
years, the sexual phase of the fungus has been placed in a number
of different genera. In de Bary's classic work, the ascosporogenous
form was known as *Peziza fuckeliana*, but this was transferred by
Fuckel to the genus *Sclerotinia* in 1869, a move acknowledged by
de Bary in 1886,[37] and in 1945 Whetzel[38] transferred it to the genus
Botryotinia, where it remained for many years[39]; since then, the
generic name has reverted to *Sclerotinia*.

Initial infection of a plant will normally be *via* conidia, which
will then germinate when conditions are appropriate. Free water,
minerals and nutrients, such as glucose and fructose, are required,
and conidia can germinate at temperatures between 1 and 30°C,
although those between 15 and 20°C are the most favourable.
Conidia may infect many different parts of the vine: young leaves,
shoots, grape flowers and berry stems, and are particularly adept at
entering stomata and opportunistically invading wounded tissue.
Primary infections usually occur around the flowering period,
followed by a period of latency, during which time the fungus is
present inside the berry without causing any external symptoms of
disease. This latency lasts until véraison when the natural resistance
of the berry is lowered. The first sign of damage to the berry
is when the cuticle begins to fissure due to pressure from within,
and this in turn leads to additional fungal infection. *V. vinifera*
cultivars vary considerably in their sensitivity to *Botrytis* infection,
those with the more compact berry bunches such as Riesling and
Sauvignon Blanc, seemingly being a lot more sensitive than those
with more dispersed clusters, such as Cabernet Sauvignon.

Once inside plant tissue, *B. cinerea* hyphae synthesise a phyto-
toxin called botrydial, which produces severe chlorosis and cell
collapse in the host and facilitates fungal penetration and colonisation. Botrydial was first detected in the ripe fruits of sweet pepper
by Deighton *et al.*,[40] and is almost certainly not host specific.
Chemically, it is a sesquiterpene, and a number of derivatives have
been synthesised from the botryane skeleton, from which it has
been ascertained that cytotoxicity is related to the presence of a 1,5-
dialdehyde functionality. Aggressive fungal strains induce 'oxygen

active species' (AOS), such as hydrogen peroxide and hydroxyl radicals, while non-aggressive strains do not. Our knowledge about the infection biology of *B. cinerea* has been considerably advanced by the discovery of a crucial role for AOS in the early stages of parasitism of this fungus.

The increased sensitivity of ripe grapes to *Botrytis* infection can be attributed to a number of factors, all of which relate to grape anatomy, physiology and biochemistry, and it is now necessary to attempt to interrelate these factors. The epidermis of the green grape skin is covered by a protective barrier, the cuticle, the thickness of which varies with *Vitis* species (10 μm in *V. coriacea*; 4 μm in *V. rupestris*; both American species). Dispersed in the cuticle are a number of minute perforations, the number of which increase as the berry matures. The cuticle is generally covered by a waxy bloom, which has a relatively uniform thickness throughout grape maturation (*ca.* 100 μg wax cm^{-2} surface). European *V. vinifera* varieties normally have a cuticle thickness in the 1.5–3. 8-μm range, with most of the very highly sensitive varieties having a cuticle thickness of less than 2 μm. In grape clusters where the berries are tightly packed, extensive berry contact results in reduced cuticle thickness. *B. cinerea* does have some cutinolytic activity, but whether it is sufficient to actively penetrate the grape cell cuticle is debatable. With some 25–40 stomata present on the surface of the average berry, these structures and the apertures in the cuticle are going to offer a more likely route for hyphal penetration. Immediately below the cuticle lies another physical barrier, the epidermis, whose cell wall thickness varies with grape variety; the most sensitive ones have the thickest cell walls. Epidermal cells also contain tannins, which exhibit weak fungistatic activity. The green grape skins (and stems) have the ability to form an insulating, corky layer directly underneath any damaged area in order to fend off microbial infection. As the grape sugar level increases during maturation, the ability to produce cork diminishes, and when it reaches *ca.* 14%, *B. cinerea* can overcome this particular defence mechanism.

The green grape skin can also synthesise anti-fungal derivatives called phytoalexins, as an immediate response to fungal infection, when normal flavonoids metabolism is diverted towards the production of stilbene derivatives (by the action of stilbene synthase, STS, for example). The chemical structure of these derivatives is

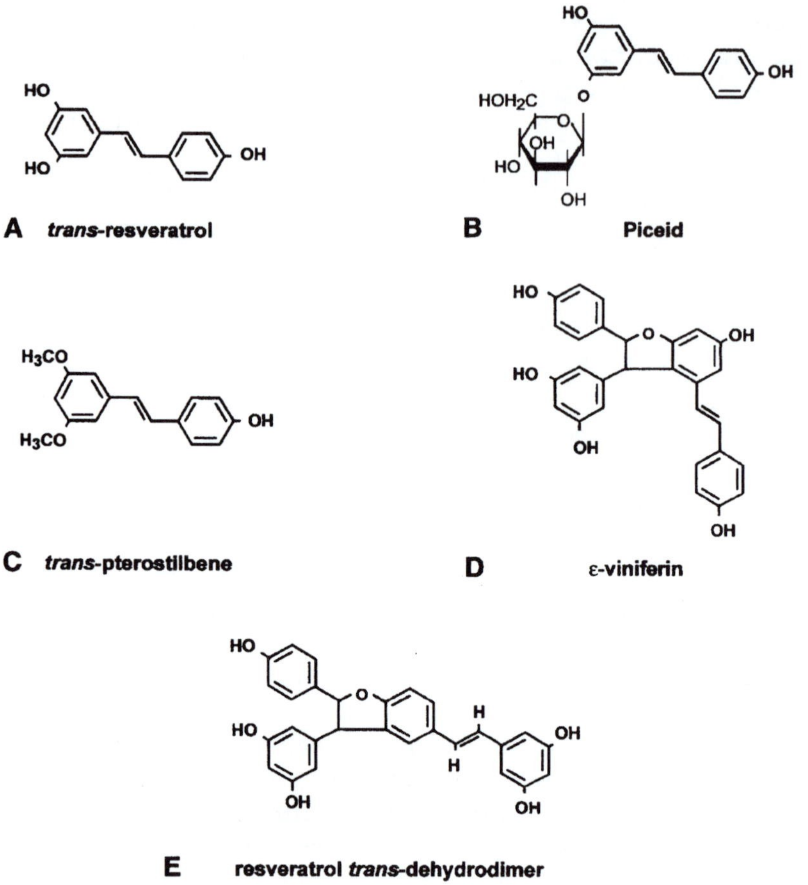

A *trans*-resveratrol

B Piceid

C *trans*-pterostilbene

D ε-viniferin

E resveratrol *trans*-dehydrodimer

Figure 4 *Stilbene stress compounds*

based on a polyphenol-type stilbene, resveratrol (*trans*-3,5,4'-trihydroxystilbene) (Figure 4A). In addition to resveratrol, the family of grapevine phytoalexins covers piceid (resveratrol-3-0-β-glucopyranose) (Figure 4B); pterostilbene, a 3,5-dimethylated resveratrol (Figure 4C); and ε-viniferin, a dehydrodimer of resveratrol (Figure 4D). The compounds are sometimes known as stress metabolites,[41] and their likely mode of biosynthesis has been documented by Jeandet and Bessis.[42] The toxicity of these phytoalexins has been elucidated, with maximum effect being shown by ε-viniferine and pterostilbene (Langcake[43]; Pezet and Pont[44]). Resveratrol, the major stress compound of the phytoalexin response, has the lowest activity.[45] As grape sugar levels increase, so does the rate of production of these stilbene derivatives, especially resveratrol, thus contributing to the enhanced sensitivity of ripe grapes to *B. cinerea*. As suggested by Bais *et al.*,[46] stilbene accumulation in young, green *V. vinifera* berries seems to stop, or at least delay *B. cinerea* infection; this defence mechanism appears to be abandoned during grape ripening. Stilbene synthesis is stimulated by polysaccharide cell wall fragments from the invading fungus, although other molecules of fungal origin may induce *de novo* synthesis of enzymes such as phenylalanine ammonia-lyase (PAL), STS and intermediary enzymes of the phenylpropanoid pathway.[47] The capacity for stilbene synthesis in plant tissues is genetically determined, and all members of the family Vitaceae can produce them, but the *B. cinerea*-resistant American *Vitis* spp. and interspecific hybrids generally show an enhanced capacity for stilbene synthesis than the more susceptible European *V. vinifera* cultivars.

The dynamics of stilbene accumulation during infection and the role of individual stilbenes are not exactly understood, but, as Van Etten *et al.*[48] have documented, many pathogens are capable of metabolising and detoxifying phytoalexins, thus encouraging vine/grape susceptibility. This is certainly true for resveratrol and pterostilbene, which are both degraded by stilbene oxidase, a laccase-like enzyme secreted by *B. cinerea*. Since the original work of Hoos and Blaich,[49] much attention has been paid to the biological significance of the laccase-mediated degradation of stilbene phytoalexins and the pathogenicity of *B. cinerea* to the grapevine. In 1998, Breuil *et al.*[50] showed that resveratrol metabolism by laccase includes an oxidative dimerisation leading to a resveratrol dehydrodimer, analogous to ε-viniferin produced by grapevines (Figure 4E), and in 1999[51], the same group characterised the pterostilbene dehydrodimer produced by *B. cinerea* laccase.

Infection of the inflorescence has long been thought of as being an important stage in the epidemiology of *B. cinerea* infection of the grape, but relatively little experimental work has been carried out on the early stages of infection. Keller *et al.*[52] studied infection patterns of the fungus in grape flowers and followed the fate of the stilbene stress metabolites. Inflorescences of field-grown Gamay vines were inoculated with a *B. cinerea* conidia suspension or dried conidia at different stages during bloom in moist weather. Conidia suspensions were also applied to various locations on flowers of pot-grown plants of the cultivars Pinot Noir and Chardonnay. The results obtained confirmed the bloom as a critical time for *B. cinerea* infection of grapes, and suggested that the most likely site of infection is the receptacle. Stilbene stress metabolites in the flowers were measured by HPLC and indicated that resveratrol accumulated mainly after pre-bloom and full-bloom inoculation, but did not prevent infection. Piceid levels did not change after inoculation, while ε-viniferin was found in necrotic tissues only, and pterostilbene and α-viniferin were not detected at all. When pot-grown plants were inoculated with conidia, it was the receptacle, but not the stigma and ovary, that produced latent infections. Stilbene synthesis was similar to the field results, with resveratrol accumulating mainly in the calyptra and receptacle area. Constitutive soluble phenolic compounds, mainly derivatives of quercetin and hydroxy-cinnamic acid, were present at high concentration in the calyptra but at low levels in the receptacle area.

Germination of grey rot conidia, and hence vine infection, is known to be stimulated by the availability of nutrients, especially sugars, with fructose being more effective than glucose and other hexoses and disaccharides.[53] This is surprising since glucose is usually the most efficient hexose, as not only a nutrient, but also a signalling compound. Recently, Doehlemann *et al.*[54] reported that they had characterised a fructose-specific transporter from *B. cinerea*, and they also documented the cloning of *FRT1*, a gene encoding for fructose transport, and found that it was similar to *FSY1*, the high-affinity, fructose-specific transporter that has been identified in the yeast *S. pastorianus*.[55]

In addition to yielding wines with highly desirable characters, botrytised grapes can cause several problems for the winemaker. For a start, they contain measurable amounts of fungal laccase (*p*-diphenol oxidase), which is far more deleterious than its grape-derived 'sister' compound, tyrosinase (notionally a monophenol mono-oxygenase), with which it has an overlapping substrate range. Laccase is more resistant to SO_2, much more stable at must and wine pH and oxidises a wider range of phenolics. It will also oxidise anthocyanins and, according to Dubernet *et al.*,[56] use ascorbic acid as a substrate. Laccases generally are remarkably non-specific as to their reducing substrate and the range of substrates oxidised. Because it is able to persist in wine, laccase can bring about serious and permanent oxidation, and most of the browning problems associated with red table wines made from mouldy grapes are largely due to this enzyme. Laccase was first discovered in the late 19th century in exudates of the Japanese lacquer tree (*Rhus vernicifera*), and was subsequently found to be a fungal enzyme by Bertrand in 1896.[57] It is one of the small group of enzymes called the 'large blue copper proteins' or 'blue copper oxidases'. The other members of this group are the plant ascorbate oxidases and the mammalian plasma protein – ceruloplasmin. The blue copper oxidases have been intensively studied, not least because they share with the terminal oxidases of aerobic respiration the ability to reduce molecular oxygen to water, the reduction being accompanied by the oxidation, typically of a phenolic subst-rate. Substrate oxidation by laccase is a one-electron reaction generating a free radical (Figure 5a), which is unstable and under-goes a second reaction.

Simple diphenols, such as hydroquinone and catechol, are good substrates for most laccases, and the diamine, phenylene diamine, is also a widely used substrate. We do not presently know the full range of laccase substrates, but syringaldazine [*N,N'*-bis(3,5-dimethoxy-4-hydroxybenzylidene hydrazine)] is considered to be uniquely a laccase substrate (Figure 4B), hence its use as an assay tool. Laccase will not oxidise tyrosine (as does tyrosinase). *B. cinerea* laccases, two of which have been purified, are extra-cellular glycoproteins, each with a molecular mass of 72 kDa and an 80% carbohydrate content, which is much higher than most other laccases of fungal origin (the typical laccase is a 60–80-kDa molecule, of which 15–20% is carbohydrate). There is much still unknown about the role played by laccases in the pathogenesis of *B. cinerea* in relation to the grapevine, but some idea of their importance may be deduced from the situation in the cucumber (*Cucumis sativus*), which is also a host for the fungus. The cucum-ber produces triterpenoid cucurbitacins, which are bitter, toxic compounds that protect the plant by specifically repressing laccase synthesis in *B. cinerea*. In basidiomycetous fungi, it is claimed that laccases are an integral part of the enzymic machinery capable of mineralising lignin, an operation that is apparently unique to that group of organisms. Conversely, in higher plants, the presence of laccase in woody tissues is thought to be a component of the lignin-synthesising system.[58] Other proposed roles for laccases in fungi include pathogenicity and pigment production. Because of their

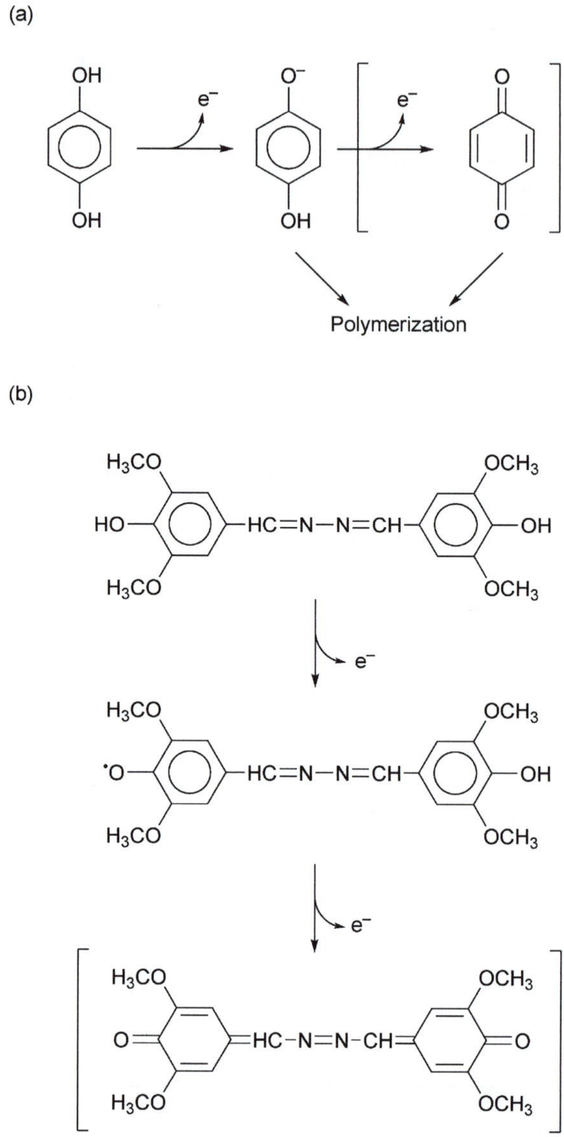

(a)

(b)

Figure 5 *(a) The typical laccase reaction, where a diphenol (hydroquinone shown here) undergoes a one-electron oxidation to form an oxygen-centred free radical. This species can be converted to the quinone in a second enzyme-catalysed step or by spontaneous disproportionation. Quinone and free radical products undergo polymerisation. (b) The laccase reaction with syringaldazine as substrate, where the initial product is a free radical. The quinone formed by a second one-electron oxidation (again by a second enzymatic step and/or by disproportionation) is deep purple in colour and not apparently prone to polymerisation*

substrate variety, laccases have been incorporated into some industrial processes such as the bleaching of textiles and pulps and the formulation of some detergents. The structure and function of fungal laccases have been reviewed by Thurston.[59]

Control of oxidation resulting from oxidase enzymes is one of the more important technical developments in modern oenology. The high laccase activity in botrytised must demands that high levels of SO_2 (*ca.* 150 g L^{-1}) are required to inhibit it, which can lead to organoleptic problems. Because of its significance, detection of laccase in wine can be important, and one of the most widely used methods is a spectrophotometric assay that employs syringaldazine,[60] but which is rather insensitive at low infection levels. Test kits for estimating laccase (and, by connotation, the level of infection) are now available.

Botrytised grapes, particularly those extracted by some degree of extended skin contact[§], are infamous for their ability to impart unwanted polysaccharides into wine. The polymers involved are mainly from the sugars glucose, mannose, galactose, arabinose and rhamnose, and most notable is β-glucan from *B. cinerea*, which has a molecular weight of around 10^6 Da and produces highly viscous wines with poor filterability. The presence of ethanol only serves to enhance the filtration problem. β-Glucan can also impair the stability of other wine components.

In addition to the above, musts made from botrytised grapes are notoriously more difficult to ferment than those derived from healthy fruit. This has partly been attributed to the fungistatic effect of an inhibitor originally called botryticin,[61] which is either partially, or completely, a mannose-based neutral heteropolysaccharide. The compound is destroyed by sulfiting and prolonged heating at 120°C, while 80% ethanol will cause it to precipitate. It is now evident that more than one antibiotic is produced by *B. cinerea*, and some of them, such as botrylactone, have now been synthesised.[62]

Fermentation of such juices is also difficult because of very high sugar levels (as high as 60° Brix is not unknown) and depleted available nitrogen levels. It is also known that *Botrytis* growth depletes the thiamine content of a must, thus discouraging yeast growth. Other problems encountered in botrytised wines,

---

[§]Often necessary because of the difficulty of removing juice from dehydrated grapes.

themselves, are high volatile acidity and increased SO_2 binding. The latter is due to elevated levels of compounds such as keto-5-fructose, keto-2-gluconic acid and diketo-2,5-gluconic acid, which, although present in non-botrytised grapes, are produced in significant amounts by *B. cinerea*. In addition, proliferating acetic acid bacteria in infected grapes are known to increase the level of compounds such as gluconic acid.

On the credit side, botrytised wines contain elevated levels of glycerol (can be as high as 30 g L^{-1}, as opposed to less than 10 g L^{-1} normally), which confer smoothness and desirable viscosity. Some glycerol is formed as a by-product of fermentation, but that which results from *B. cinerea* metabolism survives during processing and helps to give the typically 'oily' smoothness of the great sweet white wines, which rarely ferment out to more than 12–14% ethanol (leaving 5–15% sugar).

A summary of some of the other chemical changes that occur within the grape as a result of *B. cinerea* infection is as follows:

(i) citric acid content is increased, and there will be elevated levels of gluconic, glucuronic and galacturonic acids;
(ii) some esters produced during fermentation are degraded by fungal esterases;
(iii) terpenes such as linalool, geraniol and nerol are metabolised to less volatile derivatives such as β-pinene, α-terpineol and a variety of pyran and furan oxides. Over 20 such terpene derivatives have been detected from infected grapes. Muscat grapes are particularly prone to such terpene modification;
(iv) some entirely new compounds are synthesised, most important of which is sotolon (3-hydroxy-4,5-dimethyl-2(5H)-furanone), a key flavour in the French flor-sherry known as 'Vin Jaune', where it is found at high levels (120–268 μg L^{-1}) resulting from flor-yeast activity over a period of years. Sotolon, which is said to confer 'nutty/spicy' notes to botrytised wines, is also characteristic of barrel-aged port wines and can be made purely chemically by the reaction of α-ketobutyric acid and acetaldehyde. According to Rapp and Mandery,[63] in botrytised wines it is invariably found at levels above its flavour threshold (7.5 ppb). Another newly formed compound associated with botrytised grapes is the so-called 'mushroom alcohol', 1-octen-3-ol;

(v) mucic (galactaric) acid is found in rot-degraded grapes, which results in the formation of the somewhat insoluble, crystalline calcium mucate. The acid arises from the enzymic oxidation of galacturonic acid, and formation of the calcium salt is more likely if sprays containing calcium have been used in the vineyard. In botrytised wines, the precipitate normally forms shortly after bottling, so the product should be checked for calcium mucate prior to bottling.

Over the years, many attempts have been made to control grey rot of grapes with the aid of chemical fungicides, which either kill the fungus (fungicidal) or restrict its growth (fungistatic). Many compounds that have shown promising results *in vitro* have proved to be less valuable in the field because of their effect on the host plant or their propensity to induce fungal resistance. In general, most fungicides are of value only as prophylactics, which means that, to some extent, their application must rely on the anticipation of the environmental conditions conducive to the growth of fungus, a difficulty in the case of *B. cinerea*. Successful control of *Botrytis* is largely dependent on good preventive strategy; chemical control on its own is not sufficient and should go hand-in-hand with suitable viticultural practices (correct spacing of vines, rootstock selection, irrigation regime, *etc.*). Common sense has to be used; chemical control in winter is not usually a sensible option because the fungus is likely to over-winter in its resistant sclerotial stage, which fungicides cannot penetrate. Also, it is inadvisable to spray immediately prior to harvest because of the likelihood of chemical carry-over into juice/must. A problem with many fungicides is that they are rather non-specific and act on desirable members of the grape surface mycoflora, not just *B. cinerea*. Rovral® has been the 'Rolls-Royce' of fungicides for *Botrytis* control for several years, and Benlate®, while popular for many years, is now used less because of resistance problems, but apart from these, the arsenal of fungicides for *Botrytis* control is rather sparse. In the late 1990s, a new class of fungicide, the strobilurins, were developed, which have a broad spectrum of activity (including black rot and powdery mildew). They were first isolated from the wood-rotting basidiomycete, *Strobilurus tenacellus*, and have a different mode of action from sterol-inhibiting fungicides, besides possessing good residual activity. Their main drawback is their readiness to induce resistant mutants.

One of the major drawbacks of most chemical fungicides is that they are not biodegradable and have a tendency to build up in the soil. With the worldwide interest in sustainable farming, many countries are looking for 'green' ways of protecting crops, thus avoiding a dependence on chemicals. During the last few years of the 20th century, a number of biological control agents for *Botrytis* have been developed, perhaps the most notable being BOTRY-Zen®, produced by the New Zealand company, Botry-Zen Ltd., of Dunedin (formed in 2001). The product is a live spore preparation of a non-pathogenic, saprophytic fungus, which competes for the same ecological niches as *B. cinerea*. It aggressively occupies the same physical space and out-competes the rot for the nutrients in the dead and senescing floral debris in the bunches; it is a true antagonist. It is non-invasive and causes no damage to live plant tissue. With this mechanism of action, it is highly unlikely that resistance to BOTRY-Zen will develop. As the makers say, 'The active ingredient is a naturally-occurring organism. The excipients in the product are organic compounds approved in the US and Europe for use in food, and it is considered that BOTRY-Zen will be an attractive *Botrytis* control product for grape growers who want to be able to describe their crops as sustainably grown'.

In hot, dry vineyards, such as are encountered in California and Australia, environmental conditions are seldom conducive to noble rot, and it is necessary to resort to other ways of producing sweet white wines. One method is to artificially inoculate grapes with fungal spores, which will then germinate under suitable conditions. In California, grapes are placed on trays sprayed with a suspension of spores, and covered with plastic film for 24 h to attain the necessary humidity. The grape-laden trays are then stored below 75% relative humidity at 20–22°C for 1–2 weeks. Although this is 'not the real thing', grapes so treated can yield musts of around 17° Baumé, which convert to wines with at least some noble rot character.

REFERENCES

1. E.A. Bevan and M. Makower, The physiological basis of the killer character in yeast, in S.J. Goerts (ed), *Proceedings of the Eleventh International Congress of Genetics*, vol 1, Pergamon Press, Oxford, 1963, 202–203.

2. G. Philliskirk and T.W. Young, *Antonie Van Leeuwenhoek*, 1975, **41**, 147.
3. W. Magliani, S. Conti, M. Gerloni, D. Bertolotti and L. Polonelli, *Clin. Microbiol. Rev.*, 1997, **10**, 369.
4. D.R. Woods and E.A. Bevan, *J. Gen. Microbiol.*, 1968, **51**, 115.
5. D. Rogers and E.A. Bevan, *J. Gen. Microbiol.*, 1978, **105**, 199.
6. T.W. Young and M. Yagiu, *Antonie van Leeuwenhoek*, 1978, **44**, 59.
7. P. Vagnoli, R.A. Musmanno, S. Cresti, T. Di Maggio and G. Coratza, *Appl. Environ. Microbiol.*, 1993, **59**, 4037.
8. R.A. Musmanno, T. Di Maggio and G. Coratza, *J. Appl. Microbiol.*, 1999, **87**, 932.
9. H.J.J. Van Vuuren and C.J. Jacobs, *Am. J. Enol. Vitic.*, 1992, **43**, 119.
10. G.M. Heard and G.H. Fleet, *Appl. Environ. Microbiol.*, 1987, **53**, 2171.
11. R.B. Wickner, *Annu. Rev. Microbiol.*, 1992, **46**, 347.
12. R.B. Wickner, *Microbiol. Rev.*, 1996, **60**, 250.
13. A.S. Vadasz, D.B. Jagganath, I.S. Pretorius and A.S. Gupthar, *Antonie van Leeuwenhoek*, 2000, **78**, 17.
14. F. Radler, S. Herzberger, I. Schönig and P. Schwarz, *J. Gen. Microbiol.*, 1993, **139**, 495.
15. R. Klassen and F. Meinhardt, *Cell Microbiol.*, 2005, **7**, 393.
16. V.S. Javadekar, H. SivaRaman and D.V. Gokhale, *J. Ind. Microbiol.*, 1995, **15**, 94.
17. M. Ciani and F. Fatichenti, *Appl. Environ. Microbiol.*, 2001, **67**, 3058.
18. F. Comitini, N. Di Pietro, L. Zacchi, I. Mannazzu and M. Ciani, *Microbiology*, 2004, **150**, 2535.
19. A. Santos and D. Marquina, *Microbiology*, 2004, **150**, 2527.
20. R.E. Beever, E.P. Larcy and H.H. Pak, *Plant Pathol.*, 1989, **38**, 427.
21. T.W. Young, Killer yeasts, in A.H. Rose and J.S. Harrison (eds), *The Yeasts*, vol 2, 2nd edn. Academic Press, London, 1987.
22. K. Shimizu, Killer yeasts, in G.H. Fleet (ed), *Wine Microbiology and Biotechnology*, Harwood Academic, Chur, 1993.
23. D. Marquina, A. Santos and J.M. Peinado, *Int. Microbiol.*, 2002, **5**, 65.
24. N.H. Claussen, *J. Inst. Brew.*, 1904, **10**, 308.

25. J.A. Barnett, R.W. Payne and D. Farrow, *Yeasts Characteristics and Identification*, 3rd edn Cambridge University Press, Cambridge, 2000.
26. M.T.J. Custers, *Onderzoekingen over het gistgeslacht Brettanomyces* Thesis, De Technische Hoogeschool te Delft, 1940.
27. W.A. Scheffers, *Nature*, 1966, **210**, 533.
28. T.G. Phister and D.A. Mills, Technical extracts, *55th Annual Meeting*, San Diego, California, American Society of Enology and Viticulture, Davis, CA, 2004, 30.
29. C.M.L. Joseph and L. Bisson, Technical abstracts, *55th Annual Meeting*, San Diego, California, American Society for Enology and Viticulture, Davis, CA, 2004, 28.
30. A.K. Mansfield, B.W. Zoecklein and R.S. Whiton, *Am. J. Enol. Vitic.*, 2002, **53**, 303.
31. K.C. Fugelsang and B.W. Zoecklein, *Am. J. Enol. Vitic.*, 2003, **54**, 294.
32. B.W. Zoecklein, K.C. Fugelsang, B.H. Gump and F.S. Nury, *Wine Analysis and Production*, Chapman & Hall, New York. 1995.
33. P. Chatonnet, D. Dubourdieu, J.N. Boidron and M. Pons, *J. Sci. Food Agric.*, 1992, **60**, 165.
34. J. Ribéreau-Gayon, *Vitis*, 1960, **2**, 113.
35. J. Ribéreau-Gayon, *C.R. hebd. Séances Acad. Agric. Fr.*, 1970, **56**, 314.
36. A. de Bary, *Morphologie und Physiologie der Pilze, Flechten und Myxomyceten*, Engelmann, Leipzig, 1866.
37. A. de Bary, *Bot. Ztg.*, 1886, **44**, 377.
38. H.H. Whetzel, *Mycologia*, 1945, **37**, 648.
39. W.R. Jarvis, *Botryotinia and Botrytis Species: Taxonomy, Physiology and Pathogenicity*, Department of Agriculture, Monograph No. 15, Ottawa, Canada, 1977.
40. N. Deighton, I. Muckenschnabel, A.J. Colmenares, I.G. Collado and B. Williamson, *Phytochemistry*, 2001, **57**, 689.
41. P. Langcake and R.J. Pryce, *Experientia*, 1977, **33**, 151.
42. P. Jeandet and R. Bessis, *Bulletin de L' OIV*, 1989, **62**, No. 703–704, 637.
43. P. Langcake, *Physiol. Plant Pathol.*, 1981, **18**, 213.
44. R. Pezet and V. Pont, *J. Phytopathol.*, 1990, **129**, 19.
45. M. Adrian, P. Jeandet, J. Veneau, L.A. Weston and R. Bessis, *J. Chem. Ecol.*, 1997, **23**, 1689.

46. A.J. Bais, P.J. Murphy and I.B. Dry, *Aust. J. Plant Physiol.*, 2000, **27**, 425.
47. F. Liswidowati, F. Melchior, F. Hohmann, B. Schwer and H. Kindl, *Planta*, 1991, **183**, 307.
48. H.D. Van Etten, D.E. Matthews and P.S. Matthews, *Ann. Rev. Phytopathol.*, 1989, **27**, 143.
49. G. Hoos and R. Blaich, *Vitis*, 1988, **27**, 1.
50. A.-C. Breuil, M. Adrian, N. Pirio, P. Meunier, R. Bessis and P. Jeandet, *Tetrahedron Lett.*, 1998, **39**, 537.
51. A.-C. Breuil, P. Jeandet, M. Adrian, F. Chopin, N. Pirio, P. Meunier and R. Bessis, *Phytopathology*, 1999, **89**, 298.
52. M. Keller, O. Viret and F.M. Cole, *Phytopathology*, 2003, **93**, 316.
53. J.P. Blakeman, *Trans. Br. Mycol. Soc.*, 1975, **65**, 239.
54. G. Doehlemann, F. Molitor and M. Hahn, *Fungal Genet. Biol.*, 2005, **42**, 601.
55. P. Gonçalves, D.E. Rodrigues, H. Sousa and I. Spencer-Martins, *J. Bacteriol.*, 2000, **182**, 5628.
56. M. Dubernet, P. Ribéreau-Gayon, R.R. Lerner, F. Hard and A.M. Mayer, *Phytochemistry*, 1977, **16**, 191.
57. G. Bertrand, *C. R. Hebd. Seances Acad. Sci.*, 1896, **123**, 463.
58. W. Bao, D.M. O'Malley, R. Whetten and R.R. Sederoff, *Science*, 1993, **260**, 672.
59. C.F. Thurston, *Microbiology*, 1994, **140**, 19.
60. D. Dubourdieu, C. Grassin, C. Deruche and P. Ribéreau-Gayon, *Conn. Vigne Vin.*, 1984, **18**, 237.
61. J. Ribéreau-Gayon, E. Peynaud, S. Lafourcade and Y. Charpentié, *Bull. Soc. Chim. Biol.*, 1955, **37**, 1055.
62. W. Bruns, S. Horns and H. Redlich, *Synthesis*, 1995, 335.
63. A. Rapp and H. Mandery, *Experientia*, 1986, **42**, 873.

CHAPTER 10

Pests and Diseases

Ever since the vine has been cultivated, it has been subjected to the ravages of a variety of undesirable organisms, which, in the main, have involved certain fungi, bacteria, viruses, insects and nematodes. In addition, some weed species may be considered to be antagonistic towards the vine. Diseases, technically serious derangements of health, are mainly caused by fungi, bacteria and viruses, while 'damage', a different proposition, is usually the result of the presence of insects and nematode worms. Some problems of the grapevine are restricted to specific grape-growing regions, and are thus not of universal significance (*e.g. Phakopsora euvitis*, grapevine leaf rust, which at present is confined to tropical and sub-tropical areas). We shall consider a few of the more widespread pests and diseases. A compendium of grape diseases has been compiled by Pearson and Goheen.[1]

10.1 PHYLLOXERA

The grape phylloxera, *Daktulosphaira vitifoliae* (Fitch), is an aphid-like insect, often called the root-louse (Figure 1), which feeds on grape roots and leaves and is a serious pest of commercial grapevines. The insect primarily damages the root system, depriving the vine of water and nutrients, thus posing immediate management problems in the form of reduced vine growth, decreased grape yield, retarded grape maturation and lower wine quality. Like many other organisms, it has been encumbered with a number of taxonomic names, including *Phylloxera vitifoliae* Fitch and *P. vasatrix* Planchon. The latter name is interesting because it

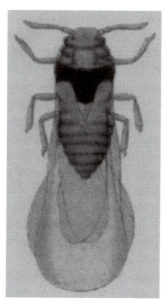

Figure 1 *Adult phylloxera*

was coined by the eminent Montpellier botanist Jules-Emile Planchon, and literally translates as, 'the dry leaf devastator'. It was Planchon, who in 1868 was the first to identify the louse as being a 'new' species, and that it was responsible for the observed grapevine damage.[2] It is native to the USA, and is widely distributed to the east of the Rockies. To the west of the Rocky Mountains (*e.g.* California) it was introduced, and, over the years, it has proved to be a serious pest in all grape-growing countries. The effects of phylloxera eclipsed all other viticultural problems when it arrived in France in the 1860s, and by the end of the 19th century, it had found its way all over Europe. The disaster in Europe came about when vines from the eastern USA were sent to France as herbarium specimens, unwittingly introducing the insect as well. By 1868, phylloxera had established itself in French vineyards, where all vines were *Vitis vinifera*, which is highly susceptible. The spread of the pest was rapid and by 1900, 75% of French vineyards were affected, something that led to much poverty and distress. Subsequently, phylloxera spread to Italy, Spain, Portugal and all other European grape-growing countries. It reached California in 1873, and later appeared in South America,

South Africa, Australia and New Zealand. It seems as though phylloxera travelled across the Rocky Mountains, from the east around 1870, only a couple of years after its appearance in Europe, and at much the same time that it must have reached Australia. European colonists to eastern North America undoubtedly brought their own vines with them, but they did not flourish because of their lack of winter hardiness and susceptibility to pests and diseases.

The arrival of phylloxera in Europe was a momentous event, but the first report of its arrival in the continent came in a wonderful example of academic understatement by the eminent Oxford entomologist, Professor J.O.Westwood, who recorded in 1869, 'In the month of June, 1863, I received from Hammersmith a vine leaf covered with minute gall-like excrescences, 'each containing', in the words of my correspondent, "a multitude of eggs, and some perfect Acari, which seem to spring from them, and sometimes a curiously corrugated coccus". Soon after the discovery of phylloxera in Europe, the possibility of using the resistance to the insect shown by American *Vitis* species was exploited by viticulturalists. Between 1885 and 1900, a tremendous effort to develop grafting stock was instigated, and so the reconstruction of phylloxerated European vineyards was underway. Prudent growers grafted their chosen variety onto a rootstock of American parentage, and it soon became apparent that the balance between the rootstock, the scion (chosen vine variety) and the soil was almost infinite. Some rootstocks proved to be so vigorous that they prompted uncontrolled vegetative growth and a concomitant vastly reduced, inferior crop. Others were too weak, and some such as 'ARGI', 'AXR' and '1202', were not sufficiently resistant to phylloxera. According to records, wild vines were common in Europe until the last decades of the 19th century, but they probably became extinct as a result of phylloxera infestation.

The life cycle of phylloxera is somewhat complicated (Figure 2), but as can be seen there are two main phases: (1) a root-feeding stage, known as 'radicicole' and (2) a leaf-feeding stage, called 'gallicole'. Principal damage to *V. vinifera* vines is caused by phylloxera living on, and feeding on, the roots. Phylloxera have sucking mouthparts, and radicoles produce characteristic lesions on grapevine roots, called 'nodosities' and 'tuberosities', the latter being essentially larger versions of the former. Nodosities are

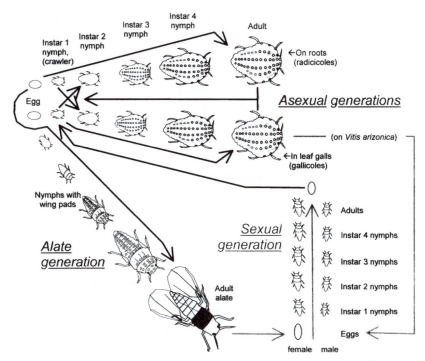

Figure 2 *Life cycle of phylloxera (by kind permission of Professor Jeff Granett)*

swellings caused by callus formation of certain sub-epidermal cells. Exactly why callus tissue is formed is still a matter of debate, but it is either due to injection of some compound by the insect (presumably a growth hormone), or maybe the insect stimulates overproduction of a phytohormone by the vine itself. Either way, the presence of the feeding louse triggers abnormal cytokinesis in the affected cells, the causal agent probably being introduced into the host *via* saliva. Nodosities, which are normally yellowish-brown, hook-shaped swellings, are galls formed on small, apical rootlets. They are generally thought to cause little damage to the vine, but they prevent further rootlet growth, which does not exactly help the plant. Tuberosities are more rounded galls formed on larger, older parts of the root system, which eventually decay and incapacitate the root, and if sufficiently abundant, may eventually result in vine death. Once the root has been compromised by nodosities or tuberosities, it is open to secondary infection by various bacteria, fungi, nematodes and other arthropods in the soil.

Root injuries severely impair the absorption of water and nutrients, and the aerial parts of the vine start to show symptoms of damage. The severity of the infection may vary according to soil conditions (drainage, *etc.*), and the variety and age of the vine. Infested vines live longer in fertile, deep, well-drained soil than in shallow soil or soil with poor drainage. Vines growing in heavy, shallow soils appear to succumb to the infestation most rapidly. Fine-textured soils, such as clays, succour the infestation more than light, sandy soils, which support vines that are almost immune to it. The reality here is that heavier soils crack when dry, and the fissures provide an ideal environment for the insect to crawl to and infest new root systems. It goes without saying that vigorous vines resist phylloxera attack better than weaker ones.

Phylloxera over-winters either on vine roots in the form of small, dark-coloured first or second instar nymphs called 'hibernants', or as a winter egg under the bark of older canes. By definition, the winter egg gives rise to the mature, wingless, female fundatrix or stem mother, which is the first generation of the gallicole form of the aphid. The fundatrix migrates to a nearby shoot tip and starts feeding, a process that elicits gall formation. As a result, the fundatrix becomes enclosed within a small, spherical gall on the underside of the vine leaf. The fundatrix is parthenogenetic and is capable of producing several hundred eggs. The first batch of eggs is usually laid early-mid summer, and two or three batches can be laid in all. First instar nymphs, called 'crawlers', emerge from the eggs and move out of their galls to feed on shoot tips, where they will initiate new galls. Over the summer, a proportion of the crawlers, either actively or passively, vacate the leaf and return to the soil surface. Once there, they find crevices in the soil and eventually reach the vine roots. Crawlers that leave the vine leaf voluntarily probably do so as a result of feeding competition, that is their population on the leaf has become too high. A moribund vine may also be the cause of their departure; otherwise they may be accidentally blown off the leaf. It is also known that crawlers can be spread *via* vineyard equipment, and so harvesting machinery, for example, should not be taken to a 'clean' site from an infested one.

Those aphids that over-winter as subterranean nymphs resume feeding in spring when sap in the vine starts to flow and soil temperature rises (*ca.* 45–65°F seems to be a critical temperature range), and in doing so become analagous to crawlers (which are technically gallicolous and found only on leaves). The radicicole nymphs begin feeding on vine roots and moult to adulthood. The adults are all females and they are very small and difficult to see with the naked eye, being some 0.7–1.0-mm long and 0.4–0.6-mm wide. Their colour varies according to their food supply: on fresh vigorous roots they appear pale green to light brown; on less vigorous roots they are brown or orange. Mature adults become brown or purplish-brown. Each adult female can lay up to 400 eggs, which when newly deposited are ovoid and lemon yellow in colour. In the absence of males, these eggs are obviously not the result of any sexual process. They develop into nymphs, which can leave the roots on which they have hatched and travel on the surface of the soil, thus dispersing themselves. They can also climb up the trunk of the vine, where they have the chance of being windblown for considerable distances. In late summer, possibly due to environmental conditions or population density, some of the nymphs develop wing buds and emerge from the ground as winged adult females (called alates). These winged females usually fly to an upright surface (*e.g.* the vine stem), and here they lay two types of eggs: a larger egg, which gives rise to female phylloxera, and a smaller one, which results in male. These male and female insects have no mouthparts and cannot, therefore, feed. Their sole function is to mate, and after they do so, the female lays a large egg, which over-winters on the bine, under the bark of older canes or trunks. As we have seen, this over-wintering egg eventually hatches into a female, and the complicated life cycle is completed. There are three to five generations of foliar phylloxera per season in eastern North America. By the end of September, nymphs begin to hibernate, and by mid-December all forms of the pest are hibernatory. The rate of development of the pest is dependent not only upon environmental factors, such as soil composition and temperature, but also on the gross structure of the vine itself.

The situation exists whereby *V. vinifera* and North American *Vitis* species have different susceptibilities to the root and leaf forms. Thus, gallicoles do not attack the leaves of *V. vinifera*, but they form galls on the leaves of many American species. These galls, which occur in the lower leaf surface, do not normally seem to cause irreparable damage to the vine concerned. High populations of the foliar form can, however, result in premature defoliation, reduced shoot growth and a reduction in the size and quality of the fruit yield. Conversely, most North American species of *Vitis* are tolerant or resistant to the radicicole form of phylloxera. On these plants, the aphid will cause nodosities but not tuberosities. The roots of *V. vinifera*, on the other hand, are highly susceptible to the radicicole form, which proliferates very rapidly and produces both nodosities and tuberosities. It was the feeding of root phylloxera on *V. vinifera* that nearly destroyed the French wine industry in the late 19th century. It should be stressed that some vines, classified as phylloxera-resistant, may 'tolerate' nodosities on their roots, and that a distinction should be made between 'tolerant' and 'resistant' when describing *Vitis* root-system phenotypes. The so-called 'resistance' to phylloxera reported for many rootstocks and wild *Vitis* species is more accurately defined as 'tolerance'. Most of the phylloxera-resistant rootstocks used in Europe exhibit nodosities on their roots and, occasionally, galls on their leaves when grown in infested soil, but they are able to withstand and overcome any potential damage. As Mullins *et al.*[3] pertinently observed,

> The use of these tolerant rootstocks, in effect, contributes to the perpetuation and spread of phylloxera. Only the exclusive use of genuinely resistant or immune rootstocks would lead to phylloxera eradication. Breeding for resistance or immunity enables the rapid screening of large numbers of progenies, because selections are made on an 'all-or-nothing' basis, but breeding for tolerance involves long-term trials of field resistance. A further disadvantage of tolerant rootstocks is the risk of breakdown of field resistance, with the appearance of more aggressive biotypes of phylloxera.

The above-mentioned differences have ultimately enabled the survival of the wine industry in Europe, for the grafting of susceptible *V. vinifera* scions onto rootstocks of resistant American hybrids has proven to be the cornerstone of European viticulture. In addition, the use of phylloxera-resistant rootstocks represents one of the first and the more successful examples of the biological control of an economic pest. The European epidemic was eventually brought under control by grafting *V. vinifera* scions onto resistant American *V. lambruscana* Bailey, rootstocks. The cultigen name '*Vitis* x *lambruscana* L.H. Bailey' is used to designate American grape cultivars having *V. lambrusca* parentage. A major resistance-breeding programme was conducted in Europe, and the resultant cultivars were commonly referred to as 'French–American' hybrids. These hybrids are important for wine producers in eastern North America, but they are susceptible to foliar phylloxera. Indeed, viticulture along the Atlantic seaboard was largely based on native American species (such as *V. lambrusca*), or on inter-specific hybrids between North American species and *V. vinifera*.

The poor winemaking qualities of the hybrid phylloxera-resistant cultivars that were bred in France in the late-19th/early-20th centuries, the so-called 'producteurs directs', has led to much prejudice against inter-specific hybrids, and hybrids in general. Complex hybrids involving *V. vinifera*, *V. rupestris*, *V. riparia*, *V. berlandieri* and other North American species were produced in France on a large scale as phylloxera-resistant cultivars. Interest in breeding for true resistance or immunity has been rekindled, somewhat by the fact that *V. cinerea* is a resistant species. Although little used commercially, muscadine grapes (*Muscadinia rotundifolia*) also have a high degree of resistance to phylloxera, and this character exhibits dominance in crosses with *V. vinifera*.

There is some variation in resistance and tolerance among American *Vitis* species, and phylloxera now occurs throughout the USA, including the south-western states of New Mexico, Arizona and California. The level of tolerance of *V. berlandieri*, for example – a species that originates from the calcareous soils of central Texas (an environment that is conducive to radicicolous phylloxera) – is much greater than species such as *V. labrusca* and *V. aestivalis*, which arose in sandy habitats in the Appalachian Mountains, an unfavourable environment for root-dwelling phylloxera.

The situation regarding phylloxera in the important North American wine-producing state of California is an interesting one and amply illustrates the complicated biology of the pest. Grape phylloxera was discovered in the Russian River Valley of Sonoma county in the year 1873, just 10 years after its detection in Europe. It is reasonably believed that infested, own-rooted vines were brought into California both from European nurseries and from sources in eastern North America, where the insect is native. At that time, there was a healthy international trade in vine cuttings, and no effective quarantine control measures were in force. The insect spread through the established grape-growing districts of California causing widespread destruction of vineyards. It was noted, even at that time, that phylloxera's destructive behaviour was heavily influenced by the soil in which the vines were growing. Soil texture, moisture content and surface temperature, all had significant effects on its ability to reproduce on vine roots, and to eventually destroy the crop. Unlike the situation in Europe, where the winged, sexual form of phylloxera allowed the insect to spread rapidly and uniformly; the movement in California was slow and sporadic. This was eventually attributed to the fact that in California, phylloxera had a simplified life cycle, being restricted to an asexual, root-infesting form. Spread of insects from a centre of infestation was by slow, seasonal movement of wingless nymphs or by distribution of infected planting stock. This difference in the biology of European and Californian phylloxera was eventually attributed to the much lower relative humidity found in Californian vineyards during the summer months.

The obvious differences between Californian phylloxera and the pest in other parts of the world, seems to me to have engendered some degree of complacency among viticulturalists (and their advisors) in that state. Even as recently as 1993, Lider *et al.*,[4] addressing the International Symposium on Viticulture and Enology in Cordoba, Spain, were wont to say, '... even today, 120 years after the first detection of phylloxera in California, only some 30% of the vineyards of the state are infested to the degree that they require phylloxera-resistant rootstocks. This is due in part to the lack of a winged form, and to the predominance of sandy soils in the Central Valley where 75% of the grape acreage exists'.

It seems that at the very time that these words were being uttered, California was being ravaged by, what is now known to be, a different, more vigorous form of phylloxera. Given the complexity of the insect's life cycle, it is not surprising that there should be different biotypes, and it has been suggested that it was a peculiarly mild version of the pest that originally arrived in California all those years ago.

Of all the hybrid rootstocks available to viticulturalists, one of them – AxR#1, a *V. rupestris* hybrid – seemed to stand above all the rest as far as Californian viticulturalists were concerned. Cynics might suggest that, in hindsight, this was inadvisable since it seems to ignore the experience and knowledge of grafting gained over

100 years by their French counterparts. The French had largely discarded many *V. rupestris* hybrid rootstocks, such as ARG1, AxR and 1202, because of their relatively low resistance to phylloxera, and had opted for *V. berlandieri* hybrid rootstocks instead. The French had known for at least 50 years that AxR#1 was not perfectly resistant, and that it would fail after about 10 or 20 years in the ground. Such information went almost unheeded, for in California, particularly in coastal vineyards, thousands of acres of AxR#1 were planted, so much so that the rootstock became unavailable at certain times. In such instances, alternative rootstocks such as the *V. berlandieri*-based SO4 were planted as a 'second choice'. Those farmers who were 'forced to use' SO4 had a lucky escape, as we shall see. Because AxR#1 had always been regarded as a panacea in California, relatively little research had been conducted into the suitability of other phylloxera-resistant rootstocks when Type B arrived with a vengeance.

Work by scientists at UC Davis has shown that there are now two types of phylloxera in the state known as 'California Biotype A' and 'California Biotype B', and that rootstock AxR#1 is resistant to the former, but not to the latter. Type B, which is far more vigorous than A, spread through California-like wildfire during the late-1980s and the 1990s, and caused havoc for many growers. Thousands of acres of AxR#1 and own-rooted (ungrafted) plants had to be grubbed up and burned at a total cost of millions of dollars (well over $1 billion). Around 80% of the vines in the Napa Valley had to be replaced. While there were plenty of other rootstocks to choose from, many growers have not been able to withstand the cost of re-planting – bearing in mind that it takes 5–7 years for new vines to produce a commercial crop of fruit.

There is some debate as to whether biotype B is a mutation, or whether there was a fresh influx of a new, more vigorous strain from the east. Those growers who had planted SO4 did not have the same problem as those with AxR#!, since *V. berlandieri* rootstocks are much more resistant to phylloxera. The trouble was that SO4 was not universally accepted by Californian growers, mainly because of its low vigour, and because it was most productive on fertile, irrigated sites. On the credit side, SO4 roots easily and is not difficult to graft. For several years, Californian growers seemed to be in a state of denial regarding the dangers of phylloxera, which hindered the development of suitable replacements for AxR#1.

They appeared more worried about changes in wine tastes and using virus-free stock, than with the assessment of suitable replacement rootstocks. Indeed, in an evaluation of rootstocks for resistance to Type A and Type B, Granett *et al.*[5] found that SO4 (together with 039-16 and 044-4) showed the best level of resistance in the field. The same group also found from laboratory evaluations that some rootstocks with partial *V. vinifera* parentage had resistance to both biotypes, while others did not.

Towards the end of the last millennium, leaf gall-forming phylloxera was found in California for the first time, which further increased the pressure on own-rooted and AxR#1 vines. This event took place almost 80 years after Davidson and Nougaret[6] published their classic monograph. East Coast wineries have always had entirely different problems with phylloxera, because while *V. vinifera* is immune to the foliar form, American–French hybrids are not. These hybrids are an important facet of the wine industry in eastern North America, and it is thought that the introduction of the foliar form into California was *via* nursery stock from the East Coast. Indeed, some native grape varieties, such as Concord, are notorious for harbouring phylloxera, even though the vines themselves do not suffer. The cultivar Concord, which is of immense importance in New York State vineyards, was once regarded as pure *V. labrusca*, but is now thought to be a natural hybrid of *V. vinifera* and *V. labrusca* (*i.e.* '*V. labruscana*'). Exactly where the foliar form of the insect was harboured on the infected material that might have 'contaminated' California is not known, but once infected, such forms can spread with the wind.

Foliar phylloxera reduces net photosynthesis of vine leaves, and gall formation causes leaf distortion, necrosis and premature defoliation. The latter, in turn, may delay ripening, reduce crop quality and render vines more susceptible to winter injury. Vines that are heavily infested with the foliar form may have the effect of root infestation exacerbated. Although research indicates that high population densities of foliar phylloxera can result in a reduction of yield and quality of the crop, it should be stressed that very high densities have to be attained before yield is affected. Despite the fact that both leaf gall-forming and soil-borne phylloxera have existed in East Coast (and European) vineyards for well over 100 years, the genetic relationship between above-ground and below-ground populations has scarcely been looked into.

Australia too has suffered, although the situation there is that some isolated areas have escaped the ravages of the insect. In 1875, phylloxera was identified at Geelong, just to the west of Melbourne and one of Australia's leading wine-producing areas. The government of the day, totally ignoring known grafting technology, ordered wholesale eradication of all infected vineyards, and when the pest arrived in Bendigo, in central Victoria, in the 1880s, the same solution was sought. This resulted in the almost immediate disappearance of two, hitherto flourishing winemaking areas, and it was over 100 years before vines reappeared. It transpires that the rationale behind these decisions was to ultimately protect the industry in north-east Victoria. This proved to be of little avail, for phylloxera reached there by the beginning of the 20th century, but by this time the lesson had been learned and grafting was used to overcome the problem. Phylloxera stopped at the Murray River and never reached the state of South Australia, mainly due to effective quarantine restrictions and its relative geographical isolation. The pest was active around both Sydney and Brisbane, but it never affected the Hunter Valley. Another anomaly in Australia, albeit one on a smaller scale, is to be found at the Château Tahbilk winery in the Goulburn Valley of central Victoria. Situated some 8 km south-west of Nagambie, Tahbilk is Victoria's oldest family-owned winery, and being founded in 1860, is one of the oldest in the country. It is the subject of a fascinating book,[7] celebrating its 125th anniversary. Intriguingly, the vineyard contains a few own-rooted Shiraz and Marsanne vines that date back to pre-phylloxera days. These vines were planted on sandy soil, which is not conducive to phylloxera, and they remain defiantly productive, surrounded as they are by replanted vines with grafted rootstocks. Having said all this, the main extant viticultural regions of Australia are in phylloxera-free regions, and most vines are grown on their own roots.

Another wine-producing country that is relatively uncontaminated with phylloxera is Chile. This is principally due to its unique geographical isolation: the arid Atacama Desert to the north; the Andes to the east; the Pacific Ocean to the west; and the Antarctic to the south! The Chileans like to claim that theirs is the only country in the world making wine from 'pre-phylloxera' vines. Although the country is free from many diseases of the vine and has often been described as a 'viticultural paradise', this statement is not strictly accurate because other wine-producing places on the planet, such as Cyprus and part of Hungary, have also escaped the pest. A very readable monograph, *Phylloxera*, has recently been written by Campbell.[8]

Even now, there is no control measure for phylloxera that actually eradicates the pest, and there is still no practical alternative to using resistant rootstocks! Pesticide application is ineffective for phylloxera control due to the vine being deeply rooted and because of the high rate of insect reproduction. Other chemical treatments, such as fumigation, do not work because they either have poor penetration, and/or they move quickly through the soil profile. Few chemicals are registered for the control of foliar phylloxera, and 'Thiodan' (endosulfan) is about the only insecticide used by commercial vine growers. Endosulfan, an organochlorine compound, is a mixture of two isomers, α- and β-endosulfan. It has a molecular formula of $C_9H_6Cl_6O_3S$ and is chemically 6,7,8,9,10,10-hexachloro-1,5,5a,6,9,9a-hexahydro-6,9-methano-2,4,3-benzodioxathiopin-3-oxide. A number of natural enemies feed on grape phylloxera, but none are commercially available for use in biological control programmes, and so at present there is no effective biological means of control. Intervention practices, such as severe pruning, additional irrigation and increased fertiliser applications, may serve to lessen phylloxera impact in the short term, but the only long-term solution involves removal of infected vines and re-plantation with more resistant rootstocks. The rate of decline of susceptible vineyards can be slowed by avoidance of water stress in the vine, and flooding has been used to control phylloxera in southern France.

As with most major catastrophes, there were many lessons to be learned, and in 1874, after the phylloxera blight had all but destroyed European viticulture, vine growers in France, Italy, Switzerland, Austria and Germany gathered in a congress in Montpellier to organise a joint search for a means to fight the pest. After 35 years, in 1908 and 1909, the blight was overcome. In its place, another insidious danger threatened European viticulture: the uncontrolled production of wine. This was an inter-continental problem because world trade became inundated with all types of inferior beverage, many improperly masquerading under the name of 'wine'. Such a situation led to widespread fraud. Also in 1908–1909, two international congresses were held, one in Geneva and one in Paris, to examine this problem. Major progress was made, thanks to the provision of the first definition of 'wine', and also by the confirmation of the principles of the Madrid Agreement of 14 April 1891, relating to repression of false indications of source. Out of all this, the Office International de la Vigne et du Vin (OIV) was born. The OIV is an inter-governmental organisation, which was founded in 1924 with representation from 33 nations. It is primarily concerned with regulation and legislation in the wine industry.

10.2 FUNGAL DISEASES OF THE GRAPEVINE

There are a number of common diseases of the grapevine that are caused by fungi, of which we shall consider downy mildew, powdery mildew, black rot, dead-arm and anthracnose. Grape disease control programme needs vary with variety, geographical location and climatic conditions, and a treatment pertinent to a French vineyard, for example, may be totally non-applicable to one in the Napa Valley in California. The grey rot organism has already been dealt with, and will not be discussed further.

10.2.1 Downy Mildew

A disease of *V. vinifera* caused by the Oomycete fungus *Plasmopara viticola* (used to be *Peronospora viticola*), which is endemic to North America (but not particularly destructive there) and now occurs in most parts of the world where grapes are grown and where rain is experienced during the summer months (*i.e.* it is not normally a problem around the Mediterranean basin). The fungus has probably been attacking the grapes native to North America for so long that a balance had long been established between fungus and vines, whereby the latter were not seriously affected. Once the fungus reached Europe *via* some of the vines brought over from America for graft material to combat phylloxera, the highly susceptible, indigenous *V. vinifera* vines were decimated (spraying, of course, was unknown in the late 19th century). All this happened just as the French wine industry was recovering from the ravages of phylloxera! The situation was to some extent ameliorated by the serendipitous discovery of Bordeaux mixture by Professor Millardet in 1882. *P. viticola* over-winters as a thick-walled, resistant oospore, which is a result of sexual reproduction. Oospores are normally to be found secreted inside the tissues of plant debris on the vineyard floor, and are carried by raindrops to the green organs of the vine. Once on the vine, oospores germinate and liberate asexual sporangia, which produce motile spores (zoospores) that can swim in a surface water film and cause secondary infections. The fungus can infect all green parts of the vine, but especially the leaves where once established, the mildew is easily visible to the naked eye as a white, downy growth on the lower leaf surface. The first characteristic signs of *P. viticola* infection are pale yellow/ greenish yellow, translucent, 'oily' lesions on the upper surface of leaves. Care has to be taken, for similar early symptoms are found in cases of powdery mildew. Examination of the underside of the leaf reveals greyish-white mycelium that subsequently becomes 'downy' when the fungus begins to sporulate. These spores can be blown to other young leaves, shoots and blossom clusters in cool, moist weather during late spring/early summer. The areas of the leaf infected by the fungus become brown and necrotic, and when the disease is severe there is much defoliation. In such circumstances, photosynthesis is retarded and this results in re-duced fruit sugar content. Other symptoms on the vine are dark-ening and death of young shoots, bending of fruit stalks and death of flowering clusters. In bouts of infection later in the year, grapes become greyish and covered in 'down'. *P. viticola* is an obligate parasite and cannot, therefore, be cultivated *in vitro*. Historically, the disease is of great interest because experiments by Millardet to control it in France led to the formulation of Bordeaux mixtures that proved effective against not only this fungus, but also potato blight (*Phytophthora infestans*) and a number of other important foliar pathogens.

The classic control method for downy mildew is to spray with Bordeaux mixture, a combination of copper sulfate and hydrated lime, which was developed in the 1880s. In recent years, Bordeaux mixture has been superceded by systemic fungicides such as fosetyl aluminium (aluminium tris-*O*-ethyl phosphonate) and phenyl-amides, and by products such as Cymoxanil (1-(2-cyano-2-metho-xyiminoacetyl)-3-ethylurea), which is a non-systemic fungicide introduced by DuPont in 1977 and specific to members of the order Peronosporales (*Phytophthora, Peronospora* and *Plasmo-para*), which are all mildews of various types. It is rumoured that in certain very well-established European vineyards, copper has accumulated to unacceptably high levels and has consequently decreased vine health and fruit yield. Traces of copper salts on fruit can cause copper-catalysed oxidations and haze formations in wine. The inheritance of resistance in the grape to downy mildew was studied some while ago by Boubals.[9] After 'oidium' and phylloxera, downy mildew was the last of North America's hostile contributions to European viticulture. It first appeared in France in 1878, and it took chemists at the University of Bordeaux only 4 years to come up with Bordeaux mixture.

10.2.2 Powdery Mildew

Also known as oidium, powdery mildew is one of the most widespread fungal diseases of the vine, and like downy mildew, seems to have originated in North America, where it was first identified as a minor problem in 1834. It was not a reported nuisance in Europe until *ca.* 1850, but soon reached epidemic proportions there with reported crop losses of around 25%. As a consequence of the origin of the disease, North American *Vitis* species are some of the most resistant species to powdery mildew. It is caused by the ascomycete, *Uncinula necator*, a member of the order Erysiphales and, although it mainly affects leaves, it is capable of infecting all green vine tissues. The fungal mycelium grows on the surface of compromised plants, obtaining nutrients by producing specialised hyphal branches called haustoria, which can penetrate host epidermal cells. Removal of nutrients by absorption results in localised necrosis of the host tissue. Infected leaves will initially have characteristic light-brown spots on their lower surfaces and will adopt a 'crinkly' appearance. Their upper surfaces then become covered with chains of asexual conidia, which first form white, powdery patches. If conditions are suitable, the fungus will often completely cover the leaf surface. Conidia are produced by surface fungal mycelium and give rise to the characteristic 'dusty' or 'powdery' appearance to the infected vine organ. Conidia germinate to form a hypha, which almost immediately produces an appressorium[*] (Figure 3) over an epidermal cell of the host leaf. A penetration peg forms below the appressorium and enters the host cell, whence

---

[*]A swollen, flattened portion of a fungal filament that adheres to the surface of a higher plant, providing anchorage for invasion by the fungus.

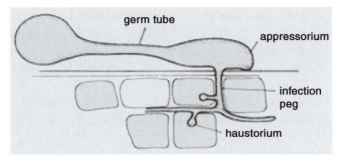

Figure 3   *Fungal infection apparatus*

a specialised structure called the haustorium is formed inside the vine leaf cell. As stated, the haustorium is responsible for the transfer of nutrients from host to external mycelium.

When infection is severe, leaves will shrivel, turn brown and fall. Infection of canes and flower clusters also occurs, and in the case of the latter the results are, fall of flowers, poor fruit set and splitting of any berries. All of this results in vastly reduced fruit yields, and splitting renders berries liable to attack from other microbes. As berries mature, they become less susceptible to attack, and when their sugar content reaches 12–15%, they appear to be immune to the invasion of the fungus. The fungus either over-winters in the vegetative form, the hyphae harbouring in dormant vine buds, or as small fructifications called cleistothecia, which are produced in autumn and are the result of a sexual reproductive process. Cleistothecia, which appear as 'black specks' on the surface of the vine, especially in older parts, can survive for considerable periods among vegetation debris on the vineyard floor. They contain ascospores, which are enclosed in an ascus (plural, asci), and it is these spores that are responsible for primary infection of the vine. Ascospores are wind-blown to green vine tissue and, once they alight, germinate to produce hyphae, appressoria and haustoria. Secondary infections are due to the aforementioned conidia, which are produced in millions and are wind-borne. Infection usually starts soon after the vine has flowered and continues on the foliage throughout the growing season. Unlike other fungal infections, *U. necator* does not require a water film or water droplets to proliferate, since conidia can germinate on a dry leaf surface as long as the relative humidity is sufficiently high. Thus, the hot, dry conditions that favour the development of powdery mildew are in sharp contrast to the warm, damp conditions necessary for the development of downy mildew.

When the fungus was first found on the vine, only the conidial (asexual) state was known, and the organism was named *Oidium tuckeri*. The organism is now correctly known as *U. necator* (Schw.) Burr; anamorph *O. tuckeri*. 'Oidium' seriously threatened the wine industry in France, but experiments aimed at controlling the disease led to the discovery of sulfur-containing fungicides. It took 10 years to arrive at an effective treatment, but once discovered, sulfur, in the form of a dust or a wettable powder, is still the most widely used fungicide. Fortunately, it has both prophylactic and curative properties, and it is the vapour phase of sulfur that is most effective in controlling *U. necator*. The time of sulfur addition is an important factor, and early application is advisable so that it will disappear by the time that fruit has set. If sulfur is applied too late in the growing season, a residue of finely divided sulfur particles can adhere to fruit, and can be introduced into must. Such material will invariably be metabolised by yeast during fermentation, to give hydrogen sulfide. It is important not to spray or dust vines with sulfur within 4 weeks of harvest.

Organic pesticides, such as strobilurines, may be used to combat powdery mildew, and these are generally more effective than sulfur. They do however induce resistance, which sulfur does not. Other compounds that are labelled for control of powdery mildew include copper fungicides, dinocap, benomyl, imidazole, mycobutanil and triadimefon. Cultural considerations (*e.g.* good canopy management) can reduce the severity of powdery mildew infections, and increase the effectiveness of chemical control agents. Canopy management is of special importance because it is known that low-light intensity favours growth of the fungus. Indeed, the first symptoms of the disease often appear on shaded parts of the vine, and this is attributable to the low-light requirement for conidia germination.

10.2.3 Black Rot

Another fungal disease, native to North America and exported to Europe, where it became a serious parasite on *V. vinifera*. It is caused by the ascomycete, *Guignardinia bidwellii* (Ellis) Viala

& Ravaz, an organism studied extensively by Donald Reddick during the early years of the 20th century. Black rot is the most widespread fungal disease of grapes, and in warm, humid weather it can elicit total crop loss. The disease is virtually unknown in regions with hot, dry summers. The fungus can affect all green vine tissues, including leaves, young canes, tendrils and petioles, but it is the fruit that is most severely damaged. Infected berries first appear light or chocolate brown, but quickly turn dark brown, with numerous black pycnidia (see below) developing on the surface. After a while, infected berries shrivel and turn into hard, black raisin-like structures called 'mummies'. Mummies are one of the main over-wintering sites for the fungus, although it can persist on other vine sites and in the soil. Mummies can remain on the vine or fall off and exist in soil. All cultivated *V. vinifera* grape varieties are susceptible, but there is a wide variety of susceptibility to this disease among native American and hybrid cultivars.

The fungus produces a mycelium that grows inside host tissue, and external symptoms of black rot first appear as small reddish-brown circular spots on the upper leaf surface. As the spots enlarge, a black border forms around their margins. When these lesions reach *ca.* 2–4-mm diameter, small fruiting bodies, called pycnidia are formed. These resemble tiny black 'pimples' and contain large numbers of asexual spores (conidia), which, when liberated, can rapidly spread the fungus. These conidia are carried by air currents and/or rain splash to tender young vine tissue, whence infection occurs. It seems as though temperatures of 60–90°F and a film of water on the vine surface are prerequisite for infection. The initial infections provide the spores that cause repeated secondary infections throughout the summer, if conditions are sufficiently damp. During the growing season of the vine, a new bout of infection will follow each outbreak of rain. Disease symptoms become evident on the plant some 10–14 days after infection. Very young vine leaves are highly susceptible to black rot, but this susceptibility decreases by the time that they have finished expanding. As a rule, berries do not become infected while the caps remain attached.

In addition to the asexual condia, *G. bidwellii* produces a second type of spore, the ascospore, which is a result of sexual reproduction.

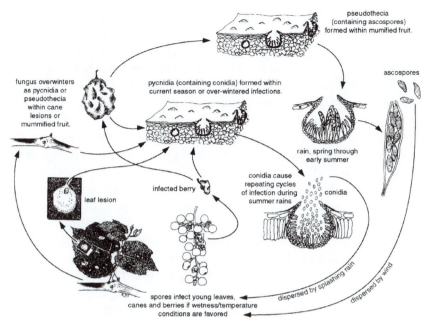

Figure 4 *Life cycle of* G. bidwellii
(Reproduced by kind permission of the New York State Agricultural
Experiment Station, Cornell University)

The ascospores, which are also capable of causing infection, are
produced in sunken, flask-shaped cavities, called pseudothecia,
which are located on mummified fruit. From mummies on the
ground, ascospores are forcibly discharged about 2–3 weeks after
bud break and continue to be ejected just after the start of bloom.
In contrast, mummies that remain on the vine will continue to
release ascospores (and conidia) right through the way to véraison.
Spring rains can also trigger the release of ascospores, which will
then be wind-blown for short distances. A simplified diagram of the
life cycle of *G. bidwellii* is shown in Figure 4.

10.2.4 Dead-Arm

If one consults a 50-year-old text on diseases of the vine, one will
learn that dead-arm disease is caused by *Phomopsis viticola*, and
primarily affects the shoots and branches (arms) of the vine,
although it will also infect fruit and leaves. Many authorities

now maintain that the name 'dead-arm' should be dropped, because Dye and Carter,[10] working with infected New Zealand vines, demonstrated that the 'disease' is actually two different diseases that often occur simultaneously and should be regarded as such. It was therefore recommended that 'Eutypa die-back', caused by the fungus *Eutypa armeniacae* (syn. *E. lata*) should be the new name for the canker- and shoot-die-back phase of what was known as 'dead-arm', and that 'Phomopsis cane and leaf spot', caused by the fungus *P. viticola*, should refer to the cane- and leaf-spotting phase of what was called 'dead-arm'. A subsequent piece of work[11] conclusively showed that it is *E. armeniacae* and not *P. viticola* that incites dead-arm symptoms on the vine, and that 'Eutypa die-back' was a sensible name for the disease. To confuse matters further, some plant pathologists recommend that 'dead-arm' should now be called 'phomopsis', especially since another four species of *Phomopsis* have been reported from vines in Australia.[12] In 1998, Phillips[13] also found the fungus *Botryosphaeria dothidea* to be associated with typical 'die-back' symptoms in some Portuguese vines. From a growers' point of view, Dye and Carter's distinction makes sound sense because *Phomopsis* and *Eutypa* cause two essentially different diseases, and their control procedures vary greatly accordingly. We shall consider 'Eutypa die-back' and 'Phomopsis leaf and cane spot' as separate entities.

The earliest symptom of Eutypa die-back, caused by the Ascomycete fungus *E. armeniacae*, is a canker that generally forms around pruning wounds in older wood of the main trunk. These cankers are usually difficult to see because they are often covered with bark, but one indication of a canker is a flattened area on the trunk. Removal of bark over the canker reveals a sharply defined region of darkened or discoloured wood, bordered by white healthy wood. Cankers may be up to 3-feet long and may extend below the soil line. When the trunk is cut in cross-section, the canker appears as darkened or discoloured wood extending in a wedge shape to the centre of the trunk (Figure 5).

The most striking symptoms of Eutypa die-back are to be found on the leaf and shoot, and they are most easily seen in spring, when healthy shoots are around 12–24-inches long. Spring shoot growth is weak and stunted above the cankered area, with very small, misshapen (cupped), chlorotic leaves. Such symptoms are difficult to see later in the season because they become immersed in healthy

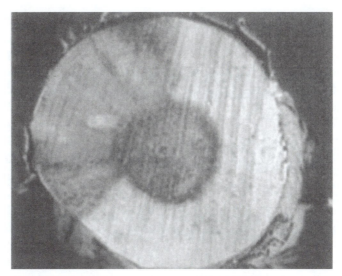

Figure 5 *Cross-section of a vine trunk infected with* Eutypa
(Reproduced by kind permission of Ohio State University Extension)

foliage. *E. armeniacae* is a very slow-growing fungus, and disease symptoms may not necessarily appear for up to 4 years after the vine has been infected. This is a worry because identification and removal of infected vines is an important management tactic. Leaf and shoot symptoms are more pronounced each year until the affected portion of the vine finally dies.

The fungus survives in infected trunks for long periods, whether as part of the in-place vine or as old, dead grape wood in the vineyard. The fungus is generally present in older wood such as vine trunks, but generally not in younger wood such as 1- or 2-year-old prunings. The fungus eventually produces reproductive structures called perithecia on the surface of infected wood, and these perithecia contain ascospores, which are eventually discharged into the air. Ascospore discharge is initiated by the presence of water, usually rainfall splash. Most spores appear to be released during winter or early spring; few are released during the summer. Unfortunately, most spores are released at about the same time when pruning is being carried out. Air currents can carry the ascospores long distances to recent wounds on the trunk. Pruning wounds are by far the most important points of infection, and it is helpful if vines can be pruned after rainfall (when there are fewer spores in the air), or as late in the season as possible. The ascospores germinate when they contact the newly cut wood, and a new infection is thus initiated. As a result of sexual reproduction, perithecia, which contain the ascospores, are formed around 5 years after the initial infection of wood, and are produced in hardened masses of fungal and bark tissue called a 'stroma', which result from cankerous growth by the host.

An important control method is the removal of infected trunks from the vineyard. The vine must be cut off below the cankered or discoloured wood, or if the canker extends below the soil line, the entire vine must be removed. If the canker does not go below the soil line, the stump can be left and a new trunk formed. Growers must remember that the best time to identify and remove infected vines is in early spring when leaf and shoot symptoms are most obvious. In addition, large wounds are less susceptible to infection during this part of the season, and fewer ascospores are present to cause re-infection. If trunks cannot be removed in the spring, they should be marked for easy identification and removal later in the growing season. All wood (especially trunks and stumps) from infected plants must be removed from the vineyard and destroyed (either buried or burned). An old infected stump or trunk lying on the ground may continue to produce spores for many years. *Eutypa* spp. are also pathogenic for apricot, cherry, almond and some other important tree crops. At present, there are no recommended fungicides available for the treatment of *E. armeniacae*. The disease cycle of Eutypa die-back is shown in Figure 6. Intriguingly, the d'Arenberg Winery in McLaren Vale, South Australia produces a 100% Shiraz wine called 'The Dead Arm' as their top cuvée. It is claimed that they use 'dead arm' fruit, *i.e.* after the disease has killed off fruit from one side of the vine, all the 'goodness', they claim, is concentrated in the fruit on the other side of the plant!

P. viticola (Sacc.) Sacc., the causative agent of leaf and cane spot, is the asexual phase (anamorph) of the Ascomycete fungus *Diaporthe perjuncta*. Cane and leaf spot is an important disease of grapes worldwide as the fungus attacks most parts of the vine, including canes, leaves, flowers, rachises (berry cluster stems), tendrils and berries, and can cause vineyard losses by (1) weakening canes, which makes them more susceptible to winter injury; (2) damaging leaves, which reduces photosynthesis; (3) infecting cluster stems, which can result in poor fruit development and premature fruit

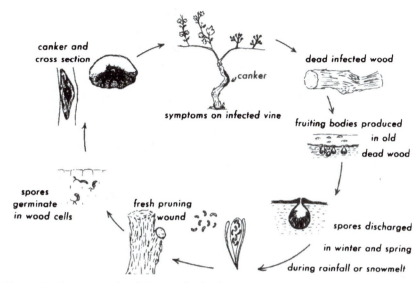

canker and cross section

canker

symptoms on infected vine

dead infected wood

fruiting bodies produced in old dead wood

spores germinate in wood cells

fresh pruning wound

spores discharged in winter and spring during rainfall or snowmelt

Figure 6 *Disease cycle of Eutypa die-back*
(Reproduced by kind permission of New York State Agricultural Experiment Station Cornell University)

drop; and (4) infecting berries resulting in fruit that reaches nowhere near ripe enough to be harvested.

Phomopsis over-winters as mycelium and resting structures (pycnidia) in the buds, barks and canes of infected vines. In spring, when pycnidia are in a moist environment (at least 96% relative humidity, for at least 10 h), they exude their asexual spores (pycnidiospores) in gelatinous, thread-like masses, which are called cirrhi. Water dissolves the cirrhi and spores are splashed or washed on to young shoots, and they enter the vine through pores on leaves or stems. Spores are not generally dispersed directly by wind, though they may be moved about in water droplets blown from infected vines. Infection causes black lesions on shoots and leaf spots. Leaf spots (brown with a yellow halo) usually appear about 21 days after an infection, while stem symptoms can take 28 days or more.

Most of the economic damage to the vine is caused by rachis and berry infection, which over the years has been studied in some detail, although there are conflicting reports as to exactly when infection might occur. In 1913, Gregory[14] reported that ripe *V. labrusca* Niagara fruit showed disease symptoms within 18 days after inoculation, indicating that fruit infection by *P. viticola* can occur in ripe fruit, while Lal and Arya[15] found that direct berry infection by *P. viticola* did not occur on ripe *V. vinifera* Thompson Seedless berries. They also reported that berries could not be infected unless they were wounded prior to inoculation. The mode of infection has produced similarly varied results. Hewitt and Pearson,[16] for example, found that most fruit infections by *P. viticola* arise from mycelial invasion from lesions on the rachis or pedicel, while Pscheidt and Pearson[17] reported that berry infection in the Concord cultivar occurs primarily during bloom and shortly after bloom, and that the infections remain latent until berries begin to ripen. As berries ripen, the fungus becomes active and fruit rot develops. They also reported that little or no fruit infection occurs after pea-sized berries are formed. Erincik *et al.*[18] indicated that both rachis and berry infections by *P. viticola* on Seyval, Catawba and Chambourcin grapes can occur throughout the growing season, from bloom right through to véraison. Thus at present, it is unclear whether the fungus can infect berries directly or if the fungus invades the fruit through infected rachis tissues. Since knowledge of the mode of infection is important in developing control measures, we need an improved understanding of the etiology of the berry infection phase of this disease.

The same Ohio workers[19] developed and validated a predictive model and a forecasting system for the disease. There are no known curative fungicides against Phomopsis leaf and cane spot of the grape, although mixtures of some fungicides with adjuvants can enhance such activity, probably by increasing uptake of the fungicide by the host plant They also attempted to evaluate the curative activity of the currently registered fungicides for 'phomopsis', such as mancozeb (a practically non-toxic ethylene bisdithiocarbamate (EBDC[†]) manganese zinc complex) and captan (1,2,3,6-tetrahydro-*N*-(trichloromethylthio) phthalimide), which are typically used as calendar-based surface applications. The Ohio team found that there was no sign of any curative effect on the disease if the fungicides were applied after the initiation of infection, whether mixed with an adjuvant or not. By incorporating their prediction model into a fungicide application programme, they envisaged that growers

---

[†] EBDCs are a group of non-systemic fungicides that have been widely used since the 1960s.

would be able to reduce the number of applications and/or improve application timing. Ideally, the fungicide used with a predictive system should have curative activity (*i.e.* control the pathogen after infection and fungal establishment has taken place), although predictive systems can also be used to optimise scheduling of protective chemicals. At the moment, however, fungicide treatments for Phomposis leaf and cane spot are preventative rather than curative.

In the past, two types of Phomopsis have been identified as being closely associated with the disease, and these were known as Phomopsis type 1 and Phomopsis type 2. Recent research, however, has shown that only Phomopsis type 2 causes the disease on grapevines, whereas type 1 is not damaging to the vine. Molecular analysis of these two types found on the vine has permitted a clear distinction into two taxa,[20] and to avoid confusion between the two types, it has been suggested[21] that Phomopsis type 1 should now be known by the taxonomical term, *Diaporthe* (*i.e.* the teleomorphic phase, *D. perjuncta*). Phomopsis type 2 retains the species name *P. viticola* and is the cause of Phomopsis cane and leaf spot, which is responsible for economic losses to the grower. Both types are capable of overwintering on infected material, and it is known that they can remain dormant for many years. Although both biovars cause bleaching of canes, *Diaporthe* does not cause leaf, shoot, inflorescence or any symptoms on grape bunches. In an extensive reassessment of *Phomopsis* on South African vines – a study that combined molecular, cultural and pathological data – van Niekerk *et al.*[22] recently described what they considered to be 15 distinct species of the fungus. This amply illustrates the problems associated with fungal taxonomy, especially the genus *Phomopsis*. When one appreciates the nomenclatural difficulties in this instance, it is relatively easy to see how 'dead arm' was misinterpreted in the past. An additional complication is that 'dead arm' was originally called 'excoriosis', a name coined by Ravaz and Verge in 1925, and one that still persists in some regions today. In those days, the causative agent was called *Cryptosporella viticola* (Reddick) Shear.

10.2.5 Anthracnose

This disease, also known as 'bird's-eye rot' or 'black spot', is caused by the Ascomycete, *Elsinoe ampelina* Shear, which has *Sphaceloma ampelinum* de Bary as its anamorph. The disease is of European origin, and most *Vitis* species are susceptible to it, although there are some resistant cultivars, such as the American Concord and Niagara. Before the 'introduction' of powdery mildew and downy mildew, anthracnose was the most damaging disease in European vineyards. It is now known from all grape-growing countries, but is mainly a problem in rainy, humid regions. Epidemics of anthracnose are sporadic but cause significant economic loss when they do occur. Infection may occur on all succulent parts of the vine, but is most prevalent on fruits and shoots.

The disease first appears obvious as dark red, circular spots (*ca.* 1/4″ diameter) on the berry, which later become sunken and greyish in colour. Even later in the season, these spots become delimited by a dark margin, which gives the typical 'bird's-eye' appearance. Fungal hyphae may extend into the pulp and cause the berry to crack. The lesions on the berry surface produce asexual conidia in small mounds, called *acervuli* (singular *acervulus*), which are characteristic of this group of fungi, and which release a pinkish mass of conidia when the moisture level is sufficient. These conidia cause secondary spread of the disease and will infect other, normally green parts of the plant when free moisture is present. The fungus, therefore, is also capable of attacking shoots, tendrils, petioles, leaf veins and young fruit stems. On leaves, the general symptoms include downward curling, laceration, chlorosis and premature leaf abscission.

The fungus over-winters as hardened, resistant sclerotia on infected canes. Sclerotia germinate in the spring, forming mycelia and then conidiophores and conidia. It has been shown that a 24-h period of wetness is necessary for sclerotium germination to occur and once this has happened, conidia are then rain splashed to any available green tissue. Ascocarps containing ascospores may also form on infected canes and shed berries from the previous year in the spring. Both conidia and ascospores serve as primary inoculum; but, whereas conidia are released passively by moisture, ascospores are actively discharged into the air and may be carried to large distances by air currents. It is only once the fungus has become well established in the host that acervuli are formed.

Fortunately, the disease is relatively amenable to being treated with fungicides, copper salts being effective. A commonly used control regime employs liquid limed sulfur as a dormant application in early spring and foliar fungicides later on in the year.

10.3   BACTERIAL DISEASES OF THE GRAPEVINE

Fortunately, there are relatively few bacterial 'bad boys' in viticulture, and we shall look at the two major examples: Pierce's disease (PD) and crown gall.

10.3.1   Pierce's Disease

Pierce's disease (PD) is a fatal disease of vines in all regions of North and Central America, which experience mild winters. It has also been reported from some parts of north-western South America. It is present in some California vineyards every year, with the most dramatic losses being in the Napa Valley and in some parts of the San Joachin Valley. The disease is named after N.B. Pierce, who was a special agent to the Secretary of Agriculture and California's first professionally trained plant pathologist. His classic publication entitled *California vine disease*, published in 1892, described all aspects of the disease that was later to bear his name. The organism responsible is *Xylella fastidiosa*, a Gram-negative bacterium related to *Xanthomonas*,[23] which inhabits the xylem of the vine cane. Once in the water-conducting elements, it proliferates massively and causes blockages, called tyloses, to be formed. These blockages severely impair movement of water in the vine, and so the symptoms of the disease are those that characterise water stress, such as scorching of leaves. Water stress begins in mid-summer and increases through to autumn. Other manifestations of PD are irregular lignification and suberisation of canes and poor development of fruit clusters, which may desiccate, shrivel and die. The first evidence of PD infection is the scorching of leaves, which in addition become yellowed (chlorotic) along their margins before drying. Typically, the leaf dries progressively leaving a series of concentric zones of discoloured and dead tissue. Scorched leaves detach from the distal end of the petiole, rather than from the base, leaving bare petioles congruously attached to canes well after normal leaf fall. About mid-growing season, when foliar scorching begins, some or all of the fruit clusters may wilt and desiccate. The bark on affected canes often matures unevenly, leaving 'islands' of mature (brown) bark surrounded by immature (green) bark, or *vice versa*.

In addition to vines, the bacterium has a wide range of other hosts,[24] including brambles (*Rubus* spp.), willows (*Salix* spp.), elderberry (*Sambucus niger*) and many grasses, and it is these plants that act as reservoirs for the infection. To move from reservoir to vine, they require the presence of vectors; these vectors are types of leaf-hoppers, mainly of the genera *Draeculocephala* and *Graphocephala*, commonly known as 'sharpshooters'. There is no vine-to-vine means for spreading of the bacterium and sharpshooters must be implicated – but once introduced into the plant, bacteria can spread internally. As a result, the best way of preventing the disease is to prevent the insects from entering the vineyard (*i.e.* removal of other host plant species). The incidence of PD in any one vineyard is often variable from year to year.

Pierce's Disease is endemic in the south-eastern states of the USA, from Florida through to Texas. In the late 19th century it swept across the USA, over the Rockies and into California, where it reached Anaheim in the Los Angeles basin (it was originally called 'Anaheim Disease') and decimated vineyards there in the 1880s. Similar outbreaks occurred in the 1930s and 1940s. The bacterium cannot survive in latitudes more northerly than are found in northern California. As with phylloxera, a fresh wave of even more rampant PD hit California towards the end of the 20th century. This outbreak was associated with a new vector, the glassy-winged sharpshooter (*Homalodisca coagulata*), which was accidentally introduced in 1989 with nursery stock from the southern USA. This insect has spread PD with astonishing speed. It is a voracious feeder and an aggressive flier, and it feeds and breeds on at least 130 host plants. Since 1998, some $65.2 million has been spent by government and industry on researching how to combat the insect. There is no known cure for PD.

In Florida and some other south-eastern states, PD has been the single most formidable obstacle to the cultivation of European-type grapes (*V. vinifera*). The disease, in the regions in which it is endemic, represents another classic example of the adaption of a host to one of its parasites. Thus, some vine species native to the south-eastern states, such as *V. coriacea* and *V. simpsonii*, are highly resistant or at least tolerant to PD. The situation regarding native North American grapes and PD is an interesting one. As has recently been reported,[25] *V. labrusca*, which is native to north-east USA where PD is absent, appears to be as susceptible to the disease as *V. vinifera*. California natives, *V. californica* and *V. girdiana*, appear to be moderately susceptible; whereas, in contrast,

M. rotundifolia and *V. arizonica*, both native to areas of severe disease pressure, appear to be very resistant. The authors also found that wild populations of *M. rotundifolia* from areas where PD is rare supported higher concentrations (up to 20 times) of *X. fastidiosa* than the populations from areas where PD is severe. Trials with wild populations of *V. girdiana* showed a similar pattern, with susceptible selections showing up to 100 times higher bacterial concentrations. These results, detailing a gradient in resistance both among and within species, are consistent with the hypothesis that PD resistance has evolved in response to disease pressure.

10.3.2   Crown Gall

Crown gall is a disease caused by the Gram-negative, non-sporing, motile, rod-shaped bacterium, *Agrobacterium tumefaciens*, an organism closely related to *Rhizobium* of nitrogen-fixing fame. The two bacteria are both classed as 'alpha protobacteria' and are in the family Rhizobiaceae. Like *Rhizobium*, soil is its natural habitat, particularly on and around root surfaces (the rhizosphere), where it survives on nutrients that have leaked from root tissue, but unlike *Rhizobium*, it is a phytopathogen. *A. tumefaciens* has been studied intensively since 1907, when as *Bacterium tumefaciens* it was identified as the causative agent of crown gall disease by Smith and Townsend.[26] The name was changed to *A. tumefaciens* in 1942.[27] The bacilliform rods measure *ca.* 1 μm × 3 μm, and bear flagella that are sub-polarly arranged around the cylindrical circumference of the cell, a condition referred to as circumthecal flagellation (Figure 7). Also present on the cell surface are short rod-like appendages called pili (singular, pilus), and after virulence has been induced, the T-pilus is formed. The organism is strictly aerobic and considered nutritionally non-fastidious. *A. tumefaciens* enters the vine root through wounds, and infection results in the formation of fleshy galls, which are caused by excessive cell divisions by vine root tissue. These galls, which are essentially neoplastic growths, can become as large as walnuts, appearing as spherical reddish-brown swellings. In nature, these tumours are formed at the soil–air junction, the so-called 'crown' of the plant. Galls can also form above ground on stems, where they are manifested as hard, black–brown lumps. Fresh crown galls are fairly hard, sturdy structures, but as they age beyond 1 year they

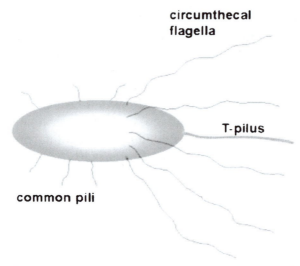

Figure 7   *Cell surface of* A. tumefaciens, *showing circumthecal flagella, pili and the T pilus*

become convoluted with cavities, more friable and easy to detach. A variety of insect life inhabits these gall cavities. It is difficult to isolate *A. tumefaciens* from aged galls, when most of the host tissue is moribund, but fresh galls are a profitable source of the bacterium. Chronically diseased root systems decrease plant vigour, and the host will exhibit symptoms characteristic of water stress. The wide variety of plants affected by *Agrobacterium*, including stone fruit and nut trees as well as the grapevine, make it of great concern to the agriculture industry. It should be noted, however, that *A. tumefaciens* is not the only, or even most common, source of galls on plants, since many are caused by insect larvae that secrete plant growth hormones and have the same effect.

As we shall see, under natural conditions, the motile cells of *A. tumefaciens* are attracted to wound sites through chemotaxis, and this is partly a response to the release of sugars and other common root solutes, and partly to the synthesis and release of novel phenolics. The bacterium is most infective when young vines are being established, and when practices such as transplanting of nursery stock are being undertaken, but it is a fact that the main means of spreading the disease is *via* infected propagating material. Vine root wounds can also be made accidentally by other cultivation practices, such as mowing and 'disking' vineyards for weed removal. In addition, tissue injured by frost is prone to infection, as are wounds made by grafting and pruning before transplantation. Once inside the vine, the bacterial infection becomes systemic, and frost injury can often cause a linear or confluent array of small tumours along the vascular system of the aerial parts of the infected plant.

In the days when phenotypic markers were of paramount importance in bacteriological taxonomy, *A. tumefaciens* isolates were originally classified into three biotypes (I, II and III) or biovars (1,2 and 3). It is biotype III (biovar 3) that is specifically confined to grapevines and molecular (DNA homology) studies, and the close relationship with *Vitis* spp. have resulted in some authorities[28] reclassifying biotype III *A. tumefaciens* isolates as *A. vitis*. One of the most valid reasons for making this change is the fact that most '*A. vitis*' strains can metabolise tartrate, which is an unusually abundant compound in the grapevine, while other biotypes cannot. The genus *Agrobacterium* has traditionally been divided into a number of species, and the taxonomy has largely reflected disease symptomology and host range. Thus, *A. radiobacter* is an 'avirulent' species: *A. tumefaciens* causes crown gall; *A. rhizogenes* causes hairy root disease; and *A. rubi* causes cane root disease.

Most of the infective capability of *A. tumefaciens* is associated with a large plasmid[‡], a covalently closed, double-stranded DNA molecule, which is supernumerary to the main core of DNA in the cell (the nuclear material or 'nucleus'). As will be explained, most of the genes involved in crown gall disease are, in fact, extrachromosomal and borne on this plasmid, which is called the Ti (tumour-inducing) plasmid (pTi). In similar fashion, most of the genes that enable *Rhizobium* strains to produce nitrogen-fixing nodules on certain plant roots, are contained on a large plasmid. Thus, from a viticultural point of view, the important biology of *A. tumefaciens* is a function of its plasmid, not its chromosome. The central role of the plasmid in crown gall disease can be simply demonstrated by 'curing' bacterial cells. Thus, if the bacterium is grown near its maximum temperature (*ca.* 30°C), the plasmid is lost

---

[‡] By definition, a plasmid is a circle of DNA separate from the chromosome, capable of replicating independently in the cell and of being transferred from one bacterial cell to another. Plasmids encode non-essential functions, in the sense that a bacterium can grow normally in culture even if the plasmid is lost.

along with its pathogenicity. If these 'plasmid-free' cells are then grown in laboratory culture, they appear to be entirely functional!

The huge Ti plasmid is approximately 200 kilobases in length and plays a fundamental role in pathogenesis and colonisation of host cells. It inserts a small segment of DNA (known as the T-DNA, or 'transfer DNA') into a host plant cell, and this fragment is then incorporated at a semi-random location into the plant genome. The Ti plasmid harbours not only T-DNA, but also many genes essential for infection and T-DNA transfer to the plant cell. The T-DNA inserted by the bacterium contains genes for the production of cytokinins and indoleacetic acid (IAA) – plant growth hormones that upset the normal balance of cell growth and initiate tumour production. The gall, in turn, provides a nutrient-rich environment for the bacterium because the DNA also contains genes encoding enzymes that cause the plant to create unique amino acid-like metabolites, called opines and agrocinopines, which act as a carbon and nitrogen source for *A. tumefaciens* but not for most other soil microbes. This phenomenon is seen as being of some ecological advantage to the gall-forming bacterium. A number of different opines and related compounds are produced and, based on the specific opine(s) produced in the tumours, *Agrobacterium* strains can be classified accordingly. There are three classes of opine: octopine, nopaline and agropine, and most common strains of agrobacteria are octopine or nopaline specific. The bacterial genes for the ability to catabolise opines are also located on Ti plasmid.

The first *A. tumefaciens* pathovar genome to be fully sequenced was that from strain C58, isolated from a cherry tree crown gall. The sequencing was simultaneously reported by Goodner *et al.*[29] and Wood *et al.*[30] in 2001. The genome of strain C58 consists of a circular chromosome, two plasmids and a linear chromosome. The presence of a covalently bonded circular chromosome is more or less universal in bacteria, but the presence of both a circular chromosome and a single linear chromosome is unique to a group in this genus. The two plasmids are pTiC58, responsible for the processes involved in virulence, and pAtC58, coined as the 'cryptic' plasmid, which has been implicated in the metabolism of opines.

It had long been assumed that elicitation of tumours provides *A. tumefaciens* with some sort of biological advantage, and it is now known that this advantage derives from the production of compounds such as opines by tumours. Opines are structurally diverse, only certain ones being produced by an individual tumour, and these are dependent on the infecting strain of the bacterium. In turn, the opines produced by a particular gall are specifically catabolised by the strain of *A. tumefaciens* that produced the gall. Furthermore, the ability to metabolise opines is closely correlated with virulence, and loss of virulence is always accompanied by the loss of ability to catabolise a specific opine. Before the identification of the Ti plasmid, these observations were the first indication that tumourigenesis involved the transfer of genetic material from bacteria to plants. It is now known that the enzymes for catabolism of specific opines are encoded on the Ti plasmid and complement the opine biosynthetic pathways encoded for on the T-DNA. Thus, agrobacteria create a favourable niche for themselves by genetic modification of a host plant cell, a process that Schell *et al.* called 'genetic colonisation'.[31] The gall provides a unique habitat wherein the bacteria are solely equipped to utilise the predominant carbon and nitrogen source (*i.e.* an opine).

Before considering the infection mechanism of *Agrobacterium* in further detail, and before discussing the consequences of this infection, it is worth outlining how the organism has achieved celebrity status in the fields of genetics, molecular biology, cell biology and agriculture. The first indication of the cellular or biochemical mechanisms involved in tumourigenesis coincided with the discovery of auxin, when, in 1941, Link and Eggers[32] showed that *Agrobacterium*-induced tumours were sources of this plant growth regulator. In the same year, White and Braun demonstrated that bacteria-free tumour tissue was capable of growth *in vitro*, and that callus could also grow in the absence of auxin. Three years after 'cytokinin'[§] was identified as a plant growth regulator, Braun[33] reported that the substance was strongly implicated in the growth of *Agrobacterium*-induced tumours. Braun had previously reported[34] that *Agrobacterium* was the source of a 'tumour-inducing principle', possibly DNA, which permanently transformed plant cells from a state of quiescence to active cell division.

With the advent of molecular techniques came the first evidence that axenically cultured crown gall tumours contained DNA of

---

[§]The discovery of the cytokinins was a direct outcome of tissue culture studies carried out over a number of years by Professor Folke Skoog at the University of Wisconsin. The first cytokinin to be elucidated was kinetin (Miller *et al.*[35]).

bacterial origin,[36] although this conclusion was exhaustively debated. Thanks mainly to Jeff Schell and other notable scientists at
the University of Ghent, Belgium, identification of the pTi followed
in 1974,[37] and this finding narrowed the search to genetic elements
derived from this plasmid, and led ultimately to the discovery of
T-DNA, a specific segment transferred to plant cells. The successful
integration of *A. tumefaciens* plasmid DNA into plants was carried
out by Mary Dell Chilton's group at Washington University,
St. Louis,[38] and it is not insignificant that a method of sequencing
DNA, and the first practical application of genetic engineering
(a human growth hormone produced by a bacterial cell) followed
later in the same year. An interesting historical review of the
early work on *A. tumefaciens*, notably studies conducted prior to
the advent techniques in molecular biology, has been written by
Braun.[39]

During the late 1970s and early 1980s, Chilton led a collaborative
study that produced the first transgenic plants – the basis for the
many significant contributions that plant biotechnology has made
to modern agriculture. In addition to Chilton's group in St. Louis,
the other major contributors were groups led by Jeff Schell at
Rijksuniversiteit in Ghent; Robert Fraley at the Monsanto Company in St. Louis; and John Kemp at the University of Wisconsin.
On the same day in January 1983, the groups led by Chilton, Schell
and Fraley announced at a conference in Miami, Florida, that they
had inserted bacterial genes into plants. In April of that year,
Kemp's group announced at a conference in Los Angeles that they
had inserted a plant gene from one species to another. These epoch-
making discoveries were soon reported in scientific journals and,
because of their importance to mainstream science, I document
them here. Schell's group had produced tobacco plants that were
resistant to the antibiotic kanamycin and to methotrexate, a drug
used to treat cancer. Their report appeared in *Nature* in May
1983.[40] This was followed, in July, by a paper by Chilton's group,[41]
who had produced cells of *Nicotiana plumbaginifolia*, a close relative of ordinary tobacco, which were resistant to kanamycin. The
following month, the Monsanto group detailed their work on
the production of kanamycin-resistant petunia plants,[42] while in
November 1983, the Wisconsin laboratory explained how they
had inserted a bean (*Phaseolus*) gene into a sunflower.[43] Genetic
manipulation was well and truly with us. The significance of the

kanamycin resistance gene (*kanR*) as a biotechnological tool is outlined elsewhere.

In an erudite article in the June 1983 edition of *Scientific American*, Chilton[44] was given to say,

> For a long time, perhaps millions of years, the common soil bacterium *Agrobacterium tumefaciens* has been doing what molecular biologists are now striving to do. It has been inserting foreign genes into plants and getting plants to express those genes in the form of proteins.

As indicated, the Ti plasmid carries two important genetic components: the *vir* (virulence) region and the T-DNA delimited by two 25-bp direct repeats at its ends. These are termed the 'T-DNA borders'. The *vir* region comprises seven major loci: *virA*, *virB*, *virC*, *VirD*, *virE*, *virF* and *virG*, which encode most of the bacterial protein machinery (Vir proteins) of DNA transport. After induction of the *vir* gene by the vine signal molecules, the T-DNA borders are nicked by the bacterial VirD2 endonuclease, generating a transferable single-stranded copy of the bottom strand of the T-DNA region (designated the T strand). The T strand does not travel alone, but is thought to associate with two *Agrobacterium* proteins, VirD2 and VirE2, forming a transport (T) complex in which one molecule of VirD2 is covalently attached to the 5'-end of the T strand, whereas VirE2, a single-stranded DNA-binding protein, is presumed to co-operatively coat the rest of the T strand molecule. The transport of VirD2 and VirE2 into the recipient cell makes biological sense.

It is known that two different classes of Ti plasmid exist: nopaline and octopine, and that they have T-DNA elements of different composition. For example, that in the nopaline Ti plasmid is a contiguous segment of ~ 22 kb, whereas the octopine-specific T-DNA is composed of three independently transported, adjacent strands: left DNA (TL), of 13 kb; centre DNA (TC), of 1.5 kb; and right DNA (TR), of 7.8 kb. TL contains the oncogenic function, while TR contains the opine synthetic genes.

The first stage of the infection process is the chemotactic response of *A. tumefaciens* to signals sent by the injured vine, and this event heralds a sequence of distinct cellular processes that culminate in the integration of bacterial DNA into the host plant. These events are summarised in Table 1. Once the T-DNA has been

Table 1   *A summary of the cellular processes involved in* Agrobacterium–
            plant interactions

Cellular process	Specific step in Agrobacterium–*plant cell* interaction	Agrobacterium *proteins involved in the* process
1. Cell–cell recognition	Binding of *Agrobacterium* to the host cell surface receptors	ChvA, ChvB, PscA, Att
2. Signal transduction	Recognition of plant signal molecules and activation of the T-DNA transport pathway	ChvE, VirA, VirG
3. Transcriptional activation	Expression of *vir* genes after phosphorylation of the transcriptional activator	VirG
4. Conjugal DNA metabolism	Nicking at the T-DNA borders and mobilisation of the transferable single-stranded copy of the T-DNA (T-strand)	VirD1, VirD2, VirC1
5. Intercellular transport	Formation of protein–DNA T-complex; formation of a transmembrane channel; export of the T-complex into the cytoplasm of the host cell	VirE2, VirE1, VirD2, VirD4, VirB4, Virb7, VirB9, VirB10, VirB11
6. Nuclear import	Interaction with the host cell NLS receptors and transport of the T-complex through the nuclear pore	VirD2, VirE2
7. T-DNA integration	Integration into the plant cell genome; synthesis of the second strand of the T-DNA	

inserted and integrated, its genes are expressed and this results in changes to the host plant phenotype.

When *A. tumefaciens* cells have been induced by vine phenolics, T-pili are generated. The T-pilus is composed mainly of processed VirB2 protein, with VirB5 and VirB7 as associated components. According to Lai and Kado,[45] T-pilus biogenesis uses a conserved transmembrane nucleoporotein- and protein-transport apparatus for the transport of cyclic T-pilin subunits to the *Agrobacterium* cell surface. T-pilin subunits are processed from full-length VirB2 pro-pilin into a cyclised peptide, a rapid reaction that is *Agrobacterium*-specific and can occur in the absence of Ti-plasmid genes.

An early step in the process of tumour formation is the attachment of agrobacteria to plant cells at the wound site, a step that occurs prior to, or concomitantly with, induction of the virulence region. The genes involved in the attachment are located in the

Agrobacterium chromosome, and include *chvA*, *chvB*, *pscA* and *att*. Intriguingly, the *att* genes are located on a cryptic plasmid, pAtC58, which is about twice as large as the Ti plasmid, and therefore likely to encode numerous bacterial functions. It has been conclusively shown, however, that the AtC58 plasmid of *A. tumefaciens* is not essential for tumour induction. Host-bacterial cell recognition is a two-step process. Firstly, the bacteria loosely bind to the plant cell surface, and, secondly, the bound bacteria synthesise cellulose filaments that stabilise the initial binding. This results in a tight association between the two. The surface receptors on the host plant cell can only accommodate a finite number of bacterial cells, and it is possible that one of the receptors may be a vitronectin-like glycoprotein. It is reckoned that up to 200 agrobacteria can attach themselves to each host plant cell. In animal cells, vitronectin is an important component of the extracellular matrix and functions as a receptor for several bacterial cells. Vitronectin-like molecules have been found on the cell surface of many plant species. To support this notion, it is known that human vitronectin, as well as anti-vitronectin antibodies block attachment of *A. tumefaciens* to cultured plant cells. In addition, *Agrobacterium* strains that are unable to bind plant cells due to mutations in their chromosomal *chvB*, *pscA* or *att* loci also show reduced binding to vitronectin. In addition to what has been said above, a plant vitonectin-like protein, other (as yet unknown) plant cell surface proteins and carbohydrates are likely to be involved in the interaction with agrobacteria.

Agrobacteria have evolved a two-component signal transduction system composed of the virulence proteins VirA and VirG. Together, these proteins sense signal molecules secreted by wounded plant cells and activate the expression of other *vir* genes, thereby initiating the process of T-DNA transport. Wounded plants secrete a characteristically acid sap (pH 5.0–5.8), which contains high levels of phenolic compounds, such as lignin and flavonoid precursors. This sap specifically stimulates *Agrobacterium vir* gene expression. The best characterised and most effective *vir* gene inducers are monocyclic phenols, such as acetosyringone (AS; 4′-hydroxy-3,5-dimethoxyacetophenone),[46] and they act as chemotactic agents that attract agrobacteria to the wound site. These molecules are not detected or are detected at minute levels in uninjured plants, but their concentration is significantly increased upon injury. The specific composition of phenolic compounds in plant exudates is thought to underline the host specificity shown by many *Agrobacterium* strains. Interestingly, many other plant–microbe interactions are initiated by specific phenolic compounds in host plant exudates (*e.g.* flavonoids, such as luteolin and chalcone induce the *Rhizobium* nodule-forming genes). When only small amounts of phenolics are secreted from the damaged plant cell, the release of sugars may help to activate the major phenolic-mediated wound signalling pathway. Monosaccharides, such as glucose and galactose, only significantly increase *vir* gene expression when AS is limited or absent. It has been shown that highly motile *Agrobacterium* strains exhibit a marked pTi-dependent chemotaxis towards phenolic compounds such as AS, which strongly induce the *vir* genes. Poorly motile strains do not show the same penchant for AS.

As a general rule, conditions that promote *vir* gene induction are not conducive to bacterial growth, whereas conditions that support adequate bacterial growth are unsuitable for *vir* gene induction. Accordingly, it has been proposed that in the plant wound environment containing low levels of opines, *Agrobacterium* stays in an optimal state for *vir* gene induction and T-DNA transfer. The high opine levels produced by mature galls restore the bacterium to a phase of vegetative growth and permit opines to be used as a carbon and nitrogen source. Although phenolic inducer molecules are required for the initiation of the infection process, most of these compounds can be bacteriostatic at higher concentrations, and it is thought that a plant-inducible locus in the *vir* region may be involved in detoxification of potentially harmful wound phenolics.

Although the signal molecules released by the wounded plant cell are recognised by the two-component (VirS/VirG) regulatory system, it is not entirely clear whether the phenolic signals are sensed directly by the sensor component, VirA, or by another receptor protein that then interacts with VirA. Sugar signals, on the other hand, have been shown conclusively to interact with a chromosomally encoded glucose/galactose binding protein, which then interacts with VirA.

Most of the events between transcriptional activation and the integration of T-DNA into the plant cell genome are beyond the scope of this book, and descriptions of the events relating to the production of the transferable T-strand, the formation of the T-complex and its transport into the host plant cell have been fully dealt with.[47–49] In a fascinating paper, Ziemienowicz[50] likens the fate of T-DNA, *i.e.* its processing, transfer and integration, to the journey of Odysseus; although, as she says, 'our hero returns from its long journey in a slightly modified form', referring to the fact that T-DNA leaves the *Agrobacterium* cell in the form of nucleic acid and returns from its 'journey' in the form of opines, derivatives of amino acids.

One aspect of T-DNA transport that deserves comment is its transmission between bacterial and plant cells, because there are parallels to be drawn here with the evolutionarily related process of bacterial conjugation that occurs in most bacteria, including *A. tumefaciens* itself. Unlike bacterial conjugation, however, the recipient cell in the crown gall system is eukaryotic, which means that there is a nuclear membrane to be penetrated before the *Agrobacterium* T-strand can be inserted into the host plant DNA.

Intercellular DNA transport requires a direct passageway between donor and recipient cells, and it is predicted that *Agrobacterium* forms a channel through which T-complexes are transferred into the cytoplasm of the host cell. The molecular mechanism by which this passageway is formed, and how it functions is still a 'biological black box', but it is likely that the *Agrobacterium*–plant cell channel is encoded by the *virB* locus, most of which is required for bacterial virulence but not for T strand production. The *virB* operon contains 11 open reading frames, 9 of which encode proteins shown to associate with bacterial membranes. Transport of T-complexes through the VirB channel is most likely an energy-dependent process, and two VirB proteins, VirB4 and VirB11, are the best candidates to provide energy for this translocation (VirB4 has a nucleotide binding site, whereas VirB11 is both an ATPase and a protein kinase, and both proteins localise to the inner bacterial membrane).

Regarding the expression of transferred genes, it should be realised that T-DNA, which encodes several proteins that lead to great changes in the host plant phenotype, can be expressed in host cells because those proteins mimic eukaryotic genes. In addition, the non-transcribed regions of each transferred gene possess many features of plant genes. One group of T-DNA genes directs the production of tumour-inducing plant growth hormones. For example, the *iaaM* and *iaaH* products, which may be regarded as oncogenes, direct the conversion of tryptophan to IAA *via* indolacetamide. A second set of transferred genes direct the production of the opines, which are formed by a condensation of an amino acid with a keto-acid or a sugar. Transferred cells synthesise and secrete significant quantities of particular opines, and the inducing bacteria typically carry genes (outside the T-DNA region, and usually on the virulence plasmid) required to catabolise the same opines synthesised by the induced tumour.

Based on the kind of opines produced in tumours, agrobacteria are classified as 'octopine', 'nopaline', 'succinamopine' and 'leucinopine' strains. At least 20 different opines are known to exist, and each *Agrobacterium* strain induces and catabolises a specific set of opines, the octopine-type Ti plasmids, for example, directing their hosts to sythesise at least 8 opines. The *ocs* gene encodes octopine synthase, which reductively condenses pyruvate with either arginine, lysine, histidine or ornithine to produce octopine, lysopine, histopine or octopinic acid, respectively. The *mas2'* product is thought to condense glutamine or glutamic acid with glucose, while the *mas1'* gene product reduces these intermediates to mannopine and mannopinic acid, respectively. The *ags* product catalyses the lactamisation of mannopine to form agropine, and both mannopine and agropine can spontaneously lactamise to form agropinic acid. In short, the tumours induced by agrobacterial strains harbouring octopine-type Ti plasmids can produce four members of the octopine family and four members of the mannityl opine family.

Agrocinopines A and B are phosphorylated opines containing nopaline. Agrocinopine A is a non-nitrogenous opine of sucrose and L-arabinose, with a phosphodiester linkage from the 2-hydroxyl of the arabinose to the 4-hydroxyl of the fructose moiety in sucrose. Agrocinopine B is the corresponding phosphodiester, in which glucose has been hydrolysed from the sucrose portion of agrocinopine A.

The DNA transmission capabilities of *A. tumefaciens* have been extensively exploited in biotechnology as a means of inserting foreign genes into plants. The plasmid T-DNA that is transferred to the plant is an ideal vehicle for genetic engineering, since, as was shown by Schell's group,[51] DNA can be introduced into plant host cells without alteration of their normal regeneration capacity. This is done by cloning the desired gene sequence into the T-DNA that will be subsequently inserted into the hosts' DNA. Interestingly, this process has been performed using the firefly luciferase gene to produce the famous 'glowing tobacco plants'. Although higher plants are the natural hosts for *A. tumefaciens*, the microbe has been made to genetically transform a wide range of other eukaryotic organisms, including yeasts and filamentous fungi. Under laboratory conditions, the T-DNA has also been transferred to human cells, demonstrating the diversity of the insertion application.[52] The mechanism by which *A. tumefaciens* inserts materials into the host cell is by a type IV secretion system, and this is very similar to mechanisms used by pathogens to insert materials (usually proteins) into human cells by type III secretion. It also employs a type of signalling conserved in many Gram-negative bacteria, called 'quorum sensing'. This makes *A. tumefaciens* an important subject of medical research as well. Originally, interest in *Agrobacterium* was sparked by the fact that it caused tumours, and research was expected to provide insight into animal tumour pathogenesis. While this goal has not yet been realised even after 100 years of research, it is evident that continued research into 'agrobiology' has more than a passing application to animal pathogenesis.

The molecular mechanisms of T-DNA transfer that underline the use of *A. tumefaciens* as a gene vector have been reviewed by Valentine,[53] who described the bacterium as being the 'David and Goliath' of modern genetics. In particular, Valentine emphasises the ethical and political aspects of the release of *Agrobacterium*-generated genetically modified plants into the environment, and discusses the attitudes of different, and often opposing, geopolitical and economic forces on such matters.

Because of its systemic nature, crown gall is a difficult disease to cure *per se*, and this is partly due to the fact that visual examination for galls has always been the primary screen for diseased material. One of the standard treatments for older vines is to excise the galls and then treat the wounds with kerosene. Other chemical treatments involve the use of creosote-based compounds, copper-based solutions and strong oxidants such as sodium hypochlorite. All of these are only transiently effective, and new galls inevitably occur. It is fortunate that in established vines, the disease is rarely fatal. The most effective method of control is to grub up affected vines and use disease-free plants for re-planting. Where re-planting is necessary, it is advisable to use land that has supported cereal crops for some time, since such soils contain reduced populations of agrobacteria. There is, of course, much interest being shown in using biotechnology to solve the problem, but the first step, the production of genetically transformed vines that express a marker gene, has yet to be unequivocally reported. Some laboratories have had encouraging results with *Agrobacterium*-mediated transformation, but no genetically transformed vines have yet been produced. To successfully engineer disease resistance into plants, the following are needed: (1) recipient cells that are capable of growing into whole plants; (2) a method to transfer resistance genes into the cells; (3) proper expression of the genes by the transformed cells; (4) a method to select the transformed cells from non-transformed cells; (5) regeneration of whole plants; and (6) evaluation of transformed plants for disease resistance. When one appreciates the difficulties associated with stage (4) above, the decision by pioneers such as Schell, to introduce kanamycin resistance into their recipient plants makes great sense. The *kanR* is one of the most widely used selectable marker genes. It specifies the information for the production of the aminoglycoside $3'$-phosphotransferase II enzyme $(APH(3')II)$ – common name, neomycin phosphotransferase II. This enzyme modifies aminoglycoside antibiotics such as neomycin and kanamycin, chemically inactivating them and rendering cells that produce the *kanR* gene product, refractory or resistant to the antibiotic. Plant cells that have received and stably expressed the *kanR* gene survive and replicate on laboratory media in the presence of kanamycin. Cells that did not take up and express the gene will be killed by the antibiotic. By linking the selectable marker gene to another gene that specifies the desired trait, scientists can identify and select plants that have taken up and expressed those genes. Once the desired plant character has been inserted, the *kanR* gene serves no further useful purpose, although it will continue to produce the $APH(3')II$ enzyme.

Success in grapevine transformation came only when researchers started using what are termed 'embryonic cultures'. These are axenically grown in artificial media in the laboratory and consist of tiny clumps of cells that are capable of growing into embryos that, in turn, can germinate into plants. These cells originate from the main body of the plant (*i.e.* are somatic cells) and not egg or sperm cells, so that each embryo is an exact replicate of the original plant (*i.e.* a clone). At present, there are around 15 laboratories working worldwide to genetically engineer grapevines, and transformed varieties are being tested in France, Australia and USA.

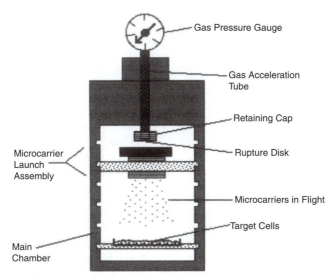

Figure 8   *Diagram of gene gun used to deliver genes into plant cells*
(Reproduced by kind permission of Cornell University)

One of the leading centres of such research is at Cornell University's New York State Agricultural Experiment Station, Geneva, NY, where they have focused on the 'biolistic' (shortened form of 'biological ballistic') method of inserting genes into the vine. In this technique, minute DNA-coated particles are used to carry foreign genes into recipient cells. DNA coding for the gene(s) of interest is coated onto tungsten micro-projectiles, which are accelerated at very high speed into the cultured recipient cells, using a 'gene gun'. The gene gun (Figure 8) was invented by John Sanford and a team of co-workers at Cornell. The device is driven by high-pressure helium gas, and when the rupture disk at the end of the gas acceleration tube is burst, a strong shock wave of gas is released, which in turn, launches the micro-carriers, which are minute tungsten particles coated with the relevant genes to be transferred. These particles penetrate the vine (target) cells, and the genes are released inside them. When used by capable hands, cell injury is minimal and the new genes have a long-term influence within the receptive cell.

Since 1973, crown gall on stone fruit and roses has been biologically controlled commercially by dipping planting material in a bacterial suspension of *Agrobacterium rhizogenes* (*A. radiobacter*) strain K84, a non-pathogenic relative of *A. tumefaciens*. This treatment proved successful, and it is now practiced worldwide. Biological control by strain K84 is primarily due to the production of two plasmid-encoded antibiotics, agrocin 84 and agrocin 434. Agrocin 84 is an adenosine analogue, while agrocin 434 is a di-substituted cytidine analogue. Another rather effective biological control agent was reported by Burr and Reid,[54] and is produced by the avirulent *A. vitis* strain F2/5. This strain produces an (as yet, unknown) antibiotic, which is lethal to many tumourigenic *A. tumefaciens* strains *in vitro*. Biological control by F2/5 is grape-specific, but is not effective against all *A. vitis* strains. Attempts are being made to modify and increase the toxicity of F2/5 towards tumour-forming agrobacteria. One successful modification was undertaken by Helache and Triplett,[55] who conferred trifolitoxin (TFX)-producing ability to F2/5. Trifolitoxin (TFX)[¶] is a peptide antibiotic active only against certain α-proteobacteria (including agrobacteria and close relatives). TFX production is expressed from the stable plasmid, pT2TFXK, which was inserted into F2/5. When the resultant modified, non-pathogenic cells were co-inoculated with virulent *A. vitis* into a suitable crown-gall sensitive host (*Nicotiana glauca*), both the number and size of tumours were significantly reduced. As these workers remarked, 'This is the first demonstration that the production of a ribosomally-synthesised, post-translationally modified peptide antibiotic can confer reduction in plant disease incidence from a bacterial pathogen'.

Agrobacteria, of course, predate the science fiction visions of 'nano-machines' performing genetic engineering and other biotechnological tasks, but they can be regarded as 21st century, microscopic, complicated pieces of machinery, which are routinely used to alter genotypes of higher plants. The use of *A. tumefaciens* is based on its unique capacity to perform, what Stachel and Zambryski[56] called 'trans-kingdom sex', *i.e.* the transfer of genetic material between prokaryotic and eukaryotic cells. Decades of research altered, augmented and greatly improved this natural capacity, and this has resulted in ingeniously modified *Agrobacterium* strains that can transfer and stably integrate virtually any gene to a variety of plant species from research model plants, such as *Arabidopsis*, to

---

¶TFX was first isolated from *Rhizobium leguminosarum* biovar. *trifolii* T24.

economically important crop plants, like rice and maize. Furthermore, the *Agrobacterium*–host cell reaction represents a unique and powerful experimental system to study a wide spectrum of basic biological processes, such as cell–cell recognition and cell-to-cell transport, nuclear import, assembly and disassembly of protein–DNA complexes, DNA recombination and regulation of gene expression.

10.4 VIRUSES

Viruses of the vine have been around for thousands of years, and descriptions that go as far back as Roman times leave no doubt as to their existence. As a rule, however, it seems as though their presence was somewhat less than totally disastrous. All this changed when grafting with American rootstocks became *de rigueur* for combating phylloxera. Many of the American rootstocks brought to Europe for this purpose were, and are, symptomless carriers of a number of viruses, and two quite destructive viral diseases, 'fan leaf' and 'leaf roll', were inadvertently spread through commercial wine-producing countries. The causal agents of many virus and 'virus-like' diseases of the vine have yet to be unequivocally identified, but it is known that none of them is transmissible by seed or pollen, yet all are transmissible by grafting. It goes without saying that vines with obvious symptoms of disease are not used for propagating purposes, but seemingly healthy material can harbour viral particles. Selection of virus-free propagating material is now much more feasible, thanks to modern diagnostic methods that include immunological techniques (*e.g.* enzyme-linked immunosorbent assay; ELISA) and nucleic acid technology, such as ds-RNA and c-DNA probes.

Plant viruses affecting grapevines are composed of an RNA core surrounded by a protein coat, the 'outer shell', and modern diagnostic methods are based on the ability of the relevant reagent to bind onto, or copy, a small portion of the virus. ELISA is one of the most common and inexpensive diagnostic methods and involves the binding of the outer shell with an antibody. In simple terms, extracts of ground vine tissue are placed in a test plate that has been coated with specific antibodies. If the relevant virus is present in the extract, it will bind to the specific antibodies on the plate, whence it can be detected by an enzyme–substrate reaction that produces a colour reaction. Several samples can be tested simultaneously and results can normally be obtained within 3 days. ELISA results should be interpreted with care because it can 'miss' infections if the pathogen is only present in minute quantity. To confirm the absence of a virus, a negative ELISA result should always be confirmed with an appropriate polymerase chain reaction (PCR), which permits the amplification (*i.e.* production of multiple copies) of viral RNA, which might be present in small amounts. Prior to PCR, viral RNA must be converted into DNA. This technique is known as reverse transcriptase PCR (RT-PCR) and requires a piece of viral RNA and primers to get the copying process started. Primers are short pieces of DNA that 'jumpstart' the genetic copying process. The polymerase (a molecule that facilitates the copying of DNA) helps produce duplicates of the original viral RNA, no matter how insignificant. The duplication is repeated many times with each copy making more copies, and so the viral RNA can be identified. Although it is expensive and laborious, PCR is now regarded as the method of choice for identifying the presence of viruses present at low infection levels and as a confirmation of 'virus-free' status.

10.4.1   Fanleaf Degeneration

Fanleaf degeneration is an important and widespread viral disease of grapevine, occurring in all vine-growing regions and probably spread by the main vector, the nematode *Xiphinema index*, and by infected plant material. It is caused by the grapevine fanleaf virus (GFLV). Apart from the symptom obvious from its name, the virus causes wide range of symptoms on the shoots of the grapevine, including shortening of internodes, zigzag growth and fasciation and acute indentation on the leaves. Chromogenic strains (yellow mosaic form) of the virus induce various patterns of yellow discolouration along the main vein of mature leaves, a condition known as 'vein-banding'. Fruit set is irregular and poor with fewer and smaller bunches, there being many aborted berries. Yield can be reduced by up to 80% in the more susceptible varieties (*e.g.* Cabernet Sauvignon), and the productive longevity of the plant is low. GFLV is a nepovirus, whose natural host range is restricted to the genus *Vitis* and whose virus particle is isometric and measures between 25 and 30 μm in diameter. In the era prior to modern diagnostics, the infection profile of certain plants, called 'indicator plants' or 'differential hosts', was used to identify and occasionally purify, the virus. Indeed the virus is mechanically transmissible to a variety of herbaceous plants, such as *Chenopodium amaranticolor* (where it can be transmitted through seeds), *Cucumis sativus* and some varieties of *Nicotiana tabacum*. Strategies for the control of fanleaf degeneration include (1) control of nematode vectors by long-term fallowing of soil and soil fumigation; (2) breeding rootstocks for resistance to the feeding of *Xiphinema* nematodes; and (3) breeding vines that are resistant to GFLV (a long-term solution). According to Mullins *et al.*,[3] 'fanleaf degeneration' is the only so-called "virus disease" of grapevines in which a viral particle is known to be the causal agent.

10.4.2 Grapevine Leafroll

Caused by the grapevine leafroll virus (GLRV), like fanleaf, grapevine leafroll is found wherever grapes are grown, infected plants being slightly smaller than healthy ones. It is especially common where *V. vinifera* varieties are grown on phylloxera-resistant stocks. Leafroll virus disease is associated with several related viruses of the Closterovirus (Greek, 'thread-like') type. The viruses cause varying degrees of disease in different grape varieties. Susceptible varieties exhibit variable symptoms. Generally, susceptible black and red-fruited varieties exhibit a red or purplish leaf colour in late summer, often, but not always, accompanied by a downward rolling of leaf margins. All or only some leaves on a shoot and some shoots on a vine may exhibit symptoms. Once symptoms show up, they recur every year. White-fruited varieties exhibit a light green to yellowish colour, again sometimes accompanied by marginal leaf rolling. Leafroll is not a fatal condition, but results in considerable reductions in fruit yield and fruit quality. Grapes on diseased vines often contain less sugar than usual because carbohydrates accumulate in the leaves, rather than the fruit. The disease slowly weakens the vine and early symptoms include delayed fruit maturity. Diseased vines eventually weaken to the point of being uneconomical to farm. Several different viruses, called 'grapevine leafroll associated viruses' (GLRaV, types 1–9) have been reported as being associated with leafroll disease, and ELISA test kits have been developed for types 1–7. The viruses and virus diseases of the grapevine have been reviewed by Hewitt[57] and Bovey and Martelli.[58]

REFERENCES

1. R.C. Pearson and A.C. Goheen, *Compendium of Grape Diseases*, American Phytopathological Society, St. Paul, MN, 1988.
2. F. Cazalis, *L'Insectologie Agricole*, 1869, **3**, 29.
3. M.G. Mullins, A. Bouquet and L.E. Williams, *Biology of the Grapevine*, Cambridge University Press, Cambridge, 1992.
4. L.A. Lider, M.A. Walker and J.A. Wolpert, Grape rootstocks in California vineyards: The changing picture, in F. Pérez Camacho and M. Medina (eds), *International Symposium on Viticulture and Enology*, ISHS Acta Horticulturae 388, Cordoba, Spain, 1995.
5. J. Granett, A.C. Goheen, L.A. Lider and J.J. White, *Am. J. Enol. Vitic.*, 1987, **38**, 298.
6. W.M. Davidson and R.L. Nougaret, *USDA Bull.*, 1921, **903**, 128.
7. E.M. Heddle and F. Doherty, *Château Tahbilk: Story of a Vineyard 1860–1985*, Lothian, Melbourne, 1985.
8. C. Campbell, *Phylloxera: How Wine Was Saved for the World*, Harper Collins UK, London, 2004.
9. D. Boubals, *Ann. Amélior. Plantes*, 1959, **9**, 5.
10. M.H. Dye and M.V. Carter, *Aust. Plant Pathol. Soc. Newslett.*, 1976, **5**, 6.
11. W.J. Moller and A.N. Kasimatis, *Plant Dis.*, 1981, **65**, 429.
12. S.J. Merrin, N.G. Nair and J. Tarran, *Australas. Plant Pathol.*, 1995, **24**, 44.
13. A.J.L. Phillips, *J. Phytopathol.*, 1998, **146**, 327.
14. C.T. Gregory, *Phytopathol.*, 1913, **3**, 20.
15. B. Lal and A. Arya, *Indian Phytopathol.*, 1982, **35**, 261.
16. W.B. Hewitt and R.C. Pearson, Phomopsis cane and leaf spot, in R.C. Pearson and A.C. Goheen (eds), *Compendium of Grape Diseases*, American Phytopathological Society, St. Paul, MN, 1990, 17–18.
17. J.W. Pscheidt and R.C. Pearson, *Plant Dis.*, 1989, **73**, 829.
18. O. Erincik, L.V. Madden, D.C. Ferree and M.A. Ellis, *Plant Dis.*, 2001, **85**, 517.
19. O. Erincik, L.V. Madden, D.C. Ferree and M.A. Ellis, *Plant Dis.*, 2003, **87**, 832.
20. D.L. Melanson, B. Rawnsley and R.W.A. Scheper, *Australas. Plant Pathol.*, 2002, **31**, 67.

21. B.E. Stummer, B. Rawnsley, T.J. Wicks and E.S. Scott, *Aust. N Z Grape grower Winemaker*, 2002, **464**, 30.

22. J.M. van Niekerk, J.Z. Groenewald, D.F. Farr, P.H. Fourie, F. Halleen and P.W. Crous, *Australas. Plant Pathol.*, 2005, **34**, 27.

23. J.M. Wells, B.C. Raju, H.Y. Hung, W.G. Weisburg, L. Mandelco-Paul and D.J. Brenner, *Int. J. Syst. Bacteriol.*, 1987, **37**, 136.

24. J.H. Freitag, *Phytopathology*, 1951, **41**, 920.

25. J.J. Ruel and M.A. Walker, *Am. J. Enol. Vitic.*, 2006, **57**, 158.

26. E.F. Smith and C.O. Townsend, *Science*, 1907, **25**, 671.

27. H.J. Conn, *J. Bacteriol.*, 1942, **44**, 353.

28. K. Ophel and A. Kerr, *Int. J. Syst. Bacteriol.*, 1990, **40**, 236.

29. B. Goodner, G. Hinkle, S. Gattung and N. Miller *et al.*, *Science*, 2001, 2323.

30. D.W. Wood, J.C. Setubal, R. Kaul and D.E. Monks *et al.*, *Science*, 2001, 2317.

31. J. Schell, M. Van Montagu and M. De Beuckeleer *et al.*, *Proc. R. Soc. Lond. B*, 1979, **204**, 251.

32. G.K. Link and V. Eggers, *Bot. Gazette*, 1941, **103**, 87.

33. A.C. Braun, *Proc. Natl. Acad. Sci. USA*, 1958, **44**, 344.

34. A.C. Braun, *Am. J. Bot.*, 1947, **34**, 234.

35. C.O. Miller, F. Skoog, M.H. Von Saltza and F.M. Strong, *J. Am. Chem. Soc.*, 1955, **77**, 1329.

36. R.A. Schilperoort, H. Veldstra, S.O. Warnaar, G. Mulder and J.A. Cohen, *Biochim. Biophys. Acta*, 1967, **145**, 523.

37. N. Van Larebeke, G. Enbler, M. Holsters, S. Van Den Elsacker, I. Zaenen, R.A. Schilperoort and J.S. Schell, *Nature*, 1974, **252**, 169.

38. M.-D. Chilton, M.H. Drummond, D.J. Merlo, D. Sciaky, A.L. Montoya, M.P. Gordon and E.W. Nester, *Cell*, 1977, **11**, 263.

39. A.C. Braun, The history of the crown gall problem, in G. Kahl and J.S. Schell (eds), *Molecular Biology of Plant Tumors*, Academic Press, New York, 1982, 155–210.

40. L. Herrera-Estrella, A. Depicker, M. van Montagu and J.S. Schell, *Nature*, 1983, **303**, 209.

41. M.W. Bevan, R.B. Flavell and M.-D. Chilton, *Nature*, 1983, **304**, 184.

42. R.T. Fraley, S.G. Rogers, R.B. Horsch, P.R. Sanders, J.S. Flick, S.P. Adams, M.L. Bittner, L.A. Brand, C.L. Fink, J.S.

Fry, G.R. Galluppi, S.B. Goldberg, N.L. Hoffmann and S.C. Woo, *Proc. Natl. Acad. Sci. USA*, 1983, **80**, 4803.

43. N. Murai, D.W. Sutton, H.G. Murray, J.L. Slightom, D.J. Merlo, N.A. Reichart, C. Sengupta-Gopalan, C.A. Stock, R.F. Barker, J.D. Kemp and T.C. Hall, *Science*, 1983, **222**, 476.
44. M.-D. Chilton, *Sci. Am.*, June 1983, **248**, 50.
45. E.-M. Lai and C.I. Kado, *Trends Microbiol.*, 2000, **8**, 361.
46. S.E. Stachel, E. Messens, M. Van Montagu and P.C. Zambryski, *Nature*, 1985, **318**, 624.
47. J. Sheng and V. Citovsky, *Plant Cell*, 1996, **8**, 1699.
48. J. Zupan, T.R. Muth, O. Draper and P. Zambryski, *Plant J.*, 2000, **23**, 11.
49. T. Tzfira and V. Citovsky, *Plant Physiol.*, 2003, **133**, 943.
50. A. Ziemienowicz, *Acta Biochimica Polonica*, 2001, **48**, 623.
51. P.C. Zambryski, H. Joos, C. Genetello, J. Leemans, M. Van Montagu and J.S. Schell, *EMBO J.*, 1983, **2**, 2143.
52. T. Kunik, T. Tzfira, Y. Kapulnik, Y. Gafni, C. Dingwall and V. Citovsky, *Proc. Natl. Acad. Sci. USA*, 2001, **98**, 1871.
53. L. Valentine, *Plant Physiol.*, 2003, **133**, 948.
54. T.J. Burr and C.L. Reid, *Am. J. Enol. Vitic*, 1993, **45**, 213.
55. T.C. Herlache and E.W. Triplett, *BMC Biotechnol.*, 2002, 2:2 (online).
56. S.E. Stachel and P.C. Zambryski, *Nature*, 1989, **340**, 190.
57. W.B. Hewitt, *Rev. Appl. Mycol.*, 1968, **47**, 433.
58. R. Bovey and G.P. Martelli, *Vitis*, 1986, **25**, 227.

Table of Wine Composition

Compound	Composition ($g\,L^{-1}$)
Acids (16 other acids detected in small concentration)	
Tartaric	2–5
Malic	tr.–5
Citric	tr.–1
Succinic	0.5–1.5
Lactic	0.4–3
Formic	0.05
Acetic	0.3–1.5
Propionic	tr.
Butyric	0.02
Pyruvic	tr.–0.13
α-Ketoglutaric	tr.–0.12
pH	3.0–4.3
Sugars (small quantities of 12 other sugars have been detected)	
Glucose	tr.–100
Fructose	tr.–100
Arabinose	0.3–1
Xylose	tr.–0.05
Alcohols (small quantities of 18 other alcohols have also been detected)	
Methyl	tr.–0.6
Ethyl	0–160

(*Continued*)

(*Continued*)

Compound	Composition ($g\ L^{-1}$)
n-Propyl	tr.–0.01
iso-Propyl	tr.–0.25
iso- and active–Amyl	0.1–0.6
2-Phenethanol	tr.–0.08
n-Hexanol	tr.–0.01
Esters, carbonyls and aldehydes (approximately 80 other esters and 7 other aldehydes have been detected)	
Ethyl acetate	0.05–0.15
Acetaldehyde	0.02–0.4
Acetyl methyl carbinol	tr.–0.08
Diacetyl	tr.–0.007
Acetal	tr.–0.01
Hydroxymethyl furfural	tr.–0.3
Polyols	
Glycerol	1–15
2,3 Butandiol	0.1–1.6
Inositol	0.2–0.7
Sorbitol	tr.–0.1
Anions (small quantities of some other anions have been found)	
Tartrate	0.5–4
Sulphate	0.1–3
Chloride	0.02–0.4
Phosphate	0.05–1
Bisulphite	0–0.3
Fluoride	tr.–0.005
Bromide	tr.–0.002
Iodide	tr.–0.0002
Borate	tr.–0.1
Cations	
Potassium	tr.–2.5
Sodium	0.02–2.5
Calcium	0.01–0.15
Magnesium	0.01–0.2
Iron	tr.–0.015
Copper	tr.–0.002

(*Continued*)

Compound	Composition ($g\ L^{-1}$)
Aluminium	tr.–0.005
Zinc	tr.–0.005
Manganese	tr.–0.001
Arsenic	tr.–0.0001
Lead	tr.–0.0005
Phenolics	
Anthocyanins	0–1
Tannins	0.2–4
Other phenolics	0.1–1
Nitrogenous compounds	
Proteins	tr.–0.04
Amino acids:	
Alanine	0.07
Aminobutyric acid	0.03
Arginine	0.02–0.1
Asparagine	0.05
Aspartic acid	0.01–0.1
Cysteine	0.01–0.1
Glutamic acid	0.09–0.4
Glutamine	0.03
Glycine	0.01–0.07
Histidine	0.005–0.04
Leucine and isoleucine	0.01–0.06
Lysine	0.02–0.07
Methionine	0.002–0.05
Phenylalanine	0.01–0.03
Proline	0.05–0.7
Serine	0.01–0.07
Threonine	0.02–0.4
Tryptophan	tr.–0.005
Tyrosine	0.005–0.04
Valine	0.01–0.08
Vitamins	*(in mg L^{-1})*
Thiamine	tr.–0.01
Riboflavin	tr.–0.3
Pantothenic acid	0.5–2
Pyridoxin	0.2–0.8

(*Continued*)

(Continued)

Compound	Composition $(g\ L^{-1})$
Nicotinic acid	1–3
Biotin	tr.–0.002
Mesoinositol	200–700
Choline	17–40
p-Aminobenzoic acid	tr.–0.2
Pteroylglutamic acid	tr.–0.004

Data as presented in Rankine (2004) represent a collation of world oenological literature and analyses of Australian wines. Here reliable and comprehensive data are available, a range of values is given, otherwise mean values are presented. (Reproduced with kind permission of Bryce Rankine and Pan Macmillan Australia Ltd.)

APPENDIX B
Density Scales

When giving density figures in the text, I have tended to quote the figures in common use for that particular wine-producing area. Density scales vary throughout the world, and the following is an attempt to interrelate them.

Four scales are used internationally for the measurement of soluble solids during winemaking, these being expressed as °Brix, °Balling, °Baumé and °Öechsle. Two other density scales exist: Plato and Klosternerburg, the latter now being almost obsolete. The Plato Scale, based on Plato's original density measurements, is still widely used in the brewing industry, as is the Brix Scale. The Balling Scale, named after its 'inventor', Karl Balling, is calibrated against the concentration of sucrose solutions at 17.5°C (*i.e.* Balling = weight percent sucrose at 17.5°C). It has been largely replaced by the Brix Scale, which was developed in the mid-19th century, when Antoine Brix recalculated Balling's results to a reference temperature of 15.5°C, but this scale has since been re-calculated again to a reference temperature of 20°C. It is the latter that is now widely in use in the fermentation industries.

Thus, the Balling and Brix scales are now identical, a solution of 20 °Brix having a specific gravity of 1.082. The Öechsle Scale (°Ö), which is used in Germany, is the most fundamental of the scales, as it relates directly to specific gravity. It is based on the difference in weight of one litre of must compared to one litre of water, and the first three figures in the decimal fraction of the specific gravity gives the Öechsle equivalent. Thus, using the example above, our 20 °Brix solution (with an SG of 1082), has a value on the Öechsle Scale of 82. In effect, this scale amplifies the density contributions of the solute over that of water, by a factor of 1000 (as does the Plato Scale). The Baumé Scale, which is widely used in Australia, is now calculated to a temperature reference of 20°C, and 1° on the Baumé Scale is equivalent to 1.8 °Brix (or Balling). Conversely, 1 °Brix is equivalent to 0.56 °Baumé. Standard tables are available for the inter-conversion of specific gravity and the major density scales.

Subject Index